CONJURING CULTURE

RELIGION IN AMERICA SERIES
Harry S. Stout, *General Editor*

A PERFECT BABEL OF CONFUSION
*Dutch Religion and English Culture
in the Middle Colonies*
Randall Balmer

THE PRESBYTERIAN CONTROVERSY
*Fundamentalists, Modernists,
and Moderates*
Bradley J. Longfield

MORMONS AND THE BIBLE
*The Place of the Latter-day Saints in
American Religion*
Philip L. Barlow

THE RUDE HAND OF INNOVATION
*Religion and Social Order in Albany,
New York 1652–1836*
David G. Hackett

SEASONS OF GRACE
*Colonial New England's Revival
Tradition in Its British Context*
Michael J. Crawford

THE MUSLIMS OF AMERICA
Edited by Yvonne Yazbeck Haddad

THE PRISM OF PIETY
*Catholick Congregational Clergy at the
Beginning of the Enlightenment*
John Corrigan

FEMALE PIETY IN PURITAN
NEW ENGLAND
The Emergence of Religious Humanism
Amanda Porterfield

THE SECULARIZATION
OF THE ACADEMY
Edited by George M. Marsden and
Bradley J. Longfield

EPISCOPAL WOMEN
*Gender, Spirituality, and Commitment
in an American Mainline
Denomination*
Edited by Catherine Prelinger

SUBMITTING TO FREEDOM
The Religious Vision of Willilam James
Bennett Ramsey

OLD SHIP OF ZION
*The Afro-Baptist Ritual in the
African Diaspora*
Walter F. Pitts

AMERICAN TRANSCENDENTALISM
AND ASIAN RELIGIONS
Arthur Versluis

CHURCH PEOPLE IN THE
STRUGGLE
*The National Council of Churches and
the Black Freedom Movement,
1950–1970*
James F. Findlay

EVANGELICALISM
*Comparative Studies of Popular
Protestantism in North America, the
British Isles, and Beyond, 1700–1990*
Edited by Mark A. Noll, David W.
Bebbington and George A. Rawlyk

RELIGIOUS MELANCHOLY AND
PROTESTANT EXPERIENCE IN
AMERICA
Julius H. Rubin

CONJURING CULTURE
Biblical Formations of Black America
Theophus H. Smith

CONJURING CULTURE

BIBLICAL FORMATIONS OF BLACK AMERICA

Theophus H. Smith

New York   Oxford
OXFORD UNIVERSITY PRESS
1994

Oxford University Press

Oxford   New York   Toronto
Delhi   Bombay   Calcutta   Madras   Karachi
Kuala Lumpur   Singapore   Hong Kong   Tokyo
Nairobi   Dar es Salaam   Cape Town
Melbourne   Auckland   Madrid

and associated companies in
Berlin   Ibadan

Copyright © 1994 by Theophus H. Smith

Published by Oxford University Press, Inc.
200 Madison Avenue, New York, New York 10016

Library of Congress Cataloging-in-Publication Data
Smith, Theophus Harold.
Conjuring culture : biblical formations of black america /
Theophus H. Smith.
p.   cm. — (Religion in America series)
Includes bibliographical references and index.
ISBN 0-19-506740-1
1. Afro-Americans—Religion.   2. Religion and culture—United States.
3. Bible—Influence—Western civilization.
4. Bible—Criticism, interpretation, etc.—United States—History.
5. Typology (Theology)—History of doctrines.
6. Magic—Religious aspects—Christianity—History of doctrines.   I. Title.
II. Series: Religion in America series (Oxford University Press)
BR563.N4S574   1994
277.3'08'08990673—dc20   93-8152

1 3 5 7 9 8 6 4 2

Printed in the United States of America
on acid-free paper

in memoriam
Ereina Christin Smith
March 2, 1978 – February 25, 1987

*For [Christ] is our peace, who has made us both one,
and has broken down the dividing wall of hostility.*
EPHESIANS 2.14

Preface

I could not conjure my God in this place, and it seemed His failure.
Surprise ... overwhelmed me.

Lorene Cary, *Black Ice* [1]

Like the author quoted above I spent a few years during my adolescence as a black student in a New England prep school. We were among the sixties and seventies generation of the "young, gifted, and black"—to recall the popular refrain of one of Aretha Franklin's songs of that period. It was a period when the African American presence in northern boarding schools felt more like a daring social experiment than something traditional. But there *was* a tradition, as Lorene Cary herself realized: "when I stopped thinking of my prep-school experience as an aberration from the common run of black life in America," and when she discovered a decades-long, ongoing "conversation" among black Americans about our pilgrimage through white academic institutions. [2] Of course, Aretha's celebratory song was not intended to represent the pathos of black Americans' more abysmal experiences. Some experiences require instead the ironic expressions articulated in the spirituals and the blues, to complement the jubilant highs of our American existence as expressed in sixties' soul music or its gospel music parallels. All those genres are needed, at any rate, to encompass my own preparatory school experience. [3]

The tradition of black Americans attending the nations' prestigious institutions of learning began in the nineteenth century, by my reckoning. I mark its inception with the matriculation of the renowned black scholar, W.E.B. Du Bois, at Harvard College in the autumn of 1888, following his graduation from the southern black institution, Fisk University. I mark it further with his subsequent studies in Germany, at the Friedrich Wilhelm University in Berlin in the early 1890s, and with his publication in 1896 of Harvard's first African American doctoral thesis. Moreover, I locate in Du Bois's Harvard and German era the inaugural period for the emergence of the character type signified by Lorene Cary's trope, "black ice." It is a trope that aptly captures the now legendary persona of a man whose acquired character armor enabled him to endure the indignities visited upon a black scholar of that day. Despite those indignities (both black-white and black-on-black), and despite his own sometimes glacial response to them, Du Bois nonetheless persisted in an unparalleled quest for intellectual excellence and humane service to his people and the nation.

The tradition thereby inaugurated, and which I inherited along with Cary and numerous others, also constitutes the matrix of this book. It is a matrix formed by fusions of the North and the South, the white and the black, and the Euro- and the Afro-American; by the fusion of double worlds of history and culture, of intellect and spirit. It is a bicultural or multicultural matrix, therefore, and the implications of such "double consciousness" (Du Bois) will be evident throughout the pages that follow. But there is a more literal "matrix" of this study of conjurational spirituality: my mother, The Reverend Josephine Jackson-Smith, in her persona as carrier and mediator of our family's folk religious convictions and sensibilities. It was in prep school that I first discovered, albeit dimly and with growing misgivings, that my mother's piety and our family's roots in black folk religion had bequeathed me an imaginative and visionary sensibility that found little or no validation in the Anglo-Saxon, Protestant, and Yankee ethos in which I sojourned. Again and again during the course of the semesters I discovered in myself what I now know to be conjurational aptitudes and impulses in the way I viewed reality and framed experience. But those inclinations appeared to be superstitious or fanciful illusions in relation to the sensibilities and behaviors of my white fellow students and teachers. My persistent and instinctive impulse, for example, to prepare for an exam by projecting myself as a masterful and triumphant performer of extraordinary feats (with and even without more mundane preparation), became cast as sheer folly as I learned to internalize the pragmatic attitudes of my peers.

A more calamitous casualty was my African American ancestral conviction that a provident God ordered my life and presided over that New England experience as its *genius loci*; as the "spirit of the place" or the "local integrity." Under assault was the conviction that my sojourn there constituted a kind of "ascension ritual"—a rite of passage conveying me 'North to freedom!'—in relation to which the school and its environs constituted my "ritual ground."[4] Such structural supports in my spiritual universe were continuously under assault by the anti-mystical, deritualized, and materialist constitution of the place.

> I was 'way down yonder by myself
> and I couldn't hear nobody pray![5]

That refrain from a slave spiritual captures, with generic eloquence, the abysmal depths of the African American passage into *terra incognita*—into white American and European institutions and cultures that grudgingly permit but typically eschew ritualized intersections between the material∠ world and our ancestrally mediated worlds of spirit. Years later, attending seminary in pursuit of theological education, I began a quest for spiritual reintegration with my folk heritage. This book on the conjurational spirituality of black North Americans is my most systematic effort in that quest.

As one among other scholars engaged in the nascent field of African American spirituality, I locate its interest somewhere between the historical and social science study of religion on the one hand, and the black theology of liberation movement on the other, spanning literary and aesthetic considerations in between. A convergence of interest in spirituality as a distinctive category of religious phenomena can be observed, first, from the religious studies side. Here I refer to the experiential focus found in the work of established scholars, including Gayraud Wilmore in his edited, interdisciplinary anthology, *African American Religious Studies* (Duke, 1989) as well as in his earlier work, *Black Religion & Black Radicalism* (Orbis, 1983). I note also the early work of Henry Mitchell, *Black Belief: Folk Beliefs of Blacks in America & West Africa* (Harper & Row, 1975). From the perspective of black theology I note the work of James Cone, particularly *the Spirituals and the Blues* (Seabury, 1972). More recent is the work of younger scholars influenced by Cone, some of whom incorporate religious studies within black (liberation) theology: Josiah Young in *Black and African Theologies: Siblings or Distant Cousins?* (Orbis, 1986); James H. Evans, Jr. in *We Have Been Believers: An African-American Systematic Theology* (Fortress, 1992); Dwight Hopkins, *Shoes That Fit Our Feet: Sources for a Constructive Black Theology* (Orbis, 1993); and Dwight Hopkins, George Cummings, and Will Coleman in *Cut Loose Your Stammering Tongue: Black Theology in the Slave Narratives* (Orbis, 1991).

Note also a convergent interest in spirituality from the side of literary studies, such as Marjorie Pryse's introduction to *Conjuring: Black Women, Fiction & Literary Tradition* (Indiana, 1985), James Evan's *Spiritual Empowerment in Afro-American Literature* (Edwin Mellen, 1987), and Michael Awkward's *Inspiriting Influences: Tradition, Revision, & Afro-American Women's Novels* (Columbia, 1989). My own contribution to this new direction in scholarship is evident in the chapter, "The Spirituality of Afro-∠ American Traditions," in the Crossroad Press encyclopedia of *World Spirituality*, Volume 18, *Christian Spirituality III* (1989), and the entry on "African American Spirituality" in the forthcoming *Encyclopedia of African American Culture & History* (Macmillan with Columbia University). In addition I have taught for several years now, at Emory and at the Harvard Divinity School, a graduate seminar on the history and phenomenology of black religion, in which the correlation of conjurational and ecstatic spiritualities with sociocultural developments and political movements is a central feature. Here I acknowledge with gratitude the colleagues, deans, and students at two institutions, North and South, who from 1987 to 1993 provided me with a community of learning and leisure without which all books are greatly impoverished.

In addition to the authors already mentioned, the reader will discover my extensive conversation in this book with Albert J. Raboteau. Raboteau's now classic text, *Slave Religion: The "Invisible Institution" in the Antebellum South* (Oxford, 1978), inspired me in the early 1980s when I was a doctoral student at the Graduate Theological Union (GTU) at Berkeley. At the time he was on faculty at the University of California, and in that capacity served on my doctoral examinations committee. He thus encountered some of the emphases in this book in their earliest, embryonic form, and has since reviewed the manuscript in its more mature development. I acknowledge with gratitude Al Raboteau's sustained attention to my career and scholarship. As usual (and of course), any excesses or defects in this study are all mine. The "ur-text" for this book was my dissertation, however, and for that project I received the providential guidance, gracious mentoring, and rigorous coaching of James Wm. McClendon, Jr., then at the Church Divinity School of the Pacific. With Jim McClendon I made a decisive turn to narrative theology, after a rich immersion in liberation theology with Robert McAfee Brown and in Tillich studies with Durwood Foster, both faculty at the Pacific School of Religion. The discerning reader will recognize this mixed patrimony at decisive moments in the following study. Among the benefits of studying at the GTU was the encouragement to practice methodological diversity in scholarship, in tandem with experiencing the cultural diversity of the Bay area. My enduring gratitude goes to Jim McClendon for throwing me a theological lifeline in the midst of that pluralist milieu; for mentoring me with Christian fraternity, and showing me his teacher's heart.

Other persons have enabled this project as well. Erica Sherover-Marcuse (1938–1988) was, throughout my dissertation phase in Berkeley, a self-avowed Jewish Marxist atheist interlocutor for my reflections. (We successfully persuaded each other that every black Christian theologian should have such an ally.) Ricky also initiated me into an emancipatory praxis that seems increasingly able, in concert with diverse religious and humanist efforts toward a "beloved community" (King), to achieve the nonviolent global communities projected in the final chapters of this book. In that connection see the section of her own dissertation, "A Practice of Subjectivity," that concludes her book, *Emancipation and Consciousness* (Blackwell, 1986). A more recent reader of my scholarship, however, persuaded me to dare represent black America's emancipatory projects in terms of the concept that was relatively understated in the philosophical and theological orientation of my dissertation. With the encouragement of my colleague and chair in the Religion Department at Emory University, Paul Courtright, I have turned the earlier project inside-out so that the theme of *conjuring culture* stands foremost in the reader's view. After months of that labor and now years since my earliest research into black America's conjurational spirituality, I am grateful to Paul for sustained exhortations (and enabling reasons) to write the book my heart wanted to write.

I owe to Paul Courtright also the introduction to my editor at Oxford University Press, Cynthia Read, whose repeatedly documented belief in the cogency of this work fortified me in the face of periodic doubts. She offered me as well the open sense of freedom and opportunity to craft a book that would be commensurate with its richly layered, fascinating, and complex subject matter. I hope that I have met that generous opportunity with some measure of eloquence, insight, and skill. Additional readers of the manuscript in progress have included my department colleagues, especially Eugene Bianchi, David Blumenthal, Martin Buss, John Fenton, and Vernon Robbins. With Paul Courtright's facilitation we met over the course of several months during my semester on leave and discussed chapter installments; such concerted effort was crucial for my timely progress and productivity. In addition at least one person besides myself has read and commented on both the dissertation and the manuscript (John Fenton), while other colleagues have reviewed chapters and consistently informed me of supplemental materials and critical perspectives. David Brown in particular, my colleague and friend in the Art History Department at Emory, gave me extensive comments on several chapters and vastly improved the ethnographic precision and coherence of the book. I wish I could have followed his suggestions more thoroughly. At any rate, all friends of conjure studies owe him a measure of gratitude for the critical service he has rendered this project. (And, as historian David Wills at Amherst has informed me in another context, every theologian needs a historian nearby to keep the theologian honest.)

The institutional support of Emory College and its library was crucial during my research and writing periods. A research grant from the Dean of the College, and the College's Summer Faculty Development Awards, provided me the necessary time for successful productivity. Emory University's Woodruff Library staff were unfailingly helpful; in particular Luella Davis helped me acquire a copy of the "race history" by my name-sharer, Theophilus Gould Steward, called *The End of the World* (1888). I am also grateful for the continuing, though unrewarded, efforts over several years of Oscar Burdick, librarian at my alma mater, the Graduate Theological Union in Berkeley, to locate a now notoriously missing nineteenth century text: David Coker's "Sermon Delivered Extempore in the African Bethel Church in the City of Baltimore, on the 21st of January, 1816."[6]

Family members conclude the circle of support and incentive I have enjoyed during this process. My mother I have already acknowledged as the "local genius" of this book, whose womanist wisdom, ritual leadership, and traditional spirituality presented me from childhood with some of the ecstatic and conjurational data this study explores. In a certain sense—Jim McClendon's sense of (auto)biography as theology—this entire project > represents my effort first to explain to my own satisfaction, and then to vindicate to others, my people's folk spirituality as mediated to me by my mother. Here I also want to remind my brother and sister-in-law, G. Winston and Elzadie Smith, that they asked me when I was a novice seminarian or graduate student to explain to them "what the Bible is all about." I promised to respond at the time, but did not or (most likely) could not. Now—some number of years, and diplomas, and research hours later—I think I have an answer. This book is my considered response, and I hope it is an accessible, illuminating one.

For his part my father, Attorney W. Harold Smith, has coaxed, exhorted, and waited to see this projected completed too. For that incentive I am also grateful, and hope that I can rest "on my laurels" for awhile now before rising to the challenge of his next parental expectation. As noted the book is dedicated to the memory of my daughter, Ereina Christin Smith (1978–1987). Her first names in Greek mean "peace of Christ," and therefore provide a fitting emblem for the transformative and conciliatory themes of this book. Moreover I promised her, during the dissertation phase of this project, that I would compensate her for all the attention she was missing from me by dedicating the book to her. Ereina's ecstatic delight at having her name in my book stands as one of my glowing memories of her. Speaking of missing attention, I now look forward to restoring to my wife, Jane Najmuddin, and our nephew, Sheryar Singha, all those portions of our house where my papers and books have proliferated during the final stages of this project. Jane's unfailing encouragement as a fellow (sister) teacher, and her gracious allowances for the rigors of this project, have calmed my soul during many weary hours of labor. Thanks be to God for her collegiality and faithful companionship!

Ascension Day   T.H.S.
May 20, 1993
Atlanta

Notes

1. Lorene Cary, *Black ice* (New York: Vintage Books/Random House), p. 24.
2. Ibid., p. 6.
3. Not only the abysmal, but also the halcyon aspects of student experience at the Phillips Exeter Academy in Exeter, New Hampshire, are represented in the Academy's remarkably candid and caring collection of alumni reminiscences, *Transitions: Exeter Remembered, 1971–1987* (Exeter, N.H.: Phillips Exeter Academy, 1990). See my poetic as well as analytic reflections in: Thee Smith, "Versions of Exeter in Rilke's Verse," pp. 49–63.

4. Robert Stepto's terminology here has several sources, as he acknowledges: "Prior to [Du Bois's *The Souls of Black Folk* (1903)], the seminal journey in Afro-American narrative literature is unquestionably the journey north. In many narratives this perilous assertion of mobility after the assault of bondage assumes the properties of an ascension, especially when the journey is a quest for literacy as well as for freedom." Robert B. Stepto, *From Behind the Veil: A Study of Afro-American Narrative* (Urbana & Chicago: University of Illinois Press, 1979), p. 67.

In addition, Stepto explains, "ritual grounds are those specifically Afro-American spatial configurations within the structural topography that are, in varying ways, elaborate responses to social structure *in this world* . . . The slave quarters are . . . the prototypical ritual ground . . . [Other] Afro-American ritual grounds are quite frequently . . . [topoi of] 'double life, with double thoughts, double duties, and double social classes' giving rise to 'double words and double ideals' that characterize what Du Bois describes to be the Negro's burden of 'double consciousness' " (pp. 68–69). Finally: "In Afro-American narrative literature, one finds as well depictions of spirit of place or *genius loci*, in which, as Geoffrey Hartman has remarked, the 'local integrities' or 'imagery of the tribes are given bounding outline' "(p. 70).

5. See the brief but insightful exposition of this verse in Lerone Bennett, Jr., "Ethos," in *The Negro Mood and Other Essays* (Chicago: Johnson Publishing Co., 1964), p. 50. See also a moving elaboration of the historical assault on Afro-American spirituality in Euro-American contexts by Dona Richards, "The Implications of African-American Spirituality," in *African Culture: The Rhythms of Unity*, edited by Molefi Asante and Kariamu Asante (Westport, Conn.: Greenwood Press, 1985), pp. 207–31: "The African-American . . . represents the embodiment of the confrontation of two divergent world views: a spiritual ethos inheriting a sacred, cosmic world view forced to adjust to a materialistic society in inhuman circumstances. What happens when the spiritual ethos of peoples of African origin is entrapped and exploited by an oppressive, materialistic society whose culture seeks to dominate them? . . . We live in the chaos of North America . . . Yet, we have not become thoroughly European . . . *We are a spiritual people, living in a profane society*" (pp. 214, 228).

6. The only currently available version of this missing document is the two page excerpt in Herbert Aptheker, ed., *A Documentary History of the Negro People of the United States*, vol. 1 (New York: The Citadel Press, 1951), pp. 67–69. The figural religious discourse evident in the extant fragment of this text, which clearly correlates the Babylonian captivity of the Jews to the black church experience of captivity by white American ecclesiastical authorities, provides a signal instance of the conjurational employment of biblical figures examined in the following pages. Coker's use of this figure, of course, echoes the sixteenth-century reformer Martin Luther's attack on the ecclesiastical hegemony of the Roman Catholic church in his classic work, *The Babylonian Captivity of the Church* (1520).

Contents

CONJURING CULTURE

Introduction:
Formulary

Out of Africa, always something new.
Pliny the Elder (23–79 C.E.)

Out of Afro-America too, "always something new." The following pages disclose something new out of Afro-America. Here the sacred text of Western culture, the Bible, comes to view as a magical formulary for African Americans; a book of ritual prescriptions for reenvisioning and, therein, transforming history and culture. In this way the veracity of Pliny's ancient proverb is confirmed. The claim 'something new out of Africa' is confirmed by a heretofore unacknowledged combination: the combining of (1) biblical interpretation, with (2) magical transformation, as practiced by the descendents of Africans in North America. Folk practices like conjure, of course, are as old as magic everywhere. What is innovative is a remarkably efficacious use of biblical figures, with a historically transformative and therapeutic intent, in the social imagination and political performances of black North Americans. The term "conjuring culture" expresses that magical intent as explicitly as possible. Once one knows what to look for, indeed, a conjurational employment of biblical figures becomes increasingly transparent—even compelling. Accordingly, throughout the following study I felt compelled not merely to explicate conjurational performances, but also to display them by mirroring, in the very style of presentation, the power and efficacy of conjuring-with-Scripture.

To display that power and efficacy calls for art as well as argument. In this regard *Conjuring Culture* argues for African spirituality as a potent source of black North American religious experience. Immediately however the data require that we also acknowledge the interplay of African-derived practices with Euro-Christian sources and traditions. While elaborating these issues the book nonetheless exceeds argumentation by way of representation: that is, through its own mimetic use of figural forms. It is evident in the Contents, for example, that the chapter titles and contents are framed by biblical titles and figures in a manner that mirrors the coded discourse, figural transformations, and conjurational performances of black culture itself. In this regard the book attempts its own performative role as well as conducting descriptive, analytical, and critical tasks in the field of religious studies and theological reflection. With this twofold intention—the combined performative-analytic presentation of a people's composite spirituality—the author seeks a more integral relationship to that spirituality. To simplify: I offer here not only analysis and critique of an Afro-Christian tradition. I also seek to extend the ongoing vitality of this conjurational tradition into future formations of history and culture. From that perspective *Conjuring Culture* is itself a conjure book. It is a conjure book that (first) critically examines the role of the Bible as a conjure book, in order (second) to advance—even correct and refine—its prescriptive efficacy for future conjurational performances.

As will become evident, this study employs the term "conjure" or "conjuration" for its versatility as a "root metaphor" (Victor Turner); a metaphor that circumscribes black people's ritual, figural, and therapeutic transformations of culture. But conjure phenomena are treated here not only as literary or cultural metaphor.[1] Conjure is fundamentally magic. It is first in consideration the magical folk tradition of black North Americans. Its practitioners have traditionally performed their craft in order to heal or harm others by the operation and invocation of extraordinary powers and processes. More concisely and comprehensively stated: conjure is a magical means of transforming reality. Here the term "magic" is best understood as one system, among humanity's more primal cognitive systems, for mapping and managing the world in the form of signs.[2] Rather than consider such phenomena merely irrational or marginal, to be relegated to the unintelligible realm of the supernatural or to the unorthodox realm of the occult, I follow contemporary scholars who regard magic as a primordial and enduring system of communication—as a form of "language."[3] But it is obviously not simply speech or expression. Rather, magic is ritual speech and action intended to perform what it expresses. As the historian of religion Lawrence Sullivan has observed, practitioners of magic "craft concrete signs that embody the inner power of will." Such signs can be correlated and can thereby create a "system of correspondences" (as in the philosophical-magical tradition of alchemy, for example). Such systems coincidentally are a principal feature of contemporary sciences like chemistry where, as Sullivan also notes, we moderns too construct a network of signs that enables us first to apprehend and then to transform reality.

> Magic reminds us that the world consists of signs, that every obvious reality is only a sign pointing to some more hidden one.... Magical diviners have

organized the outward signs that disclose the inner structures of all existence into various systems of correspondences, much in the same way that chemists have mapped the variables that organize the relations among the elementary structures of matter into the periodic table of elements.... [Thus magic] can make present and subject to human control the powers hidden within the dynamic structures that create arbitrary order in culture and cosmos.[4]

However, when we turn to the black American conjuror we find not only a magician but also as a kind of doctor. Such conjurors also claim the power to heal by means of their prescriptions—their pharmacopeia of herbs and roots, human artifacts, and similar materia medica. In the following pages I propose that this unconventional representation of conjure be extended to include other cultural performances that involve curative transformations of reality by means of mimetic operations and processes. To be precise: I do not employ the notion of conjure only as a metaphor for such transformations. In addition to demonstrating the fitness of the metaphor, I also analyze the internal principles of conjure as a folk practice and then show how these principles are operative as well in larger cultural contexts—predominantly in Afro-American (and to some degree in Euro-American) social history.

The word "conjure" comes from Middle English and Old French forms of the Latin *conjurare*: to swear together (as in "jury"). Whereas the notion of "conspiring" is now obsolete, conjure can still be used in the sense of "abjure"—to entreat solemnly or constrain someone by oath. More familiar and pertinent here are the three related meanings: (1) to invoke or summon (up) a spirit, as in sorcery; (2) to effect by the use of "magical" arts; and finally (3) to summon up an image or an idea as an act of imagination. At the same time that we note the familiarity of these connotations we should acknowledge their tendency to efface the more varied African American folk meaning of the term. Notably omitted from conventional references is conjure as the use of natural and artificial materials for medicinal and quasi-medicinal purposes—that is, conjure as a *pharmacopeic tradition* of practices. This African American meaning, conjure as folk pharmacy, appears prominently in the traditional designation of the expert practitioner as a "conjure doctor." This prominence sharply differentiates black American usage from popular use of the word to mean only imagination on the one hand, or only sorcery, witchcraft, or occultism on the other. It also indicates a point of differentiation between African American and English or European conjure traditions. In this regard, one commentator warns against a 'Eurocentric' assimilation of conjure to mean only witchcraft. "The role and function that conjurers served in the efforts of enslaved Africans to build a culture ... cannot be understood in terms of an ethnocentric equation between conjuration and witchcraft."[5]

On the contrary, a pharmacopeic emphasis in conjure studies is capable of linking magical and *supernatural* elements, on the one hand, with medicinal practices and *natural* processes on the other.[6] On the other hand, an emphasis on the healing practices of conjure need not (and must not) exclude other referents. For example, I will argue further in the following chapters that reflection on conjure in its full ethnographic and phenomenal reality requires that we hold in concert both its therapeutic or benign referents and its occult and malign attributes. To repeat: all the conjurational elements are readily observed in black American experience, including sorcery or malign occultism, the most prosaic deception and manipulation, as well as herbal healing and other pharmacopeic practices. The claim here is a matter of emphasis. It is an emphasis that leads to a reconstruction of the word "conjure" in its North American context to include the prominence of pharmacopeic operations and intentions. The new emphasis balances conventional references to occultism and witchcraft with a comparable emphasis on curative and pharmacopeic referents.[7]

In addition, this study proposes a view of conjure performance that encompasses social-historical transformations as well as folklore practices. I refer to such transformations as instances of *conjuring culture*, specifically where I find (1) ritually patterned behaviors and performative uses of language and symbols (2) conveying a pharmacopeic or healing/harming intent and (3) employing biblical figures and issuing in biblical configurations of cultural experience. This third, biblical and figural criterion for identifying such cultural transformations also has a basis in folklore practices. Perhaps the most explicit reference for this basis is found in the folklorist Zora Neale Hurston's catalogue of the "paraphernalia of conjure," in which the final, culminating entry in her list is the Bible. In addition, the folklore collector Harry Hyatt has corroborated the use of the Bible as a talisman or amulet, on the one hand, and as a source of conjurational spells, incantations, or prayers on the other.[8] I will examine the field work of each of these researchers in more detail later, but already the indications suggest a folk use of the Bible as a 'conjure book': as a kind of magical formulary for prescribing cures and curses, and for invoking extraordinary powers in order to reenvision, revise, and transform the conditions of human existence.

The chapter titles of this book correspond almost completely to the titles of biblical books. That schema is informed both by the black preaching tradition and by a tradition of biblical interpretation called figuralism or typology. The rationale for the particular ordering of my chapter headings can be derived from the biblical hermeneutic (interpretive tradition) of typology presented by Northrop Frye in *The Great Code* (1981). Frye's own title is taken from the literary theory of the poet William Blake, in which the "Old and New Testaments" are presented as "the great code of art." Frye extends Blake's claim by his treatment of the Bible as a source of representation and of cultural meaning in Western civilization and history. In Frye's presentation, as one commentator notes, the Bible as a whole functions as "the code, or perhaps key to the code, of Western culture . . . as a kind of microcosm within which the imaginative experience of all of Western civilization is contained."[9] As his "key to the code," Frye himself refers to biblical figuralism or typology. "The typological way of reading the Bible is indicated too often and explicitly in the New Testament itself for us to be in any doubt that this is the 'right' way of reading it."[10]

Most crucial for the purposes of this study are the typological codes that predominate in Afro- and Euro-American cultural history. Frye himself observes seven major codes, or "phases," operating within the Bible and constituting a typological system as in the following schema:

Old Testament

Genesis (creation)   Wisdom
Exodus ("revolution")   Prophecy
Law

New Testament

Gospel   Apocalypse

It is evident, Frye observes, that the first five of these phases have their "center of gravity" in the Old Testament and latter two in the New. Moreover, typological conventions or relationships operate between the figures, so that "each phase is not an improvement on its predecessor but a wider perspective on it." Thus the sequence or order of figures is itself another aspect of biblical typology, in which each phase is a "type" of the one following it and an "antitype" of the one preceding it.[11] The type-antitype relationship of these codes is evident if one considers, for example, Genesis as a (proto)type of Exodus, and Exodus as an (anti)type which recapitulates or fulfills Genesis. The most elegant statement of this relationship is the one formulated by Erich Auerbach: the first type signifies the second, while the second comprehends or fulfills the first.[12] To be explicit: in the Genesis-Exodus dyad the creation of a world ex nihilo (Latin: out of nothing) prefigures the Exodus creation of a people, Israel, out of Egyptian bondage. Conversely, the Exodus story of the formation of a people from a band of former slaves and refugees completes the patriarchal history and destiny of Abraham's descendents promised by God in Genesis.

A more recent treatment of the typological relationship between Genesis and Exodus has been provided by the literary historian Werner Sollors. Sollors's succinct expression "typological ethnogenesis" offers a useful neologism in the field of literary criticism.[13] Typological ethnogenesis is the formation of peoplehood through the hermeneutic of biblical typology. With this term Sollors refers to the premier instance of figuralism that has influenced and shaped both Afro- and Euro-American history and culture. Each people or *ethnos*, black and white, has envisioned and revised its existence in terms of characters and events found in the Exodus story. By means of figural participation in biblical narrative each has engendered or regenerated its corporate identity as an antitype of ancient Israel. Such identity formation in North America has focused most clearly on the

Exodus themes of liberation and promised land in biblical narrative. Both black and white Americans identify their New World existence with ancient Israel's transition from bondage and oppression to freedom and a promised destiny. In Chapter 2, I explore in more detail this figural relation between (ethno)genesis and Exodus recapitulations in North American social history.

In looking beyond the two initial figures of Genesis and Exodus, it may be tempting to reduce the number of correlations between biblical narrative and African American social history. Are not five of Frye's figural phases sufficient for understanding major formations in black social history, instead of his full schema of seven? This conjecture seems plausible when comparing the density and duration of biblical and Western history, on the one hand, to the less extended history of black America on the other. On the contrary, I deem it useful and necessary to add two more figures to Frye's seven, thus making a total of nine figural elements for comprehending black experience in biblical terms. To Frye's seven phases I add "spirituals" and "praxis," the former corresponding to the Book of Psalms in the Hebrew Scriptures and the latter to the Book of Acts (Greek: *praxis*) in the Christian Scriptures. The prominence of music generally in black culture, and the significance of slave songs or spirituals in particular, justify adding the category of "spirituals" to my typological schema. The importance of radical activism, and of critical social theory in understanding and advancing black political struggle, justifies the addition of "praxis."

In the following I present my complete schema, which is designed to display most adequately both the figural and the conjurational dimensions of black cultural experience. Observe that six of these figures are contextually located in the Hebrew Scriptures and three in the Christian Scriptures. It will also be evident that the schema corresponds directly to the book's Contents. In this regard the rationale for the ordering of chapter headings is internally consistent with the subject matter of the study.

Hebrew Scriptures

Genesis (cosmogony)	Spirituals (psalms; aesthetics)
Exodus (conjuring-God-for-freedom)	Wisdom (proverbs, world view)
Law (curing violence I)	Prophecy (oracles; vocation)

Christian Scriptures

Gospel (curing violence II)	Apocalypse (judgment;
Praxis (acts/activism)	revelation)

Three further observations may be useful with regard to this schema. First of all it preserves the typological conventions that Frye has indicated, with his observation that each figure is a (proto)type of the one following it and an antitype of the one preceding it. In Chapter 3, for example, I show how figural and conjurational aspects of the role of law in black social history constitute a type of the Exodus phenomenon examined in Chapter 2. Typically, black Americans have employed legal processes and strategies as instruments for achieving, first, manumission or emancipation, and then civil rights. That is, law has been transformed (conjured) from its conventional role as an instrument for legitimating and enforcing bondage, to a countercultural role as an 'Exodus' device for setting captives free. In this regard the emancipatory signification of law has, sometimes, been successfully realized or fulfilled in the Exodus configuration in black experience. Law (treated in Chapter 3) has served to secure access to a "Promised Land" of freedom and citizenship (the focus of Chapter 2).

Similar resonances operate between Chapter 6 and Chapter 7, to take another example. In Chapter 7 on gospel formations black America's prophetic quest, for deliverance from racist oppression and violence, finds a figural resolution of issues raised in the preceding discussion of prophecy. In this connection the title of Chapter 6 could well be *Prophesy Deliverance!*, the related title of Cornel West's Afro-Christian and revolutionary treatise. The seventh chapter is thus a pivotal one for both the theological and the emancipatory interests pursued in this study. It follows the prophetic configuration in black social history with the gospel configuration and attempts to distill, in pharmacopeic terms, a definitive statement of the Afro-Christian project of curing racism and racist violence. Chapter 7 is succeeded in turn by a related discussion of social activism and praxis.

Here it is instructive to note that the sequence of chapter headings is not a linear, chronological system. Rather, biblical configurations of experience can be partially, proleptically, or retrogressively occurring at any period in American social history. For example, early nineteenth century black slaves sang spirituals that strongly displayed Exodus figuration. At the same time more literate freepersons (Ethiopianists), as we shall see in Chapter 2, were writing with prophetic expectation of a messianic deliverer from their "Babylonian captivity" (Coker). Whereas Exodus is early in biblical Israel's chronological history, the Babylonian exile and other experiences of Diaspora are late occurrences. Nonetheless interpretive performances using both sets of terms may configure the same (prophetic) era in Afro-American history. Conversely, two widely separated historical epochs (for example, the 1860s emancipation of the slaves and the 1960s civil rights movement) may be configured by the same biblical figure (Exodus, of course, in this example). In this regard we must bear in mind a configural feature that the theologian Harold Dean Trulear calls the "flexibility in the Black story." Trulear insists that "conceptualizing the Black story ... does not mean that it is plagued by the linear rigidity and teleology of western philosophy of history."[14] Nonetheless it will be necessary, in the chapter on Apocalypse, to address the eschatological trajectory of all the preceding figures toward their narrative consummation in the closing text of the Christian Scriptures: the Book of the Apocalypse.

Finally, I have reorganized the preceding schema into three parts in order to accommodate diverse scholarly and disciplinary concerns. Ethnographic and ethnohistorical considerations predominate in Part I (although such considerations also recur in later chapters). For, the opening chapters on Genesis, Exodus, and Law seem an appropriate place to locate considerations of ethnic (African American) appropriations of biblical narrative, in view of the fundamentally ethnocentric (that is, Hebraic) character of the biblical corpus itself. In Part II, by contrast, I shift to other disciplinary concerns, drawing implicitly upon a phenomenology of conjure construed as an African American spirituality. I also address hermeneutic and aesthetic, cognitive and performative issues. More conducive to these disciplinary interests is my reformulation of the biblical series of Psalms-Proverbs-Prophets, to read instead, Spirituals-Wisdom-Prophecy. To conclude this overview, I have relegated the most explicitly theological perspectives to the end of study in Part III. The New Testament sequence of Gospel-Acts-Revelation (minus the Epistles) appears here as Gospel-Praxis-Apocalypse—where Gospel and Apocalypse are irreducibly theological in focus and intent.

It is now evident that this book bears another mimetic feature in addition to its organization by biblical themes and titles: that is, its multicultural discourses, methods, and approaches. Here I intend to mirror the multicultural character of African American culture itself, which is not monocultural in terms either of Afro- or Euro-American referents. A useful analogy is provided here by the practice of weaving, specifically the operations of 'warp and woof.' In the following study a 'warp' of conjurational performances comprising biblical figures and social historical configurations is threaded against a 'woof' of disciplinary approaches and methods drawn from the human sciences or cultural studies. The disciplines involved include the history of religions and social history, literary criticism and critical theory, ethnography and ethnomusicology, and phenomenology and biblical hermeneutics.[15] (Coincidentally, this concatenation of interests involves a complex interdisciplinary task for which there are few if any precedents in those disciplines. Whereas European or white American scholars like Auerbach, Frye, Sollors, and Sacvan Bercovitch have treated the biblical hermeneutic of figuralism or typology, none of them was concerned to correlate that hermeneutic with folk magical traditions or spiritualities like conjure. From another direction, black scholars and others working in the field of black studies have not treated conjure in a way that allows its engagement with biblical hermeneutics to be displayed to clear view.) By default, such a convergence of culture-specific data and Western academic disciplines is peculiar to the present study.

I do not, in this respect, regard my task here as unique however. Rather, I consider that I am extending into new domains a tradition of black American letters and scholarship that includes such well-known writers as W.E.B. Du Bois, James Weldon Johnson, and Alain Locke. In our own time this tradition is exemplified by scholars as diverse as St. Clair

Drake in history and sociology, Charles Long in the history of religions, Henry Louis Gates, Jr., in literary theory, and Cornel West in critical theory and philosophy of religion and culture. For writers in this long-standing tradition, the engagement between black studies materials and Euro-American scholarship and disciplines is axiomatic. They all presuppose that the goals of black intellectual emancipation are best advanced by diverse forms of bicultural parity and engagement between African American culture and the European American civilization within which it is socially and historically located. Such engagement, indeed, is one that black people have long deserved and yearned for—from our slave ancestors, legally prohibited from learning to read, to their most literate and cosmopolitan descendents. For Johnson and Locke a bicultural proficiency or 'multiculturalism' was an express ideal of the Harlem Renaissance: to combine both Afro- and Euro-American practices and materials, perspectives and traditions (not only in music, but also in the arts and letters generally). In the following correlations, for example, between black religious experience and the theory of violence developed by the French American scholar, René Girard (see Chapters 7 and 8), I am schooled by Du Bois's incantatory declamation in *The Souls of Black Folk* (1903):

> I sit with Shakespeare and he winces not. Across the color line I move arm in arm with Balzac and Dumas . . . I summon Aristotle and Aurelius and what soul I will, and they come all graciously with no scorn nor condescension. So, wed with Truth, I dwell above the Veil. Is this the life you grudge us, O knightly America? Is this the life you long to change into the dull red hideousness of Georgia? Are you so afraid lest peering from this high Pisgah, between Philistine and Amalekite, we sight the Promised Land?[16]

Du Bois's declaration may be read as that of a conjurational 'summoner' and biblical figuralist in his own right. It is emblematic of scholarship that still has in sight the black intellectual tradition's yearned-for "Promised Land." In this study, to update Du Bois's references, so-called Eurocentric scholarship does not condescend to illuminate Afrocentric contents—nor vice versa. Rather, "wed with Truth," both world-communities of scholarship and experience are positioned here in relation to one another. With Du Bois I aim to "dwell above the Veil" of alienation that continues to deny us the promise of illuminating each other's worlds from our differing perspectives. That analogy leads to a final introductory comment.

By implication this book presents conjure not only as a specific phenomenon in the field of African American religious studies, but more broadly as a new conceptual paradigm for understanding Western religious and cultural phenomena generally.[17] In this regard the fact that the project is interdisciplinary means that my excursions into disciplines outside my own fields of religious and theological studies will require, inevitably, corrections and reformulations. Moreover the merits, limitations, and implications of such an interdisciplinary venture are not (and, in the nature of things, cannot yet be) fully apparent to me. For these reasons I can not yet justify thoroughgoing, systematic formulations based on 'conjurational modes of perception,' for example; or a fully programmatic, black theological agenda based on a 'conjurational spirituality'. I certainly think that the future of black religious studies and black theology as academic disciplines will be significantly advanced, if not redirected, by the interdisciplinary approach pursued here. However, what this project in its current state of development requires first of all, before considerations of systematic development or theological extensions, are sympathetic-critical readers and collaborators. Here I call upon interested colleagues and students to assist in clarifying and exploring the category of 'conjuring' as a heuristic concept in black studies specifically, and in Western cultural studies generally. I propose the concept as a new paradigm in religious studies, and as an indigenous spirituality with implications for revitalizing the next generation of black studies and black theology. On the other hand, I am more profoundly seeking to participate with others in the multiple, "omni-American"[18] traditions of faith and empowerment that this study is designed to illumine and celebrate.

Notes

1. This ambivalence or, rather, multivalence between conjure as literary metaphor and conjure as folk practice is evident as well in *Conjuring: Black Women, Fiction and Literary Tradition*, ed. Marjorie Pryse and Hortense Spillers (Bloomington: Indiana University Press, 1985). See especially the introductory essay by Marjorie Pryse, "Zora Neale Hurston, Alice Walker, and the 'Ancient Power' of Black Women," pp. 1–24. I am indebted to my colleague in the Religion Department at Emory University, Vernon Robbins, for calling this cognate text to my attention.

2. In this regard, my efforts converge with those of "historians of religion and other scholars" whom Gayraud Wilmore describes as having engaged, in recent years, in "uncovering a form of belief and practice in the religious institutions of Afro-Americans that may be categorized as 'black religion' . . . [the knowledge of which] may help scholars understand something about the most elemental religious consciousness which humans have expressed." Gayraud S. Wilmore, *Black Religion and Black Radicalism* (Maryknoll, N.Y.: Orbis Books, 1983), p. 2.

3. "The tendency now is to understand magic in terms of communication, as a kind of 'language'." Walter Burkert in Robert Hamerton-Kelly, ed., *Violent Origins: Ritual Killing and Cultural Formation* (Stanford: Stanford University Press, 1987), p. 161.

4. Lawrence Sullivan, ed., *Hidden Truths: Magic, Alchemy, and the Occult* (New York: Macmillan Publishing Co., 1987/1989), pp. x–xi. As Sullivan makes clear, magic allocates to the objects, beings, and processes that arise in the most fundamental contexts of human experience (in nature, in society, in the soul or psyche) a corresponding system of signs. It then employs those signs as instruments for affecting the corresponding items as they are experienced in the world. Thus another commentator, David Pocock, observes that "what we call 'magic' . . . is only one of, and of the same order as, many symbolic actions which overcome the discrepancies of thought."

However, despite the fact that magic performs a cognitive function similar to that performed by those other human activities, it is nonetheless "perpetually opposed to 'religion' and 'science' in our literature." David Pocock in Marcel Mauss, *A General Theory of Magic* (Boston: Routledge and Kegan Paul, 1972), pp. 2, 4. For a brief discussion by Pocock of the modern "dissolution of 'magic' as a category" see pp. 1–2 of his foreword to Mauss's text. Cf. Amanda Porterfield's expression "the post-modern picture of reality as cross-referencing network of signs" in her essay "Shamanism: A Psychosocial Definition," *The Journal of the American Academy of Religion* 55:4 (Winter 1987): 736. Also cf. Morton Marks's suggestion that the gods or the *orisha* of the Yoruba people of West Africa represent chemical or "pharmacological categories," in his "Exploring *El Monte*; Ethnobotany and the Afro-Cuban Science of the Concrete," in Isabel Castellanos and Josefina Inclán, eds., *En Torno a Lydia Cabrera* (Miami: Ediciones Universal, 1987), pp. 229, 231. I am grateful to my colleague David Brown, in the Art History Department of Emory University, for informing me about this intriguing claim.

Finally, what distinguishes magic from science as a cognitive and transformative system is its heightened, intensive reliance on ritual performance and mimetic (imitative) efficacy. Thus in the African American magical tradition of conjure, practitioners effect transformations of reality by performing imitative operations upon natural and artificial substances. Those substances include herbs and nail clippings, potions and clothing, grave dirt, and so forth. Mimetic operations also attend the use of biblical themes and figures in black America's conjurational tradition, as we see below.

5. John W. Roberts, *From Trickster to Badman: The Black Folk Hero in Slavery and Freedom* (Philadelphia: University of Pennsylvania Press, 1989), p. 70. Although Roberts's referent for his term "ethnocentric" is ambiguous, I take him to mean 'Eurocentric,' rather than refering to African American ethnicity (a more obfuscating construal).

While the specific term, "conjure," may be conventionally limited to sorcery or witchcraft, English magical traditions also exhibit a broader spectrum of practices that similarly includes healing and divination. See the highly regarded historical study of magical and conjuring practices in Keith Thomas, *Religion and the Decline of Magic: Studies in Popular Beliefs in Sixteenth and Seventeenth Century England* (New York: Charles Scribner's Sons, 1971). In that context the terms "cunning men" and "wise women" connote a wider performance role than our conventional designations for a conjuror. "Healing was only one of the magic functions performed by the cunning men and wise women. Of the others the most common seems to have been the detection of theft and the recovery of stolen goods . . . employing one of several possible methods of divination" (pp. 212–13).

6. In contrast, there is evidence that some sectors of African American folk tradition sharply distinguished between 'medical' and 'magical' conjure, as this apparently ironic disclaimer from a former slave interviewee attests: "I am a great Christian. . . . I don't believe in conjurers because I have asked God to show me such things—if they existed—and He came to me in person while I was in a trance and he said, 'There ain't no such thing as conjurers.' [But] I believe in root-doctors because, after all, we must depend upon some form of root or weed to cure the sick." *The American Slave: A Composite Autobiography*, vol. 19, *God Struck Me Dead*, edited by George P. Rawick (Westport, Conn.: Greenwood Publishing Co., 1972), p. 76. "Root-doctor" is an alternative name for a conjuror or conjure doctor, but this speaker's conviction is self-contradictory without a tradition of distinction

between "conjurers" (as antithetical to Christianity) and "root-doctors" (as compatible with being "a great Christian").

The existence of such a distinction has been further corroborated by the anthropologist Melville Herskovits and is even correlated with gender distinctions (cited in Chapter 1, n. 65). Nonetheless, in this study I have not attempted to generalize from such data or render the distinction as sharply as these sources. Certainly a consistent and coherent distinction between 'medical' and 'magical' conjure would strengthen my case, but I have learned to be suspicious of such dichotomies when treating African and African American folk materials. In any event, I judge that if I am successful in stating the case for conjurational processes in cultural formation irrespective of the distinction, my hypotheses will be all the more compelling if the medical or pharmacopeic element in such processes is elsewhere shown to be a privileged or preferential factor in Afro-Christian strategies of social transformation.

7. See for example: Faith Mitchell, *Hoodoo Medicine: Sea Islands Herbal Remedies* (N.p.: Reed, Cannon & Johnson Co., 1978).

8. Zora Neale Hurston, *Mules and Men* (New York: Harper and Row, 1990), p. 280. Hyatt's corroborations are worth quoting in some detail. Note the magical features involved in reading scripture—either repetitively or backwards.

Under the subheading, Bible and Prayer, an informant from Snow Hill, Maryland (entry 344), reported: "I heard you have to go to a runnin' stream whare the water runs clear an' go thare nine mornin's, an' read a certain chapter somewhare, the Bible, on ever' mornin'. An' when you go the ninth mornin' you meet the devil thare on the ninth mornin' an' he'll give you a test. An' if you kin stand the test after you git thare that he puts on you, says you kin do anything you want to an' he'll defend yo'. You kin go out an' kill a person almost an' he'll clare you." Another informant, from Brunswick, Georgia (entry 345), reported that "Dey tell me dat chew take a Bible an' go to de crossroads on a Sunday night at twelve a'clock an' yo' read, but chew have tuh read it backwards—an' when yo' read one verse, dey say yo' hear te voice of de next verse somewhere in de woods; an' when yo' finish it up, dat chew have *sold yoreself to de devil*. Dat's all ah know 'bout dat." Harry Middleton Hyatt, *Hoodoo, Conjuration, Witchcraft, Rootwork*, Volume One (Hannibal, Missouri: Western Publishing, Inc., 1970), p. 103 (cf. pp. 123–24).

On the combined use of a medicinal concoction (ammonia and soda) and reading a Psalm in order to effect a cure see entry 965 under the subheadings, Folk Medicine: Principles of Healing—Witchcraft and Bible (p. 362ff.). A defensive or counter-conjurational use of the Bible was reported by an informant from Memphis, Tennessee: "Whenevah ah'm afraid of someone doin' me harm ah read the 37 Psalms [Psalm 37] an' co'se *ah leaves the Bible open with the head of it turned to the east as many as three days*. An' den sometimes ah read it an' ah have read it an' *left it open nine days*—each mawnin' *until aftah twelve* at eatin' an' [then] ah closed it. An' ah found that it re'lly did help me" (p. 417). This feature of nine days recurs in a conjure formula for regaining an estranged friend (entry 2895; Little Rock, Arkansas informant): "Well, yo' take [some] candles an' yo' take the Bible an' yo' read it an' yo' prays, an' *pray oyah that candle while it's burnin'*, see, an' yo'll do that ah think fo' 'bout nine days an' nights, an' yo'll gain friendship wit de party that mistreats you" (p. 834).

In conclusion, Hyatt's Introduction highlights in particular the magic recitation of the Book of Psalms (and he was delighted to discover a similar use of the Book of Common Prayer). But he also admits that "most material of this sort

I never recorded" (p. TX). His theological (and Eurocentric?) sensibilities seem to have been offended in this regard, if I correctly infer from this reference to his divinity school education (circa 1916) in the 'higher criticism' of the Bible: "[M]y one great gift from Chicago did originate in the seminary—a civilized view of the Bible and religion. . . . From [Mercer, Easton, Frazer, and Schweitzer] I learned that the Bible is not a book of magic, or a crystal ball revealing present and future events . . . but a collection of texts to be studied by scientific methods and judged by the time in which each was written. This knowledge was the greatest preparation I carried into the folklore field" (p. XVII). This knowledge evidently conflicted with the view of the Bible as magical exhibited by Hyatt's informants, and presumably accounts for his censoring many more instances of Bible conjuration than he recorded.

9. Lynn Poland, "The Secret Gospel of Northrop Frye," review of *the Great Code: The Bible and Literature*, by Northrop Frye in *The Journal of Religion* 64:4 (October 1984): 513. A more comprehensive study of representation in Western culture, including biblical with classical texts and converging with the related interests of this study, is Erich Auerbach's *Mimesis: The Representation of Reality in Western Literature* (Princeton: Princeton University Press, 1953).

10. Immediately Frye explicates the term "right" with the notion of intention: " 'right' in the only sense that criticism can recognize, as the way that conforms to the intentionality of the book itself and to the conventions it assumes and requires." He then continues in the declarative mode with wry effect: "Naturally, being the indicated and obvious way of reading the Bible, and scholars being what they are, typology is a neglected subject, even in theology, and it is neglected elsewhere because it is assumed to be bound up with a doctrinaire adherence to Christianity." Northrop Frye, *The Great Code: The Bible and Literature* (New York: Harcourt Brace Jovanovich, 1981/1982), pp. 79–80.

11. Ibid., p. 106.

12. Erich Auerbach, "Figura," in *Scenes from the Drama of European Literature*, ed. Wald Godzich and Jochen Schulte-Sasse (Minneapolis: University of Minnesota Press, 1984), p. 58.

13. Werner Sollors, *Beyond Ethnicity: Consent and Descent in American Culture* (New York: Oxford University Press, 1986), pp. 50, 57.

14. Harold Dean Trulear, "The Lord Will Make a Way Somehow: Black Worship and the Afro-American Story," *The Journal of the Interdenominational Theological Center* XIII:1 (1985): 101.

15. I am indebted to the feminist theologian Sharon Welch for this useful analogy.

16. W.E.B. Du Bois, *The Souls of Black Folks,* in *W.E.B. Du Bois: Writings*, The Library of America (Cambridge, England: Press Syndicate of the University of Cambridge, 1986), p. 428. The passage is found in the final paragraph of chapter 6, "Of the Training of Black Men."

17. I am especially grateful to the Indologist Paul Courtright for reiterating this possibility, and for encouraging me to invite the critical review of interested and informed readers by stating the case as explicitly as possible at this early stage of a new venture in religious studies.

18. See Albert Murray, *The Omni-Americans: New Perspectives on Black Experience and American Culture* (New York: Outerbridge & Dienstfrey, 1970).

I

ETHNOGRAPHIC PERSPECTIVES

The model of post-modern ethnography is not the newspaper but that original ethnography—the Bible.

Stephen A. Tyler, "Post-Modern Ethnography: From Document of the Occult to Occult Document"[1]

In Part I I survey the black religious uses of the Bible that correspond to distinct formations in North American social history. The three chapters on Genesis, Exodus, and Law treat two interrelated phenomena in synchronic rather than diachronic relationship: biblical interpretations by black Americans occur in coincidence with (not necessarily in causal relationship to) biblical formations of social history.[2] The most familiar instance of biblical formation in black social history is the Exodus configuration. This formation features in recurring generations the transfer of people from oppression to freedom under the leadership or inspiration of Moses figures like Abraham Lincoln, Harriet Tubman, and Martin Luther King, Jr.

But numerous other biblical types or figures have also informed ethnic experience, resulting in other social-historical configurations, including Ethiopia and Egypt, Promised Land and Wilderness, Captivity and Diaspora. Furthermore, close examination of these Afro-American configurations shows their interrelationship with Euro-American social formations—formations that are also based on traditions of biblical interpretation. Most notable in relationship to black experience is the Puritan figuration of New England as New Israel or Canaan Land. Central for many ethnic and social groups throughout United States history is the country's religious and nationalist configuration as "promised land"; as a New World of freedom and opportunity over against the Old World of European feudalism and class oppression. A crucial task for the integrity of this project is to show the relationship between African and Puritan

American biblical typologies, and their mutual grounding in an ancient, Christian hermeneutic or interpretive tradition.

But it is insufficient to account for black religious figuration in terms only of the influence of Puritan sources and the Christian appropriation of biblical typology. Indigenous African and extra-Christian elements were also at work in the success and character of that appropriation. In particular I refer to African American folk traditions of conjure or conjuration. The first chapter in Part I introduces conjure as a major folk practice in black religion and introduces black American uses of the Bible as varieties of conjurational performances. Conjure is construed here as a pharmacopeic (or healing-harming) practice and features a performative use of speech and ritually patterned or imitative behaviors. I show how conjurational performances at the level of social history employ biblical figures with a curative intention (as materia medica), and for the purpose of reenvisioning and transforming lived experience and social reality through mimetic or imitative operations. Ethnographic data are presented to support this construal, and in general the perspective throughout Part I is ethnographic and ethnohistorical.

However, I do not intend this work to be read as a project of ethnography or history, but of religious studies. Although the approach here is descriptive at some points and analytic at others, it cannot, by the nature of the subject matter, be convincing only as a type of social-scientific or historical treatise. Rather we are concerned with an intersubjective dimension in religious experience for which a phenomenological attitude is necessary: see whether this display of the phenomena is intuitively cogent and compelling.[3] In addition to historical excavation and folklore studies, therefore, the task is to display the interior aspects of a collective ethnic experience: black America's experience of biblical figuration and conjurational processes operating throughout its social history. In short, the study is a project in religious studies and features some ethnographic and ethnohistorical implications for theological studies as developed in the final chapters.

Notes

1. Stephen A. Tyler, "Post-Modern Ethnography: From Document of the Occult to Occult Document," in *Writing Culture: The Poetics and Politics of Ethnography*, ed. James Clifford and George E. Marcus (Berkeley: University of California Press, 1986), p. 127.

2. That is, I do not require a diachronic theory, or a cause and effect relationship, between African American biblical hermeneutics (interpretive traditions) and social-historical formations. The tradition of Exodus interpretations of black American experience, for example, need not be construed as directly generating exodus patterns in black social history. Causal relations may be an appearance arising from multiple instances of convergence between circumstance, on the one hand, and people's interpretive predisposition or post facto explanations on the

other. It is sufficient to regard such appearances as arising in a synchronic manner, as concurrent or coincident, rather than as diachronic phenomena occurring in linear sequence. More than this requires a theory of "consciousness as causal." See Willis W. Harmon, "The Postmodern Heresy: Consciousness as Causal," in *The Reenchantment of Science: Postmodern Proposals*, ed. David R. Griffin (Albany: State University of New York Press, 1988), pp. 115–28. Or else it requires a theology of providential operations in history. Either of these is beyond the scope of the present study (although either could be brought into fruitful relationship with this study). It is evident, therefore, that biblical patterns may occur and recur in social-historical experience without regard to linear chronology. The case of individual figuration is perhaps more obvious in this regard: persons who perform and display Mosaic roles and characterizations can appear both early and late in Afro-American history, irrespective of the sequence of biblical chronology or the hermeneutic disposition of interpretive communities.

3. "The test of a phenomenological description is that the picture given by it is convincing, that it can be seen by anyone who is willing to look in the same direction, that the description illuminates other related ideas, and that it makes the reality which these ideas are supposed to reflect understandable." Paul Tillich, *Systematic Theology*, vol. 1 (Chicago: University of Chicago Press, 1951), p. 106.

1

Genesis

And God stepped out on space,
And he looked around and said:
I'm lonely—
I'll make me a world.

And far as the eye of God could see
Darkness covered everything,
Blacker than a hundred midnights
Down in a cypress swamp.

James Weldon Johnson,
"The Creation"[1]

These verses announce the Genesis theme in James Weldon Johnson's collection of sermonic poems, *God's Trombones (1927)*. Easily recognizable, as the poem proceeds, is the fact that "The Creation" is Johnson's retelling of Genesis chapters 1–2. It is also, we learn in the preface, the seminal composition in this classic work of black religious expression. Each of the eight chapters in *God's Trombones* is exquisitely crafted and presented in its provenance: black America's folk preaching tradition, with its rich appropriation of biblical narrative and symbolism. The book as a whole, moreover, aims to circumscribe a broad span of biblical themes with selections that extend from "The Creation" to "The Judgement Day." By virtually encompassing the Hebrew and Christian Scriptures within the range of a single work, *God's Trombones* offers a rare, even a unique, literary production. At the same time, black culture is replete with kindred impulses and projects. Many observers have noted this enduring cultural intention: to represent black experiences and perspectives by means of a diverse selection of biblical texts and themes. In this study I use the concept of "conjuring culture" as the vehicle for analyzing some of those representations, and the role they play in the historical projects and cultural strategies of black North Americans. However, *Conjuring Culture* not only provides an analysis of such representations, it also embraces and advances their underlying, conjurational intent.

Johnson himself was a celebrated poet and novelist of the Harlem Renaissance, the 1920s emergence of black arts and letters. In the book's preface he tells us that most of the poems in *God's Trombones* are based on the memory of sermons heard in his childhood. But the "immediate stimulus" for the work in general, and for "The Creation" poem specifically, was a sermon heard by the mature poet. With care Johnson describes a lecture trip to Kansas City during which he spoke at several church meetings. Late one evening he found himself in the unexpected situation of listening to a series of preachers. In that context he heard the particular sermon that inspired the book. The speaker was a "famed visiting preacher" who attempted at first "to preach a formal sermon from a formal text." But the congregation "sat apathetic and dozing"— understandably so, having already endured three short sermons. To his credit the preacher, sensing "that he was losing his audience and his opportunity . . . suddenly . . . closed the Bible, stepped out from behind the pulpit and began to preach. He started intoning the old folk-sermon that begins with the creation of the world and ends with Judgment Day."

The result was "electric," Johnson tells us. He watched the congregation, himself included, grow immediately attentive and spellbound, then responsive to every turn of the preacher's art.[2] I return to that art momentarily. But first, let us examine in detail this multicultural narrative of world-creation—this Afro-Hebraic cosmogony.

In its earliest transmission through oral forms and varients, the sermon may have preceded Johnson's literary rendition by as much as two centuries. That oral tradition reached him first in "boyhood," as he sat listening in church, to

> sermons that were current, sermons that passed with only slight modifications from preacher to preacher and from locality to locality. Such sermons were "The Valley of Dry Bones," which was based on the vision of the prophet in the 37th chapter of Ezekiel; the "Train Sermon," in which both God and the devil were pictured as running trains, one loaded with saints, that pulled up in heaven, and the other with sinners, that dumped its load in hell; the "Heavenly March," which gave in detail the journey of the faithful from earth, on up through the pearly gates to the great white throne.[3]

Somehow through this process of oral transmission the tradition conveyed to the Harlem poet its own energy and intention. Regarding that intention, Zora Neale Hurston has claimed that "even the Bible was made over to suit our vivid imagination."[4] How has the Genesis cosmogony, with Johnson as a poetic medium of folk expression and transformation, been "made over" in this work?

Mimetics

Happily for my purposes Johnson described in some detail the preaching event from which he crafted the text of "The Creation." In particular he discussed the "art" by which the preacher was able to evoke, through dramatic performance, the embodied presence of the God he preached. It was a mimetic (imitative) art, as Johnson's description suggests, for the preacher performed as a theurgic mime.

> He was wonderful in the way he employed his conscious and unconscious art. He strode the pulpit up and down in what was actually a very rhythmic dance, and he brought into play the full gamut of his wonderful voice, a voice—what shall I say?—not of an organ or a trumpet, but rather of a trombone, the instrument possessing above all others the power to express the wide and varied range of emotions encompassed by the human voice—and with greater amplitude. He intoned, he moaned, he pleaded—he blared, < he crashed, he thundered. I sat fascinated; and more, I was, perhaps against my will, deeply moved; the emotional effect upon me was irresistible.

Johnson admitted the compelling force of the preacher's performance on himself as audience and as artist: "Before he had finished I took a slip of paper and somewhat surreptitiously jotted down some ideas for the first poem, 'The Creation.'"[5]

By choosing "God's Trombones" as the overarching title for his collection of poems, Johnson showed implicit awareness of the mimetic dimensions in his work. More explicitly, his preface emphasizes the preacher's voice as a mime or simulator of that musical instrument, the trombone. In turn, as Johnson emphasized, the trombone itself is one of the few instruments capable of 'miming' the human voice. Like the violin, it can render the full range of harmonic sound values achievable by the human voice and appreciable by the human ear. Referring to a standard dictionary entry, the poet extolled the trombone as "the only wind instrument possessing a complete chromatic scale enharmonically true, like the human voice or the violin."[6] In fact, a double mimesis is at work here: the preacher's voice imitates a trombone which imitates the human voice. Next, moving from the preacher's mimicry to the sermonic poems, a more significant simulation appears.

More profoundly than the preacher's voice, the poems themselves become *God's* "trombones" to the extent that they simulate divine rather than human instruments. The poems come to represent (in Johnson's phrase) God's "wide and varied range" of theurgic expression: that is, God's cosmic, historical, and transhistorical activity. One aspect of this representation of range is evident in the way in which the sermon titles combine to achieve a virtual, rather than literal, encompassing of the entire Bible. Of course the order, selection, and extent of Johnson's themes do not strictly circumscribe or replicate the books of the Bible. Rather, he

evokes the impression of comprehensiveness by an alternation and pro-
gression of patterns. The patterns seem to alternate (1) ritual and liturgical
forms, on the one hand, with (2) biblical selections proceeding from the
earlier Hebrew to the later Christian Scriptures, as follows:

Listen, Lord—A Prayer	ritual petitions; a sermonic introduction
The Creation	Hebrew narrative (Genesis 1–2)
The Prodigal Son	Christian parable (Luke 15)
Go Down Death—A Funeral Sermon	ritual invocations; a eulogy
Noah Built the Ark	Hebrew narrative (Genesis 6)
The Crucifixion	Christian narrative (Mark 15)
Let My People Go	Hebrew narrative (Exodus 6ff.)
The Judgment Day	Christian apocalyptic sermon; and ritual warning/exhortation to sinners (Revelation 20)

This strategy for simulating biblical range and comprehensiveness may
have been suggested to Johnson by the preaching tradition he encountered
in his youth. We have Johnson's reference, already noted, to "the old folk-
sermon that begins with the creation of the world and ends with Judgment Day." As elaborated in the preface, this was "a stereotyped sermon"
that

> began with the Creation, went on to the fall of man, rambled through the
> trials and tribulations of the Hebrew Children, came down to the redemption
> by Christ, and ended with the Judgment Day and a warning and an exhortation
> to sinners.[7]

Perhaps as important as the biblical range of the sermon was the opportunity it afforded for displays of the preacher's individual virtuosity—his
or her range of oratorical and expository expertise. Significant in this
connection is Johnson's comment that the "framework" of the sermon
granted the preacher "the widest latitude that could be desired for all his
arts and powers." Those powers rendered the preacher "a master of all the
modes of eloquence . . . [one who] often possessed a voice that was a
marvelous instrument, a voice he could modulate from a sepulchral whisper
to a crashing thunder clap."[8] In this regard we may observe that the poetics
or 'sermonics' of Johnson's text achieve a "trombone" effect, first at the
level of the preacher's oral performance, by extension at the level of the
poet's literary craft, and finally at the interpretive level of religious meaning.
We may call this last level, borrowing Amos Wilder's term, the "theopoetic" level of meaning.[9]

At this theopoetic level there are also divine simulations at work; a
theurgy is in operation that displays the "trombone" effect even in the
Creator's cosmogonic performance. (Compare here Plato's *Timaeus*, in
which the "demiurge" must, by metaphysical necessity, create the world
through a series of mirrored images of higher, more divine realities—

through *simulacra*.) Remarkably, here we may draw upon the German verse of another poet, Rainer Maria Rilke, to elaborate this dimension of Johnson's poem. Using Rilke's verse we may say that Johnson intended his verse—his theopoetic "trombones"—to sound

> like many, many intervals
> in his might and melody
> . . . .
> As though God went with his wide
> sculptor-hands through the pages
> in the dark book of first beginning.[10]

Like the "many, many intervals" and the "wide sculptor-hands" of God in Rilke's poem, what is represented by Johnson's theopoetics is the divine instrumentality itself. Representation is another word for mimesis, and in this regard Johnson's verses may be said to 'mime' divine creativity. But this divine creativity is itself, as represented by Johnson, mimetic in nature. Like a musical instrument that simulates human experience, God's gestures simulate their intended effects on physical materials and on the sentient beings of the universe. Beginning with material productions, we see in "The Creation" that God's smile suffices to achieve the appearance of light. That marvel occurs presumably by way of the heightening (brightening) of the divine countenance: an obviously mimetic or mirror effect. (Compare the expression "the light of thy countenance," as used in Psalms 4.6; 80.3, 7; 89.15; and 90.8.) Or again, God's smile actuates the existence of the rainbow by modeling a graceful, curvilinear form. Here a further mimetic effect is expressed by the felicitous way in which the newly created rainbow "curled itself around [God's] shoulder," like the prized bow of an archer.

Such mimetic elements in Johnson's *literary* expression are corroborated by transcriptions of *oral* expression in many folk sermons. We may review only a few examples here, of course, and even these specimens have been edited for publication. They are nonetheless sufficient to indicate the genuine folk character of Johnson's theopoetics. Two such examples were included in *The Book of Negro Folklore*, edited and published in 1958 by two other celebrated Harlem Renaissance authors, Langston Hughes and Arna Bontemps. The first sermon is quite brief—a sermon fragment really—and its title consists of the simple theopoetic referent, "God." The first line bears similarities to the first line of Johnson's creation poem ("And God stepped out on space, / And he looked around"). This sermon poem begins, "I vision God standing / On the heights of heaven." Subsequently, its three stanzas of eleven lines each conclude with the same five line refrain: "His eye the lightning's flash, / His voice the thunder's roll. / Wid one hand He snatched / The sun from its socket, / And the other He clapped across the moon." Mimetic effects are obvious here: God's eye simulates the flash of a lightning bolt, his voice the thunder, and his hands the eclipse of heavenly luminaries. Such mimesis continues in the subsequent lines:

"I vision God wringing / A storm from the heavens; / Rocking the world / Like an earthquake," and finally "I vision God standing . . . Blowing His breath / Of silver clouds / Over the world."[11]

Remarkably parallel mimetic features appear in another folk sermon included in *The Book of Negro Folklore* under the title "Behold the Rib!" The sermon was originally reported by Zora Neale Hurston and collected in her *Mules and Men* (1935). The following lines are selected for their exemplary expression of divine mimesis in the vernacular language of black folk religion. Full appreciation requires the reader to envision the preacher's mimed body gestures, to imagine an intoned voice style, and to hear an exclamatory effect wherever "hah!" or "Oh" or "I can see!" punctuates the delivery.

> High-riding and strong armded God
> Walking acrost his globe creation, hah!
> Wid de blue elements for a helmet
> And a wall of fire round his feet
> . . . .
> And Oh—Wid de eye of Faith
> I can see him
> Standing out on de eaves of ether
> Breathing clouds from out his nostrils,
> Blowing storms from 'tween his lips
> I can see!
> Him seize de mighty axe of his proving power
> And smite the stubborn-standing space,
> And laid it wide open in a mighty gash—
> Making a place to hold de world
> I can see him—
> Molding de world out of thought and power
> And whirling it out on its eternal track,
> Ah hah, my strong armded God![12]

This "strong armded God," as a preceding verse makes clear, operates in the world not only mimetically but also magnificently: "God shook his head / And a thousand million diamonds / Flew out from his glittering crown / And studded the evening sky and made de stars."

In concluding this section on mimetic elements in certain traditional Afro-American sermons, it is crucial to observe as well their expression of a magical worldview. Indeed magic itself is essentially mimetic, inferred the nineteenth century historian of religion James G. Frazer, in his magnum opus, *The Golden Bough* (1900). With his concept of "sympathetic magic" Frazer assigned to magic generically, the element of mimesis that I have highlighted with reference to African American folk expression in particular. Among religion scholars he thereby inaugurated a mimetic theory of magic. On this view magical arts consist in first discerning, and then performing an operation based on, the imputed affinity that one thing has for another. Such affinity consists in the perception of two things as similar: similar on the basis of appearance, function, prior experience of contact, or in some other way. Seizing upon such an affinity or 'sympathy', the practitioner of magic devises an effective means of turning to human advantage the perceived similarity or habitual proximity of the two objects.

> From the first of these principles, namely the Law of Similarity, the magician infers that he can produce any effect he desires merely by imitating it: from the second he infers that whatever he does to a material object will affect equally the person with whom the object was once in contact, whether it formed part of his body or not.[13]

Frazer's larger theory has been criticized on several counts by contemporary anthropologists. On behalf of primal cultures we now acknowledge that many principles of natural process—the dynamics of stimulus and response, growth and decay—are readily available to the observation and reasoning powers of humans in all ages and cultures.[14] Untenable now is this corollary to Frazer's nineteenth century theory of progress (as science and religion) versus primitivism (as magic and ritual): the corollary that "magic is the direct coercion of natural forces by man; religion is the propitiation of divinities by the believer."[15] A still popular misconception, this claim is specious on both counts: from the side of what is called magic we find the supplication and petitioning of deities, and from the side of so-called religion we see coercive efforts to manipulate God, nature, and human society.[16]

Moreover, as René Girard has argued, mimetic manipulations of reality persist throughout the history of religions and civilizations. From Girard's perspective it is mimetic conflict that links religion and violence in the ongoing development of cultures, from our prehistory as a species ("hominization") to the present complexity of world civilization. Indeed, Girard sees a specieswide identification of "violence and the sacred" that extends from prehistory to our own time; from the primitive discovery of ritual homicide as a prophylactic against the devastation of internecine violence to the contemporary threat of nuclear holocaust. In our contemporary conjunction of *nuclear* violence and the sacred, "once again, violence prevents violence from breaking out . . . [since] nuclear armaments alone maintain world peace."[17] On this view it can hardly be demonstrated that the later stages of human history and cultural development transcend the primitive employment of surrogate or mimetic constructions designed to manipulate and control reality.

Despite its excesses and inadequacies, then, Frazer's theory is inviting as a heuristic for exploring mimetic elements in black America's magical tradition of conjure or conjuration. In particular, a useful point of departure for analyzing conjure is the term that Frazer offered as an alternative in discussing sympathetic magic, namely, "homeopathic magic." Homeopathy denotes phenomena in which 'like' affects 'like'—in which a substance similar to another can affect that other for good or ill, in order to benefit or harm. It is a most intriguing term in the context of this study. For it renders magical operations with a concept that is more applicable to the kind of pharmacopeic phenomena highlighted in subsequent discussion.[18] As we will see, the term "homeopathic" answers to the philosophical and hermeneutic need to articulate a concept of negation that is continuous with both pharmacopeic performance and the rational-analytic requirements of black American critical thought. In subsequent discussion homeopathic structures will provide a means for describing not only folk magical consciousness, but other cognitive dimensions of black existence as well.

Theopoetics

In this section I examine more closely the theopoetic dimension in Johnson's poem. "The Creation" introduces us, first of all, to a "lonely" God. "And God stepped out on space," the poet begins, and for emphasis isolates a startlingly nontraditional, two-word divine self-declaration: "I'm lonely." Here we discover a God who yearns to coinhabit a world of other beings. Immediately this representation poses a theological dissonance: this God or *theos* (Greek) is not the God of Greek philosophy or of Hellenistic Christianity. The God of classical Christian theology, in his aseity or self-sufficiency, requires the existence of no other beings. It is otherwise with the God of "The Creation." This deity speaks out of the immensity of his aloneness (his "space" or "darkness"[19]). Rather than thought-thinking-itself, as in Aristotle's metaphysics, or the God of the Hebrew tetragrammaton who declares, "I am that I am" (Exodus 3.14), this God says instead: "I'm lonely—I'll make me a world." Another contrast follows immediately. In creating light and separating it from darkness, the divine fiat (Latin) or command, "let there be," is not accomplished by speaking a word—a *dabar* (Hebrew) or a *logos* (Greek). Here we do not find that "in the beginning was the Word" (John 1.1), but rather that

> . . . God smiled,
> And the light broke,
> And the darkness rolled up on one side,
> And the light stood shining on the other,
> And God said: That's good![20]

Thus a nonverbal theurgy or god-work is represented early and then sustained in the poem. Unlike the God of the biblical cosmogony, whose divine speech initiates the creation of heavenly luminaries and separates the heavens from the earth (Gen. 1.6–10), this God performs the cosmic thaumaturgy—the 'wonder-works' of the universe—through bodily gestures and physical actions.

> Then God reached out and took the light in his hands,
> And God rolled the light around in his hands
> Until he made the sun;

And he set that sun a-blazing in the heavens.
And the light that was left from making the sun
God gathered it up in a shining ball
And flung it against the darkness,
Spangling the night with the moon and stars.
Then down between
The darkness and the light
He hurled the world;
And God said: That's good![21]

Of course, biblical literature also employs anthropomorphic metaphors for divine action. The Psalmist, for example, declares that "the heavens are the work of thy hands" (Psalm 102.25). The Book of Job declaims, "Have you an arm like God? or can you thunder with a voice like him?" (Job 40.9). Second Isaiah queries, "Who has measured the waters in the hollow of his hand, and meted out the heavens with his span?" (Isaiah 40.12). Also notable in this regard are the figurative resonances between "The Creation" and Psalm 104 (vss.2–4). This Psalm addresses God directly as "thou" who

> coverest thyself with light as with a
>   garment,
> who hast stretched out the heavens like a tent,
> who hast laid the beams of thy chambers
>   on the waters,
> who makest the clouds thy chariot,
>   who ridest on the wings of the wind,
> who makest the winds thy messengers,
> fire and flame thy ministers.

Thus Johnson's poem reflects, but also selects and highlights, a shared aspect of the Bible in its full textual diversity. Nonetheless we may consider the cosmogonic narrative in Genesis as a genre distinct from the Psalms and other biblical texts. If we do so, then the verbal and disembodied nature of its God contrasts significantly with the embodied and theurgic performance of Johnson's deity. The contrast is insistent and even conspicuous. Indeed, we may wonder whether it reflects a distinctive bias in the folk tradition behind the poem.[22] Consider for example the following verses: in "The Creation."

> And God walked, and where he trod
> His footsteps hollowed the valleys out
> And bulged the mountains up.
> . . . .
> And he spat out the seven seas—
> He batted his eyes, and the lightenings flashed—
> He clapped his hands, and the thunders rolled—
> And the waters above the earth came down,
> The cooling waters came down.
> . . . .

And God smiled again,
And the rainbow appeared,
And curled itself around his shoulder.

It is notable that even when this theurge does speak in command—
in order to say more than "That's good!"—the words themselves are punc-
tuated by his expressive body movements.

Then God raised his arm and he waved his hand
Over the sea and over the land,
And he said: Bring forth! Bring forth!
And quicker than God could drop his hand,
Fishes and fowls
And beasts and birds
Swam the rivers and the seas,
Roamed the forests and the woods,
And split the air with their wings.
And God said: That's good![23]

Immediately after this verse God is depicted walking among the Creation, inspecting "all that he had made." But eventually the deity realizes,
"I'm lonely still," and sits down "where he could think; / By a deep, wide
river he sat down." Finally he conceives (what we expect in due course)
the idea of creating human beings. God thinks, "I'll make me a man!"
And there, by the riverbank,

... the great God Almighty
Who lit the sun and fixed it in the sky,
Who flung the stars to the most far corner of the night,
Who rounded the earth in the middle of his hand;
This Great God,
Like a mammy bending over her baby,
Kneeled down in the dust
Toiling over a lump of clay
Till he shaped it in his own image.... [24]

The poem ends immediately after this point. The poet has gracefully linked
an earthy and ethnic metaphor, the southern plantation "mammy" bending
over a child, with the primordial birthing of humanity by "This Great
God." Returning then to the text of Genesis (vss. 2.7), Johnson depicts
God breathing into sculpted clay "the breath of life" whereby the man
becomes "a living soul." The word "Amen" concludes the poem in affirmation of this theurgic performance, and again a final "Amen."
    It is possible, of course, to end our analysis here and regard "The
Creation" as a simple rendition of the first chapters of Genesis that, coincidentally, uses the folk idioms of black America. With more probing,
> however, we will see how these idioms radically transform the nature of
the Genesis cosmogony, rendering it African American as well as Hebraic
in character. Admittedly, the forms of representation that effect this trans-

formation of Genesis into an Afro-Hebraic text remain subliminal to most readers. Indeed, some forms of representation may not have been fully transparent to the artist himself. Although Johnson's employment of folk styles and expressions was highly reflective and deliberate, he may nonetheless be regarded (like all artists) as a mediator of cultural intentions that exceed any conscious crafting of materials.[25] On the one hand, accordingly, we have seen how the poetics of *God's Trombones* skillfully and exultantly displays the folk ethos and its spirituality. On the other hand, the theopoetics of Johnson's text shows traces of deeper intentions than were explored by the author.

What remains unexplored at the theopoetic level of "The Creation" is the deity's performance as a type of conjuror in the African American folk tradition. The "mammy" figure working with clay to create a divine facsimile—God "in his own image"—bears correspondence to another figure pervasive on southern plantations: the figure of the conjure woman and conjure doctor, the hoodoo man or root worker. In each case thaumaturgy is effected by means of (1) assembling or aggregating suitable materials (clay, roots, or herbs, for example) and then (2) transforming reality by performing efficacious operations on those materials. However, in contrast to traditions in which "conjuring" means transformation by means of witchcraft or sorcery, in black America a conjuror is also a folk doctor. More explicitly the "conjure doctor" is a quasi-medical practitioner who transforms reality by means of prescribed operations involving a repertory of efficacious materials. The conventional materials, so-called materia medica, include substances like herbs and roots, powders and grave dirt, insect and animal products like cobwebs, human body effects such as hair or fingernails, as well as clothing, fluids, and other paraphernalia.[26] Thus conjure, with its repertory of substances and prescriptions for their use, is intensively imitative or mimetic and medicinal or pharmacopeic. Concisely stated, conjure is not only sorcery or witchcraft but also a tradition of healing and harming that transforms reality through performances and processes involving a mimetic use of medicinal and toxic substances.

As we have seen, Johnson's poem highlights an indigenous preference in black folk tradition for embodied (versus purely verbal) transformations of the world. The poem does not explicitly, however, represent such transformations as conjurational performances. It is understandable that Johnson did not do so, if we grant that he was intent on representing biblical and typically Christian elements in the folk preaching tradition. In any case, it was left to another Harlem Renaissance artist to indicate directly the extrabiblical aspects of the tradition: The folklorist and novelist Zora Neale Hurston. Among Renaissance writers Hurston was probably the most uninhibited in representing folk tradition in folk terms. Most fortuitous for this study, she reported in some detail how the tradition "made over" the Bible into a conjure book.

Mosaica

The way we tell it, hoodoo started way back therebefore everything. Six days of magic spells and mighty words and the world with its elements above and below was made. And now, God is leaning back taking a seventh day rest. When the eighth day comes around, He'll start to making new again.

Zora Neale Hurston, *Mules and Men*[27]

In this passage Zora Neale Hurston reported an innovative development in African American folklore: the 'making over' of the Bible's Creation story into a *conjure* story.[28] Conversely stated, Hurston reported that black folk tradition had extended the field of conjurational performances from its marginal social context, on the one hand, to a cosmic scale of reference on the other. Ironically, her literary affirmation of this folk strategy—the strategy to locate conjure phenomena in the most primal past—"took conjuring a great leap forward."[29] Hurston advanced conjure in this regard precisely by pushing its origins as far *backward* as possible: all the way back to the biblical Creation story. "The way we tell it," she wrote with implied corroboration of her sources, "hoodoo started way back there before everything." Furthermore, the biblical God was the primal practitioner of this "hoodoo." Indeed it was the Creator God of the Bible, the claim goes, who performed as a conjuror by making the world through "six days of magic spells and mighty words." The exalted nature of this claim can best be appreciated by comparison with another celebrated claim for conjure. It is a claim which finds wider representation in folklore sources: conjure as a Mosaic tradition.

In Afro-American folk culture Moses is best understood as a figure who represents the conjurational nature of the cosmos and of divine creativity. Moses-the-conjuror, on this view, is mimetically related to God-as-conjuror and to reality as conjurational. What is the basis of that *mimesis*? Hurston herself highlighted the semidivine aspect of the Moses figure in folk tradition by emphasizing Moses as 'maker'. Unlike other human beings, she explained, Moses was privileged to discover God's "secrets" of original Creation and therefore to perform at a divine level of creativity himself. In continuing the passage, the author (combining her skills as folklorist and storyteller) explained that

Man wasn't made until around half-past five on the sixth day, so he can't know how anything was done. Kingdoms crushed and crumbled whilst man went gazing up into the sky and down into the hollows of the earth trying to catch God working with His hands so he could find out His secrets and learn how to accomplish and do. But no man yet has seen God's hand, nor yet His finger-nails.[30]

No man has seen God, that is, except Moses (albeit he saw only God's "back" in Exodus 33.23–34.6). Despite an absolute disclaimer that human beings have direct access to the divine presence, power, and performance, Hurston proceeded to express the most exalted claims for Moses as a divine intimate, indeed as a source of transformation and generativity in his own right. Moses alone was capable of performing as a maker on the scale of God's making:

> Moses was the first man who ever learned God's power-compelling words and it took him forty years to learn ten words. So he made ten plagues and ten commandments. God gave him His rod for a present... he lifted it up and tore a nation out of Pharaoh's side.... Moses talked with the snake that lives in a hole right under God's foot-rest.... The snake had told him God's making words. The words of doing and the words of obedience. Many a man thinks he is making something when he's only changing things around. But God let Moses make.[31]

In this passage Hurston previewed themes which would appear four years later in her novel *Moses: Man of the Mountain* (1939). In the novel's introduction she sought to prepare readers for her fictional representation of Moses as "the finest hoodoo man in the world."[32] Thus the introduction elaborates two themes in particular: Moses as a semidivine wonder-worker—a thaumaturge, and Moses as beneficiary of snake-wisdom. The author was fully conscious of departing from "the common concept of Moses in the Christian world" and sought to acclimate readers to alternative conceptions based on African or African American cultural referents. It was through her dual voice of ethnographer-novelist[33] that she transmitted and mediated those referents.

> All across the continent [of Africa] there are the legends of the greatness of Moses, but not because... he brought the laws down from Sinai. No, he is revered because he had the power to go up the mountain and to bring them down.... Anyone could bring down laws that had been handed to them. But who... has the power to command God to go to a peak of a mountain and there demand of Him laws with which to govern a nation?[34]

This passage goes on to celebrate Moses as the kind of thaumaturgic performer that we saw earlier in the creator God of *God's Trombones*. Unlike Johnson, however, Hurston is not concerned to render orthodox representations of thaumaturgy: "Who else has ever commanded the wind and the hail? The light and darkness? That calls for power, and that is what Africa sees in Moses to worship. For he is worshipped as a god." With this reference to worship the author supplies a concluding heterodox and ethnographic note. She correlates the biblical association of Moses and serpents (Exodus 4.3f., 7.8–12; Numbers 21.6–9; John 3.14) with the Haitian worship of Damballah, a transatlantic deity associated with serpents. "In Haiti, the highest god in the [Vaudou] pantheon is Damballa Ouedo... and he is identified as Moses, the serpent god. But this deity did not originate in Haiti. His home is in Dahomey and [he] is worshipped

there extensively. Moses had his rod of power, which was a living serpent."
I return to West African influences on black North American representations of Moses in the next section.[35]

Other sources corroborate Hurston's claim for the exalted stature of
Moses across a number of contemporaneous African cultures.[36] Indeed
⟩ the exaltation of Moses is a transatlantic phenomenon, linking Africa to
the Americas by way of biblical figuralism. In corroboration of Hurston's
research in the United States during the 1930s we may cite the contemporaneous research of Harry Middleton Hyatt. In 1970 Hyatt published
his massive compilation of verbatim field notes, *Hoodoo-Conjuration-
Witchcraft-Rootwork*. One entry in particular records the statement of an
informant labeled "Hoodoo Book Man," so called because he supported
his statements by referring to a book of cabalistic lore, *The Sixth and
Seventh Books of Moses* (so named only because it contains Hebrew names
and phrases derived from the Cabala).[37] As the reader observes immediately, Hyatt transcribed his informants' colloquial English with little or
no editing. Most usefully, Hoodoo Book Man's remarks evince some of
the same elements found in Hurston's passage: "You see, *hoodooism* was
given down from de foundation of Moses long years back. . . . *Hoodooism*
started way back in de time dat Moses days [*sic*], back in ole ancient times,
nine thousand years ago."[38] Like Hurston, this informant was concerned
to indicate the great antiquity of hoodooism. Additionally the high regard
for Moses operates here as well. There is an obvious contrast, however:
the time frame of Hurston's passage stretches further back than "Moses
days": back to the Creation of the world.

The same contrast is less marked but nonetheless evident in another
source, found in the now-standard collection in folklore studies *Drums
⟩ and Shadows (1940)*. This collection was commissioned by the federal
Works Project Administration in the 1930s to report on field work conducted along the Georgia and South Carolina seacoast. Describing their
research in Yamacraw, Georgia, the authors gave details of their conversation with one aged informant, Thomas Smith. Smith instructed them
that "the same magic power that Moses had used when he turned his rod
to a snake before Pharoah still exists today among Negroes." Moreover,
he declared: "Dat happen in Africa duh Bible say. Ain dat show dat Africa
wuz a lan uh magic powuh since duh beginnin uh histry? Well den, duh
descendants ub Africans hab duh same gif tuh do unnatchul ting."[39]

For this informant, it appears, the representation of Africa as "a land
of magic power since the beginning of history," rather than Moses, is the
focal point of conjure tradition. Like that of Hurston, but stopping short
of her cosmogonic context for conjure, this informant's perspective pushes
beyond the Mosaic period to an earlier history of "magic power." But
where he identifies that power with Africa, and with the descendents of
Africans 'who have the same gift to do unnatural things', Hurston's passage
identifies conjure first with the biblical God and then with Moses as God's
initiate. These variations—Moses-the-conjuror, God-as-conjuror, Africa as

primal land of conjure—make variability and indeterminacy themselves increasingly obvious elements in this matter of conjure origins. That is, the focal point of interest seems to shift intrinsically so that only for some purposes is the focus on Moses as semidivine (Hurston). From other points of interest the focus is more precisely Moses as Egyptian/African (Thomas Smith), or simply Moses as founder of "hoodooism" (Hoodoo Book Man). It may be that variability is a constitutive element even within a single perspective, so that (for example) Hurston's semidivine Moses may simultaneously, by inversion, signify God-as-Mosaic. In this last representation, perhaps the most radical of all, the biblical Creator becomes a 'Mosaic double'.[40] God and Moses are doubles in the sense that the category "Mosaic" signifies a superlative which is retrojected onto the divine nature. On this view, to be superlative is to be "the finest hoodoo man in the world" (Hurston). God as superlative becomes God-the-conjuror by (mimetic) assimilation to Moses the supreme conjuror.

By way of corroboration Lawrence Levine also provides evidence of a Mosaic assimilation of the divine nature. In his study *Black Culture and Black Consciousness (1977)*, Levine discusses Moses symbolism in his treatment of the slave spirituals. He focuses particularly on songs that were sung and heard during that crucial period of the Civil War transition from slavery to freedom.

> Judging from the songs of his black soldiers, [United States] Colonel Higginson concluded that their Bible was constructed primarily from the books of Moses in the Old Testament and of Revelations in the New: "all that lay between, even the life of Jesus, they hardly cared to read or to hear." "Their memories," he noted at another point, "are a vast bewildered chaos of Jewish history and biography; and most of the great events of the past, down to the period of the American Revolution, they instinctively attribute to Moses." [41]

The attribution by slaves and freed persons—of "the great events of the past ... to Moses"—is clearly continuous with the attribution of conjure foundations to Moses many decades later as attested by Hurston, Hyatt, and other sources. Even more exalted were representations of Moses that rivaled Christ in eminence, as Levine observes further in the same passage: "Shortly after his arrival in Alabama in 1865, a northern army chaplain wrote of the slaves, 'Moses is their *ideal* of all that is high, and noble, and perfect, in man,' while Christ was regarded 'not so much in the light of a *spiritual* Deliverer, as that of a second Moses'." This folk type, Christ as "second Moses," represents a figural parallel to the assimilation of God as Mosaic and reinforces the need for further inquiry. What is the impulse driving black Americans across two different centuries to invest such extraordinary claims in the figure of Moses?

Pharmacosm

The term ashé, *the "power-to-make-things-happen," may be the Afro-Cuban way of referring to a plant's chemical constituents, its magico-*

> *medicinal properties . . . [and] the "orishas" may be pharmacological*
> *categories.*
>
> Morton Marks, "Exploring *El Monte*: Ethnobotany and
> the Afro-Cuban Science of the Concrete"[42]

⟩ In this section I posit a transatlantic parallel between the function of Moses
in black North American traditions and that of gods and ritual specialists
in West African traditional religions. In order to substantiate this homology I refer to the (poly)theistic and the magical-cosmological traditions
deriving from two African peoples, respectively: the Yoruba of Nigeria
(West Africa) and the Kongo, or Bakongo, people of the Congo-Angola
region (Central and South Africa).[43] Yoruban influence in the New World
is more observable in the Caribbean and South America (most notably
Cuba, Haiti, and Bahia in Brazil). Kongo-Angolan ritual and cosmological
influences in North America in particular have been reviewed recently by
the historian Sterling Stuckey in his 1989 study, *Slave Culture: Nationalist
Theory and the Foundation of Black America*, and by the art historian Robert
Farris Thompson in *Flash of the Spirit: African and Afro-American Art and
Philosophy* (1984).[44] While acknowledging real differences among the diverse African cultures caught in the Atlantic slave trade, these and other
historians, for example, the historian of religion Albert Raboteau, allow
for some degree of generalization based on a number of common characteristics:

> Similar modes of perception, shared basic principles, and common patterns
> of ritual were widespread among different West African religions. Beneath
> the diversity, enough fundamental similarity did exist to allow a general de-
> scription of the religious heritage of African slaves, with supplementary in-
> formation concerning particular peoples. . . . A common religious heritage
> then resulted from the blending and assimilation of the many discrete religious
> heritages of Africans in the New World.[45]

⟩   From this perspective we may trace, in black North American culture,
general *continuities* of tradition that derive from elements common to a
broad number of West African peoples. Moreover we may trace such
continuities even where we are not able to ascertain discrete "survivals"
or "retentions" from traditions specific to a particular people. "It is the
continuity of perspective that is significant," as Raboteau claims, "more
so than the fact that the cults of particular African gods . . . have been
transmitted to the New World."[46] Two broad areas provide the ethnographic basis for such continuities; these areas constitute the fundamental
data of religious experience: beliefs and practices. In this instance the beliefs
concern spiritual beings, and the practices comprise magical and ritual
performances. The first area (1) not only involves beliefs per se but includes
of course communications and relationships with divine or divinized
beings. Raboteau's catalogue of such beings comprises (a) a "High God"
or "Supreme Creator," alongside (b) a pantheon of lesser deities or sec-

ondary gods, and also (c) a world of spirits and (d) a cult of ancestors.
The second area (2) constitutes a system of practices involving (a) magic
so-called, which in Africa is intensively related to (b) medicine, and also
involves (c) witchcraft and counterwitchcraft, and (d) divination. Inte-
grating those practices and augmenting them are (e) ritual performances
that incorporate such acts as drumming, singing, chanting or 'incanting,'
dancing, and—most conspicuously—spirit possession and ritual sacrifice.[47]

Turning now to Afro-American representations of Moses, I advance
my hypothesis on the basis of continuities first in area (1): that is, with
respect to Yoruba *beliefs* regarding specific divine figures or characterizations of deity. Since Moses is also represented by the tradition as a type
of conjuror, I subsequently take up continuities in area (2): with regard,
that is, to Kongo *practices* involving charms and sacred medicines as a
source for black American conjure traditions. In alternating here from
Yoruba deities to Bakongo charms I follow Robert Farris Thompson, who
compares the complexity of the Yoruba pantheon to a correlative complexity among the Bakongo: "they have, instead, a complex system of
'minkisi' or sacred medicines." Raboteau also notes that "the religious life
of the peoples of the Kongo focused not upon a pantheon but upon a
large range of *minkisi*, or sacred medicines, embodying spirits who could
harm or cure."[48] In each case I could of course consider a larger number
of West African traditions, with their diverse pantheons and magical practices, on the view that slave culture constituted a blend or compound of <
influences drawn from disparate African sources, and that one should
account for as many of them as possible. Most conspicuously absent
in the following treatment, for example, is the continuity between African American figuration of Moses and the deity Damballah of the Fon
people.[49]

The limited scope of this study, and the ethnographic clarity of de-
scriptions of the Yoruba pantheon, recommend the following effort to
trace Afro-American characterizations of Moses from the compounding
of two or three *orisha* alone. The purpose of such exploration is not to
achieve a level of ethnographic precision. I am not concerned to ascertain
or claim the formation of specific Afro-American beliefs or traditions based
on precisely delimited West African beliefs or traditions. The quest for
such precision I have already relinquished (rightly so, given the indeter-
minate nature of our data and the inchoate nature of these cultural transformations) in preference for Raboteau's broader category of "continuity
of perspective."

Orisha *Moses*

*It was not possible to maintain the rites of worship, the priesthood, or the
"national" identities which were the vehicles and supports for African
theology and cult organization. Nevertheless, even as the gods of Africa
gave way to the God of Christianity, the African heritage of singing,*

> *dancing, spirit possession, and magic continued to influence Afro-*
> *American spirituals, ring shouts, and folk belief.*
>
>   Albert J. Raboteau, Slave Religion[50]

Unlike developments in the Caribbean and South America, the conditions
of deculturation and repression in North America ensured the atrophy of
every African pantheon of gods among the generations of native born
black people. Nonetheless, three *orisha* roles appear to be continuous with
folk representations of Moses: (1) the diviner (the orisha Ifa or Orunla);
(2) the herbalist (the orisha Osanyin or Osain); and (3) the messenger/
trickster (the orisha Eshu or Elegba). The lord of divination, Ifa, corresponds to Moses as seer and prophet, whereas Moses the hoodoo doctor
or root worker corresponds to Osanyin the herbalist. Finally, Eshu as
messenger and trickster (compare the Greek god Hermes) corresponds to
Moses as God's spokesperson and wonder-worker. Here recall the Exodus
narrative in which Moses performs magical (conjuring) tricks before Pharaoh's court, including the transformation of Aaron's staff into a serpent
(Exodus 7.10; cf. 4.3). Taken together the three roles of diviner, herbalist,
and messenger/trickster reflect distinct elements in Moses' composite persona as represented either in biblical or black folk tradition or in both.
Remarkably, the art historian Robert Farris Thompson observes a nexus
of interrelationships among precisely these same orisha. To begin with,
the orisha are not high gods or self-sufficient beings but demiurges: they
need each other's varied and distinctive powers. Thus, the other orisha
need both Osanyin's herbal powers, on the one hand, and Eshu's ener-
gizing creativity on the other. Furthermore, Ifa himself needs Osanyin's
knowledge of cures since, according to both mythical and practical understanding, to divine can also mean to prescribe. (This correlation is
narratively configured in the myth which tells the story of Osanyin's initial
opposition to Ifa, his withholding of herbs from the other orisha, and his
eventual reconciliation to Ifa and to his fate after the experience of being
punished by maiming.) Similarly Osanyin needs Ifa's divinatory powers,
presumably in order to 'pharmacognize' (compare pharmacognosy) which
herbs are efficacious: that is, to discern their healing and harming properties in specific cases.[51]

Further substantiating my hypothesis are the evident correlations between the orisha Ifa, as seer or diviner, and Moses the biblical prophet.
The prophetic dimension of Mosaic figuration is treated more specifically
in the discussion of Exodus (Chapter 2). Here the figure of Moses as
Osanyin, as herbalist and pharmacist, is more pertinent. One researcher
has drawn for us a portrait of Osanyin with remarkable correspondences
to the Afro-American conjuror as an herbal doctor or 'root worker'. An
additional element of this portrait is the representation of Osanyin as a
homeopath (a specialist who employs a form of a disease in order to cure
that disease).[52]

Osanyin is the chemist or the pharmacist of the Orisha. . . . He represents the two sides of herbal power. The same thing that can cure you, if used improperly can kill you. Osanyin is representative of balance in nature. . . . He is, on the one hand, the forest that gives the herbal tonics that can help you, and on the other hand, the poison mushrooms that can kill you. . . . Osanyin can be used as a weapon of war, because of his knowledge of witchcraft, potions and charms, and can conversely be used to disarm witchcraft and heal illness. Osanyin is the greatest of witchdoctors. . . . [53]

Several points of correlation, both obvious and subtle, exist between this portrait and Moses symbolism in Afro-American folk tradition. The most obvious correspondence is the superlative valuation of Osanyin, on the one hand, as "the greatest of witchdoctors," and the superlative valuation of the Afro-American Moses, on the other, as "the finest hoodoo man in the world" (Hurston). There are more deep-structural correspondences, however, between Afro-Mosaic representations and the medicinal and pharmacopeic elements in this portrait. In order to examine those elements it is useful to turn to the sacred medicines of the Bakongo people. Before leaving this discussion of Moses as an orisha, however, I emphasize that the demise of the Yoruba pantheon in North America—"the death of the gods" (Raboteau)—need not mean the complete atrophy of Yoruba theistic personalities or principles.[54] On the contrary, in regard to spirit possession and conjure particularly it is worth considering "that some elements of African religion survived in the United States . . . as aspects hidden under or blended with similar European forms" (Raboteau).[55]

I have stated the case with respect to the Yoruba pantheon; later I discuss it with respect to Bakongo *minkisi.* Here I contend that a covert transmission of African spiritual principles or personalities lies hidden under the conventional Euro-Christian forms found in black religious traditions. Such a claim is conducive to my hypothesis, that the extraordinary black North American interest in Moses derives from his figural representation as a blend or compound of otherwise long forgotten sacred figures.

Minkisi *of Conjure*

Versions of some of the ritual authorities responsible
for Kongo herbalistic healing and divination
appeared in the Americas and served as avatars of
Kongo and Angola lore in the New World. . . .
Those in the United States were known largely as
"conjurors" and "root-persons."

Robert Farris Thompson, *Flash of the Spirit*[56]

The exalted status of Moses in African American folk tradition can be attributed to the covert survival, through the mediation of this biblical figure, of otherwise extinct African gods. Here I acknowledge, in passing, that such figural mediation operates parallel to African continuities in the Caribbean and South America. In these other sectors of the African diaspora the figures of African gods are "hidden under or blended with" (Raboteau) the veneration of Catholic saints. This mediation of orisha by Christian figures is particularly evident in the phenomena of spirit possession in Brazilian Candomble, Cuban Santería, and Haitian Vaudou. But in North America, in addition to mediating an extinct or extinguished (poly)theism, the Moses figure is continuous with folk practices of conjure. In this regard, however, Moses as ritual specialist assumes the role not so much of an orisha as that of a 'medicine man.' In both roles Moses is a titular figure for conjure practices biblically configured as 'Mosaic' practices. The materials employed in those practices, moreover, fall into two areas: folk forms of materia medica; and—crucially for a traditional oral culture encountering a literate culture—literary materials in the form of biblical texts. Accordingly, the figuration of Moses as a conjuror mediates between Bakongo sacred medicines, on the one hand, and biblical texts construed as "books of Moses" on the other.

Kongo charms or sacred medicines—*minkisi* (singular: *nkisi*)—are containers within which are placed materia medica that, in combination with the living entity or spirit of the charm, create its phenomenal power. The word *nkisi* is related etymologically to words used in other Central African cultures to mean "spirit." The container itself may be made of leaves or cloth, shells or ceramic, sculpted wood or statuettes, or may form various kinds of packets, bags, or sachets (compare the conjuror's bag). These are lifeless objects until they host the 'medicines' that activate the power of the nkisi. The most crucial medicinal materials are grave dirt, kaolin or riverbed clay, or perhaps a funerary relic (for example, a bone) of an ancestor, priest, or witchcraft victim. Robert Farris Thompson refers to these objects as "spirit-embodying" materials because they 'ground' the living presence of the nkisi within its otherwise inert container. A second class of materials, such as seeds or herbs, stones or chicken claws (vegetable, mineral, or animal materials), is described by Thompson as "spirit-directing" because they instruct the spirit in its task. The ethnographer Wyatt MacGaffey observes that, although the pharmacopeic properties of drugs (purgative effects, for example, or tranquilizing) are intrinsic to some Kongo materia medica, "most of them function as material metaphors that describe the powers attributed to the particular spirit." MacGaffey draws this intriguing formulation from Luc De Heusch's description of minkisi as composed of "spirits of the dead metonymically caught in a metaphorical trap." Metonym and metaphor are highly suggestive terms for understanding literary applications in black North American conjure.[57]

These are related figures of speech, of course, and each constitutes a type of mimesis in which one thing is correlative to another on the basis of a perceived likeness. A metonym is a term used *figuratively* to name or signify another according to that term's attributes or associations (for example: the wealth possessed by *the Crown*). A metaphor is a term whose *literal* meaning denotes a likeness or analogy in a second item (for example: a ship *plows* the sea with its prow). The combined use of metonyms and metaphors is evident (and sometimes overlapping) in the functioning of minkisi. On the one hand the contents of the charm are "verbal" and "visual metaphors" for the intended operation of the embodying spirit. "The first [verbal metaphor] is a punning principle by which an element, by its name, is made to evoke a particular concept." For this principle MacGaffey gives the example of a snail shell, the name of which rhymes with a Kongo verb meaning "to be strong" and the spiral shape of which symbolizes long life. Such shells, accordingly, are placed in minkisi designed to support childbirth or to strengthen the body. "The second [visual metaphorical] principle . . . [is] relatively specific to the kind of function the *nkisi* performs." For example: to promote divination a nkisi may contain the head of bird whose cry predicts the future; a snake head is placed in nkisi intended to punish evildoers; a round stone in *nkisi* designed to heal tumors.[58]

Moreover, as the art historian David Brown observes, such metonymic and metaphorical principles are correlative to Thompson's "spirit-embodying" and "spirit-directing" functions, respectively. The "spirit-embodying" function is evident in the metonymic principle whereby an object can signify the nature or character of a charm's indwelling spirit. For example (with respect to the title of Thompson's book, *Flash of the Spirit*), "such objects as mirrors or pieces of porcelain attached to the exterior of the nkisi may also signify power—the flash and arrest of the spirit." Coincidentally, MacGaffey notes that this metonymic relation can also extend beyond the charm's spirit to the object of its operation, "in that elements associated with the client or victim were put into the nkisi and that elements of the nkisi were put into him, too, in the form of a potion that he drank and of materials that were rubbed on his body or attached to it."[59] But most pertinent for my purposes is the fact that black North American conjuration operates in continuity with the metonymic and metaphorical principles of Kongo *minkisi*. An indication of that continuity is available in an obvious parallel between definitive statements regarding the two traditions. Compare De Heusch's description of a nkisi as a spirit "metonymically caught in a metaphorical trap" with the claim of a conjure informant as recorded by the researcher Harry Hyatt: "To catch a spirit, or to protect your spirit against the catching, or to release your caught spirit—this is the complete theory and practice of hoodoo."

Combining these two formulations, conjure operations may be said to consist in the use of mimetically encoded materials—some encoded as metonyms that name or signify the spirits involved, and some as metaphors that signify operations performed by or upon those spirits. The following formulation by David Brown elaborates this combined formulation. Observe Brown's correlation of spirit-embodying functions with metonymy in items (1) and (3), and the correlation of spirit-directing functions with metaphor in item (2).

1) Spirit-power harnessed from the "other world" through metonymy (graveyard dirt, the bone or body of a dead person or animal—e.g., the spider, the "black cat bone") is given direction ("told" what to do) through

2) objects that signify intention metaphorically or mimetically (wordplay is often involved) and then is

3) linked metonymically, in turn, to a specific person upon whom or object upon which the conjure is supposed to "work" through something of the person's body or clothing that has been collected.[60]

This "working" of conjure, Brown proceeds to explain, is like constructing a train of forces or a "circuitry" in which energy is first (metonymically) "embedded" in appropriately efficacious or conducive materials, and then (metaphorically) directed according to the intention of the practitioner. Here it is instructive to recall the contemporary view of magic as a kind of communication system, and as comparable in intent to science as a system of signs and correspondences. In this connection Brown adopts Lévi-Straussian language and describes conjure as

> a bricolage of diverse, symbolically resonant materials constructed according to a set of principles that Claude Lévi-Strauss . . . has aptly named the "science of the concrete." Like Lévi-Strauss's *bricoleur*, the conjurer "interrogates all the heterogeneous objects of which his treasury is composed to discover what each of them could *signify* and so contributes to the definition of a set which has yet to materialize."[61]

To sum up: conjurors materialize their "set" of significations, first, through the metonymic use of (spirit-embodying) substances, and then through the (spirit-directing) intentions mimetically represented in "material metaphors" (MacGaffey), thus performing formulaic operations within a "science of the concrete" (Lévi-Strauss).

Conjure Book

The Bible. All hold that the Bible is the great conjure book in the world.

Zora Neale Hurston, "Paraphernalia of Conjure"[62]

In addition to his treatment of Kongo minkisi, Robert Farris Thompson highlights the importance of vision and balance in Bakongo cosmology. In precolonial tradition Bakongo cosmograms served to effect "an ideal balancing of the vitality of the world of the living with the visionariness of the world of the dead." One means for achieving this cosmic balance or wholism was pharmacopeic, however. "Charms and medicines were constantly produced in the search for the realization of a perfect vision in the less than perfect world of the living." In contrast to a pantheon of divinities, Bakongo culture featured a roster of religious practitioners in fealty to "God Almighty," Nzambi Mpungu, "whose illuminating spirit and healing powers [were] carefully controlled by the king (*mfumu*), the ritual expert or authority (*nganga*), and the sorcerer (*ndoki*)." Thompson then details a division of labor, among the intermediate class of these practitioners, for achieving that cosmic balance and vision. Some of the ritual experts healed with charms, some with roots and herbs, and others through divination. In addition, he proceeds to distinguish the *beneficent* intentions of herbalists and diviners, over against the *maleficent* orientation of a third class of practitioners: sorcerers. "Now the king and the ritual experts controlled mystic powers for the common good. By contrast, the *ndoki*, or sorcerer, was a caricature of the individualist who, lacking social conscience, bases his security . . . in utterly selfish action devoid of mercy. . . ."[63]

Less sharply differentiated roles may be indicated, however, in the case of black practitioners in the Americas. Even where they are not regarded as sorcerers or practitioners of "malign occultism,"[64] conjure practitioners can be dually oriented—both beneficent and malignant. Thus Raboteau stresses the "dual potential," in the knowledge and performances of Afro-American conjurors, for both help and harm. In subsequent chapters it will be useful to refer to this duality in morally neutral terms such as Raboteau's alternative expressions: conjure help is "defensive protection" on behalf of the client; conjure harm is "offensive aggression" on behalf of the client. In this regard Raboteau indicates that sorcerers are socially distinguished by the claim that they specialize in offensive and aggressive operations predominantly. But here we should be careful not to collapse the defensive and offensive distinction into the moral dichotomy of good versus evil. That is to say, offensive conjure can be good as well as defensive conjure, and either can be malign. Melville Herskovits, for example, joined other researchers in differentiating between 'medical' and magical practitioners as two separate classes of conjurors, but without attempting to distinguish the classes in terms of good and evil magic.[65]

With this caveat in mind I return to Thompson's typology of religious roles among the Bakongo, in which social dysfunction and cosmic imbalance are attributed to the sorcerer, whereas the intentions of the king and the ritual expert are represented as aiming toward social harmony and cosmic balance. The king, acting as "supreme authority," achieves this intention by the impartial exercise of justice—specifically by rendering judgments and authorizing punishments against criminals and state enemies and by "decorous" living. For their part, the means employed by ritual experts for achieving the ancestors' vision of a perfect cosmos are talismanic, pharmacopeic, and divinatory; consisting, that is, in the use of minkisi as charms or medicinal and divination instruments. In this regard we may represent the black North American folk version of Moses on the model of a Bakongo ritual expert (*nganga*), who employs a combination of charms, medicines, and divination to realize an ideal world as envisioned by the ancestors and as represented in cosmograms. Replacing both ancestors and cosmograms in black North American culture (as a surrogate), by hypothesis, are biblical figures and texts that have provided the dom-

inant New World sources for envisioning and revising existence in terms
of an ideal world.

On this view the biblical world is reconfigured as a *pharmacopeic cosmos*.
More lucid and concise is the term 'pharmacosm', which designates a world
capable of hosting myriad performances of healing and harming, in accordance with an ideal world as envisioned by ancestral sources. Two
instances of conjurational lore from Zora Neale Hurston show how the
Bible has reinforced, and provided new (surrogate or diasporan) resources,
for the reconfiguring and maintenance of a 'pharmacosmic' worldview
among black Americans. Notice, in this folk sermon collected by Hurston,
correspondences between the world as a magical container and a nkisi
charm that contains the materials for God's world transformation.

> So God A'mighty, Ha!
> Got His stuff together
> He dipped some water out of de mighty deep
> He got Him a handful of dirt
> From de foundation sills of de earth
> He seized a thimble full of breath
> From de drums of de wind, ha!
> God, my master!
> Now I'm ready to make man
> Aa-aah![66]

Like a conjuror "God A'mighty . . . / Got His stuff together": that is,
the 'spirit-embodying' and 'spirit-directing' materials of world Creation.
The cosmos as God's 'conjure bag' is the representation found in this
sermon, and the cosmogony in the biblical text of Genesis provides its
figural basis. The "stuff" of charmed materials for the divine Creation of
humanity consists of dirt (compare grave dirt in conjure lore) in combination with a fluid—typical prescriptions in the paraphernalia of conjure.
In addition this combination is activated by wind, a reference to the divine
breath of the Genesis cosmogony. These *materia sacra* combine to create
humanity as a central element within a pharmacosm or, to use the folk
vernacular, as creatures inhabiting a "hoodoo" world. Moreover, it is a
world that permits and fuels our own conjurational performances as well:
a world in which it is axiomatic that "spirits or powers may be embodied
in material objects or charms, and may be manipulated by the knowl-
edgeable to harm people or to protect them."[67]

Another instance of the use of biblical referents in conjure lore pertains
to the text itself as an artifact. I have already cited the popularity, in conjure
lore documented by Hurston and Hyatt in the 1930s and still in print,
of a book of cabalistic spells, "seals," or amulets. For her part Hurston
replicated such a text in her fiction, in which she too represented a mystic
book as a source of knowledge for conjure performances—as a formulary
or conjure book. Like *The Sixth and Seventh Books of Moses*, however, this
fictive book was not biblical but magical: the book of the Egyptian god
of writing, Thoth. The book is also represented in Hurston's allegorical

novel of American slavery, *Moses: Man of the Mountain*, as a 'book of Moses'. The legendary book of Thoth is also a book of Moses because, through his heroic acquisition of it in the course of the novel, Moses acquires its "mystic powers" and appropriates its "symbols, seals and syllables" for performing his thaumaturgy.

> There is a book which Thoth himself wrote with his own hand which, if you read it, will bring you to the gods. When you read only two pages in this book you will enchant the heavens, the earth, the abyss, the mountain, and the sea. You will know what the birds of the air and the creeping things are saying. You will know the secrets of the deep because the power is there to bring them to you. And when you read the second page, you can go into the world of ghosts and come back to the shape you were on earth.[68]

Elsewhere, in presenting a cosmic origin for conjure lore, Hurston was concerned to indicate an antiquity for the tradition far greater than that of textual artifacts. "Belief in magic is older than writing," she declaimed in the passage immediately preceding her conjurational cosmogony in *Mules and Men*. But in *Moses Man of the Mountain* she shifted strategies and linked the stature of conjure to a mythic book in Egyptian antiquity—as if to lay claim not only to cosmic but also to historical preeminence on behalf of the tradition. However, this strategy effectively transforms the Bible as a cultural artifact. For, by representing the source of Moses' wisdom and power as a formulary of magical spells and incantations, Hurston in effect configured the Bible too as such a formulary. (Recall here that the Bible of some African American slaves consisted of ≤ the Book of the Apocalypse and the books of Moses. "All that lay between, even the narratives of the life of Jesus, they hardly cared to read or to hear."[69]) Indeed the Bible, as substantially "the books of Moses," becomes magical and Egyptian by the mediation of Moses as a magical figure. In this way Hurston derived a conjurational provenance for the Bible, by representing Moses' literary sources as magical and Egyptian (rather than in the orthodox terms of divine revelation and Jewish monotheism). In this way, too, the figure of the conjure book becomes formative[70] for subsequent developments in black religious and folk culture.

Notes

1. James Weldon Johnson, *God's Trombones: Seven Negro Sermons in Verse* (New York: The Viking Press, 1969), p. 17.

2. On the art of black preaching see Gerald A. Davis, *I Got the Word in Me and I can Sing It, You Know: A Study of the Performed African-American Sermon* (Philadelphia: University of Pennsylvania Press, 1985); William H. Pipes, *Say Amen Brother! Old-Time Negro Preaching* (New York: William-Frederick Press, 1951); Bruce Rosenberg, *The Art of the American Folk Preacher* (New York: Oxford University Press, 1970); Henry H. Mitchell, *Black Preaching* (Philadelphia: Lippincott, 1970).

3. Johnson, *God's Trombones*, p. 1.

4. Zora Neale Hurston, *Mules and Men* (New York: Harper & Row, 1990), p. 3.

5. Johnson, *God's Trombones*, p. 7

6. Ibid.

7. Ibid., pp. 1–2.

8. Ibid., p. 5.

9. See Amos Niven Wilder, *Theopoetic: Theology and the Religious Imagination* (Philadelphia: Fortress Press, 1976).

10. Rainer Maria Rilke, "The Angels," *Translations from the Poetry of Rainer Maria Rilke*, trans. M. D. Herter Norton (New York: W. W. Norton & Co., 1966), p. 45.

11. "God," in *The Book of Negro Folklore*, ed. Langston Hughes and Arna Bontempts (New York: Dodd, Mead, 1958), pp. 250–51. Reprinted from E.C.L. Adams, *Nigger to Nigger* (New York: Charles Scribner's Sons, 1928).

12. Ibid., p. 234–35. Cf. Hurston, *Mules and Men*, pp. 139–40. Consider the figural resonance between lines 10 through 12, "I can see! / Him seize de mighty axe of his proving power / And smite the stubborn-standing space," and representations by the Yoruba people in West Africa of their orisha or thunder god, Shango, whose thunderstones typically form the shape of a double axe. See Robert Farris Thompson, *Flash of the Spirit: African and Afro-American Art and Philosophy* (New York: Vintage Books, 1984), pp. 84–93.

13. James G. Frazer, *the Golden Bough: A Study in Magic and Religion*, vol. 1 of 2 (New York: Macmillan and Co., 1911), p. 52. Cf. the debate between René Girard and Jonathan Z. Smith on the notion of imitation in Frazer's concept of "sympathetic magic," in Robert Hamerton-Kelly, *Violent Origins: Ritual Killing and Cultural Formation* (Stanford: Stanford University Press, 1987), pp. 223–24.

14. As Bronislaw Malinowski pointed out, Frazer himself acknowledged as much at the end of *The Golden Bough*, thus contradicting his view of 'primitives' as incapable of reasoning in terms of causality. Bronsilaw Malinowski, *A Scientific Theory of Culture* (Chapel Hill: University of North Carolina Press, 1944), p. 196. Cf. Malinowski's discussion of "Rational Mastery by Man of His Surroundings" in his *Magic, Science and Religion and Other Essays* (Garden City, N.Y.: Doubleday and Co., 1948), pp. 25–36. Cf. also William James's comment: "Quite apart from the fact that many anthropologists—for instance, Jevons and Frazer— expressly oppose 'religion' and 'magic' to each other, it is certain that the whole system of thought which leads to magic, fetishism, and the lower superstitions may just as well be called primitive science as called primitive religion." William James, *The Varieties of Religious Experience* (New York: Penguin Books, 1985), pp. 30–31.

15. Malinowski, *A Scientific Theory of Culture*, p. 200; cf. p. 196.

16. Consider the "church magic" sanctioned by ecclesiastical authorities in medieval Europe, and researched by Keith Thomas in *Religion and the Decline of Magic: Studies in Popular Beliefs in Sixteenth and Seventeenth Century England* (New York: Charles Scribner's Sons, 1971).

17. René Girard, *Things Hidden Since the Foundation of the World*, with J.-M. Ourgoulian and G. Lefort (Stanford: Stanford University Press, 1987), pp. 283, 255; Girard, *Violence and the Sacred*, trans. Patrick Gregory (Baltimore: The Johns Hopkins University Press, 1977), p. 258.

18. Frazer himself gives several examples of curative or medical uses of magic, but without correlating such instances in any special way to the term "homeopathic." Frazer, *the Golden Bough*, pp. 78–84. Nor does Frazer seem to have been aware of Afro-American conjure traditions, despite his diverse interests and variety of examples drawn from African, native North American, and South American cultures. Finally, Frazer wrote that he considered using the term "imitative" or indeed "mimetic magic." But for reasons described previously as ideological, he was loathe to impute deliberate, conscious, or rational agency to practitioners of magic (an imputation that he considered to be implicit in the notion of imitation).

19. On African American theopoetics of darkness, see the quite different treatment in Howard Thurman, *The Luminous Darkness* (New York: Harper & Row, 1965), and Eulalio R. Balthazar, *The Dark Center* (New York: Paulist Press, 1973). Thurman (1900–81) was a mystical philosopher in the black American spiritual tradition. Balthazar articulates a process theology in which 'blackness' as mystery supersedes reason as the symbolization of supreme reality, and in which faith as a saving 'darkness' saves also from the religious roots of racism embedded in conventional Western forms of rationalist theology.

20. Johnson, *God's Trombones*, p. 17.

21.Ibid., p. 18.

22. In this connection the Nigerian philologist Modupẹ Oduyọye, in his "Afro-Asiatic" exegesis of Genesis 1 and 2, argues that "most demythologizing ends in remythologizing." Modupẹ Oduyọye, *The Sons of the Gods and Daughters of Men: An Afro-Asiatic Interpretation of Genesis 1–11* (Maryknoll, N.Y.: Orbis Books, 1984), pp. 34, 10–11. His commentary serves our purpose by suggesting how black folk tradition may display an interest in reversing the strong verbal aspect of deity in the Genesis *mythos*, simply in order to "remythologize" Genesis as an Afro-American text. Indeed, Oduyọye argues that such transpositions of mythic imagery are an embedded feature of the Hebrew text itself.

> The Yahwist writers of Genesis 2 and the Priestly writers of Genesis 1 were demythologizers of ancient [Near Eastern] myths. To demythologize, you effect a drastic change in imagery. If *mythology says* that the creator is a male-female couple . . . producing by sexual intercourse [as in the "copulating gods" of West African as well as Near Eastern traditions], *the Yahwist says*: 'No. The Creator is a single plastic artist using fine dust of the soil as plaster or clay to mold man, woman, and beasts' (Gen. 2.4ff), and *the Priestly writer says*: 'No. The creator is a spirit who merely blows the spirit from his lips as words of life and power: and sun, moon, stars, sea, land, and fish appear. The creator creates by words of power—words of command' (Gen. 1:1ff.). . . . One reason why the Priestly writer (ca. 450 B.C.) found it necessary to change the metaphor of the Yahwist (ca. 950 B.C.) is that those who believe that God is an image-maker may be inclined themselves to make images. This activity had been banned in Israel. God was therefore presented as a commander: the spirit (voice) proceeding from his lips brings things to life. . . . (Ibid.)

The preceding argument, however, needs to be qualified as a matter of emphasis: the priestly emphasis on God as commander includes the strong reference to God as making an image ("Let us make man in our image," Gen. 1.26), on the one hand, and the Yahwist depiction of God as forming humanity from dust refers also to the animating breath of the divine spirit (Gen. 2.7). Despite overstatement, Oduyọye's comments are especially illuminating in view of the predominant emphasis on verbal command in Genesis 1, as contrasted with Johnson's predominant use of divine imaging, simulation, or mimetics in his African American rendering.

23. Johnson, *God's Trombones*, p. 19.

24. Ibid., p. 20.

25. It is true that a signal effort of Harlem Renaissance artists—James Weldon Johnson and Zora Neale Hurston prominent among them—was to invigorate Afro-American arts by means of a deliberate retrieval of folk culture and traditional forms of expression. That mandate was brilliantly proclaimed by Alain Locke in his 1925 manifesto, *The New Negro*. But Locke's programmatic motivations gave direction to cultural imperatives that both preceded and survived the artists of the 1920s. Those imperatives—toward free expression of black experience, toward cultural parity and artistic recognition—had been abused and exploited by the minstrel tradition of the nineteenth century, on the one hand, and have most recently reemerged in the post-1960s Black Aesthetic movement on the other. On the black American minstrel tradition see Sam Dennison, *Scandlize My Name: Black Imagery in American Popular Music* (New York: Garland Publishing, 1982). On the more recent Black Aesthetic or Black Arts movement see Houston Baker, Jr., *Blues, Ideology and Afro-American Literature: A Vernacular Theory* (Chicago: University of Chicago Press, 1984), and Addison Gayle, Jr., ed., *The Black Aesthetic* (Garden City, N.Y.: Anchor Press/Doubleday, 1972). The perseverance of such imperatives indicates the degree to which Harlem Renaissance artists were only partly the creators of black culture; they were also its servants.

26. See the "Paraphernalia of Conjure" in the appendix to Zora Neale Hurston, *Mules and Men* (New York: Harper & Row, 1990), and also the multiple listings of conjure materials in Albert J. Raboteau, *Slave Religion: The "Invisible Institution" in the Antebellum South* (New York: Oxford University Press, 1978).

27. Zora Neale Hurston, *Mules and Men* (New York: Harper & Row, 1990), pp. 183–84. Some terminological clarifications may be helpful here. Hurston and other sources use the term "hoodoo" to describe practices that are also called conjure, conjuring, conjuration, or conjury, and sometimes root work or (more rarely) witchcraft. Such terms as "Voodoo" or "sorcery" are less appropriate, however, despite the linguistic and syntactical similarities operating in the homonyms, "hoodoo" and "voodoo," and in the synonyms, conjuration and sorcery.

Here "hoodoo" and "conjure" or "conjuration" are the preferred vernacular terms. They designate (a) a North American variant of the African religious 'survivals' found in the New World. That variant is properly distinguished from (b) traditional West African religions, on the one hand, and (c) Caribbean and South American developments on the other. A brief illustration: when referring to African survivals among black folk communities in North America it is more appropriate to speak of practitioners than worshippers. Such practitioners focus typically on the employment of spells in combination with materia medica. They are much less likely to engage in the worship or summoning of spirits or deities. Divine worship and spirit communication are more typically elements of Haitian Vaudou or other Caribbean and South American religious traditions: Santería in Cuba, Shangó in Trinidad, and Candomble and Macumba in Brazil are the major examples.

To reiterate: black North American conjure traditions can also be distinguished from the conventional senses of sorcery or witchcraft. An unqualified application of the term "conjure" is its Latinate senses can obscure the pharmacopeic and medicinal elements that distinguish African American hoodoo from its related traditions. So distinctive is this element that one commentator has emphasized the vernacular use of the pharmacopeic referent 'root work': "While its precise definition is unknown, it probably has to do with the preparation of potions from roots." Ronald M. Wintrob, "Hexes, Roots, Snake Eggs? M.D. versus Occult," *Medical Opinion* 1 (1972): 52. See also n. 54 below.

28. On "conjure discourse" or "conjure narrative" as an Afro-American oral communications genre see the discussion in David Brown, "Conjure/Doctors: An Exploration of a Black Discourse in America, Antebellum to 1940," *Folklore Forum* 23:1/3 (1990): 26ff.

29. Marjorie Pryse, "Introduction: Zora Neale Hurston, Alice Walker, and the 'Ancient Power' of Black Women," in *Conjuring: Black Women, Fiction and Literary Tradition*, ed. Marjorie Pryse and Hortense Spillers (Bloomington: Indiana University Press, 1985), p. 11

30. Hurston, *Mules and Men*, p. 184.

31. Ibid.

32. Zora Neale Hurston, *Moses Man of the Mountain* (Urbana and Chicago: University of Illinois Press, 1984), p. 147.

33. Hurston's biographer, Robert Hemenway, observes that "her fiction represented the processes of folkloric transmission, emphasizing the ways of thinking and speaking which grew from the folk environment." Robert E. Hemenway, *Zora Neale Hurston: A Literary Biography*, with a foreword by Alice Walker (Urbana and Chicago: University of Illinois Press, 1980), p. 242. Hemenway's comment is also reviewed in Pryse and Spillers, *Conjuring: Black Women, Fiction and Literary Tradition*, p. 12. Cf. recent discussion of the ethnographer as creative writer in *Writing Culture: The Poetics and Politics of Ethnography*, ed. James Clifford and George E. Marcus (Berkeley: University of California Press, 1987).

34. Hurston, *Moses Man of the Mountain*, p. xxi

35. Elsewhere, in her 1938 study of Haitian Voodoo, *Tell My Horse*, Hurston further elaborated the symbolic identification of Moses with an African traditional deity. There she also emphasized Damballah as "something of a creator, if not actively, certainly The Source." Zora Neale Hurston, *Tell My Horse* (Berkeley, Calif.: Turtle Island, 1983), p. 142. Robert Farris Thompson elaborates instead the Haitian identification of Damballah the serpent deity with St. Patrick, who in Roman Catholic hagiography drove the serpents out of Ireland, but without reference to a cosmogonic aspect. Thompson, *Flash of the Spirit*, pp. 176–79. Albert Raboteau corroborates both representations of Moses: "Damballa-wedo . . . is sometimes identified with Moses because of the miracle of the brazen serpent and sometimes with St. Patrick, pictured driving snakes from Ireland." Raboteau, *Slave Religion*, pp. 23–24.

36. See, for example, Vittorio Lanternari, *The Religions of the Oppressed: A Study of Modern Messianic Cults* (New York: Alfred A. Knopf, 1963), pp. 13, 39–41.

37. This work is an English translation of a nineteenth century German book

containing "wonderful magical and spirit arts," originally published in 1849 and claiming to derive from "rare old Mosaic books of the Talmud and Cabala," as stated in *The New Revised Sixth and Seventh Books of Moses and the Magical Uses of the Psalms*, ed. Migene Gonza'lez-Wippler (Bronx, N.Y.: Original Publications, 1991), p. 5. With its caballistic collection of spells, amulets, Hebrew names, and alphabetical characters, the folk usage of this book constitutes a clear form of European influence on African American conjure traditions. Finally, on the basis of his informants, Hyatt claimed that the book "is part of the Bible for most hoodoo believers." Harry Middleton Hyatt, *Hoodoo-Conjuration-Witchcraft-Rootwork: Beliefs Accepted by Many Negroes and White Persons, These Being Orally Recorded among Blacks and Whites*, vol. 2 of 5 (Hannibal, Mo.: Western Publishing Co., 1970), p. 1755.

38. Hyatt, *Hoodoo*, vol. 2, p. 1758.

39. *Drums and Shadows: Survival Studies among the Georgia Coastal Negroes*, Savannah Unit, Georgia Writers' Project, Works Projects Administration (Athens: University of Georgia Press, 1940), p. 28.

40. Cf. René Girard's concept of "mimetic double" as discussed in Chapter 7.

41. Lawrence W. Levine, *Black Culture and Black Consciousness: Afro-American Folk Thought from Slavery to Freedom* (New York: Oxford University Press, 1977), p. 50.

42. "Afro-Cuban herbalists knew the orisha owners, ritual applications, and curative powers of hundreds of trees, roots, barks, grasses, herbs, vines and flowers. In their exploration and classification of the Cuban forests and savannas, they were undoubtedly guided by the cognitive categories anthropomorphized as the 'orishas', which could comprise philosophical, aesthetic, anatomical, botanic and even chemical dimensions." Morton Marks, "Exploring *El Monte*: Ethnobotany and the Afro-Cuban Science of the Concrete," *En Torno a Lydia Cabrera*, ed. Isabel Castellanos and Josefina Inclán (Miami: Ediciones Universal, 1987), pp. 229, 231.

43. "Spelling Kongo with a *K* instead of a *C*, Africanists distinguish Kongo civilization and the Bakongo people from the colonial entity called the Belgian Congo (now Zaïre) and the present-day People's Republic of Congo-Brazzaville, both of which include numerous non-Kongo peoples. Traditional Kongo civilization encompasses modern Bas-Zaïre and neighboring territories in modern Cabinda, Congo-Brazzaville, Gabon, and northern Angola." Robert Farris Thompson, *Flash of the Spirit: African and Afro-American Art and Philosophy* (New York: Vintage Books, 1984), p. 103.

44. Sterling Stuckey, *Slave Culture: Nationalist Theory and the Foundation of Black America* (New York: Oxford University Press, 1987). Stuckey's first chapter highlights the prominence of ritual circles and circular dances in discussing "Slavery and the Circle of Culture," pp. 93–97. See also *Thompson's Flash of the Spirit* and his study coauthored with Joseph Cornet, *The Four Moments of the Sun: Kongo Art in Two Worlds* (Washington, D.C.: National Gallery of Art, 1981) and, for primary sources, John M. Janzen and Wyatt MacGaffey, *An Anthology of Kongo Religion: Primary Texts from Lower Zaire* (Lawrence: University of Kansas Press, 1974). Thompson's earlier art history study of Yoruba *orisha* is *Black God's and Kings* (Bloomington: Indiana University Press, 1976).

45. Raboteau, *Slave Religion*, pp. 7–8.

46. Ibid., p. 16

47. Ibid., pp. 8–16. For a critique of the term "survivals" and a preference for the term "transformations" see Levine, *Black Culture*, p. 4.

48. Thompson, *Flash of the Spirit*, p. 107. Raboteau, *Slave Religion*, p. 11.

49. Cf. n.35.

50. Raboteau, *Slave Religion*, p. 92.

51. "Each *orisha* is served by its own sacralizing herbs, taken with permission from Osanyin, just as each deity is accompanied by his or her Eshu, source of individualizing power and vitality." Moreover: "[Osanyin] needs Ifa ... the knowledge of the leaves must be shared. But the obverse is equally true: Ifa needs Osanyin. Without the lord of leaves and his many medicines, Ifa's effectiveness would be seriously diminished. And so the diviner herbalist and the herbalist come to share the *osun* staff." Thompson, *Flash of the Spirit*, pp. 42, 45.

52. "Rootwork is a curious word. While its precise definition is unknown, it probably has to do with the preparation of potions from roots. In a broader sense rootwork can be defined as a highly organized system of beliefs shared by blacks who were raised in the Southeastern U.S. or who retain close ties there with family and friends. ... [They consult] a rootworker, known also as a conjurer, rootdoctor, or hoodoo man. The rootworker is believed to possess power to cast spells as well as remove them. The person who comes under the spell of a rootworker is considered rooted, mojoed, criss-crossed, hoodooed, or conjured." Wintrob, "Hexes, Roots, Snake Eggs?" pp. 54–55. Wintrob continues with a discussion of ethnographic attribution: "It is usually agreed that rootwork beliefs and practices are derived from three sources: African witchcraft, Haitian voodoo, and European witchcraft. But anthropologists engage in lively debate as to which source is most influential." See Raboteau, *Slave Religion*, pp. 48–60, 85–86, regarding the Frazier-Herskovits debate on this precise issue.

53. Gary Edwards and John Mason, *Black Gods—Orisha Studies in the New World* (Brooklyn, N.Y.: Yoruba Theological Archministry, 1985), pp. 29–30.

54. I am indebted to my colleague David Brown, in the Art History Department of Emory University, for this precise formulation.

55. Raboteau, *Slave Religion*, p. 59.

56. Thompson, *Flash of the Spirit*, p. 107.

57. Luc de Heusch, *Why Marry Her? Society and Symbolic Gestures* (Cambridge and New York: Cambridge University Press, 1971), p. 182; Wyatt MacGaffey, "Complexity, Astonishment and Power: The Visual Vocabulary of Kingo Minkisi," *Journal of Southern African Studies* 14:2 (January 1988): 190; Thompson, *Flash of the Spirit*, pp. 117–18.

58. MacGaffey, "Complexity, Astonishment and Power," pp. 192–93. Cf. John M. Janzen, *The Quest for Therapy in Lower Zaire* (Los Angeles: University of California Press, 1978), pp. 44–53.

59. Thompson, *Flash of the Spirit*, pp. 117–18. MacGaffey, "Complexity, Astonishment and Power," pp. 190–91. Brown, "Conjure/Doctors," p. 20.

60. Brown, "Conjure/Doctors," pp. 20–21.

61. Brown, "Conjure/Doctors," p. 20, quoting Claude Levi-Strauss, *The Savage Mind* (Chicago: University of Chicago Press, 1966), pp. 18, 21.

62. Hurston, *Mules and Men*, p. 280

63. Thompson, *Flash of the Spirit*, p. 107.

64. Norman E. Whitten, Jr., "Contemporary Patterns of Malign Occultism among Negroes in North Carolina," *Journal of American Folklore* 75:298 (1962): 311–25.

65. Raboteau, *Slave Religion*, p. 33. I resist, as a prematurely theological imposition, the impulse to make this distinction a strict moral dichotomy between good and evil. A somewhat related distinction has been emphasized by Melville Herskovits, who cited, and extensively quoted from, the unpublished study by Vivian K. Cameron, "Folk Believes Pertaining to the Health of the Southern Negro," master's thesis, Northwestern University, 1930 (pp. 36–37):

> [T]wo groups of practitioners are known and recognized not only by themselves but also by their particular clienteles, as distinct from each other. One deals in what may be termed "medicine," that is, roots, herbs, barks and teas, while the other is composed of those who work by means of magic. So clear cut is this feeling of difference between the members of these two groups that there is reason for deep insult if a practitioner of the medical type is mistaken for one of those who practice magic.

Herskovits himself endorsed Cameron's differentiation of 'medical' from magical conjurors by referring to distinctions of dress and gender between the two—women predominating as pharmacopeic practitioners and men as magicians, a distinction that he found corroborated in Newbell Niles Puckett's early study, *Folk Beliefs of the Southern Negro* (Chapel Hill: University of North Carolina Press, 1926). Herskovits then stated his own corroboration of the distinction with reference to West African indigenous cultures, in his classic study, *The Myth of the Negro Past* (New York: Harper and Brothers Publishers, 1941), p. 241:

> [C]ritical assessment of the magico-medical complex of the American Negro ... [indicates] point after point at which African tradition has held fast. In West Africa those who deal in herbs, roots, and other curatives are invariably differentiated from those whose cures for illness and other less mundane evils come from their supernatural powers.

However, Herskovits immediately cites Puckett and others in order to stress the moral neutrality of these distinctions.

> The conviction held by American Negroes that no dichotomy exists between good and evil in the realm of the supernatural, but that both are attributes of the same powers in terms of predisposition and control, is characteristically African.... The immediate African parallel... [is] striking because it contrasts vividly with the European habit of separating good and evil so strongly that the concept of the two as obverse and reverse of the same coin is almost nonexistent. (p. 243)

More recent discussion of this issue can be found in John W. Roberts's folklore study *From Trickster to Badman*. In addition to the pharmacopeic and the magic performances of the conjuror, Roberts emphasizes that "the conjurer's characteristic actions complemented those most often associated with the trickster ... a trickster possessed of spiritual power." In this regard, the criteria for good and evil can be referred to the slave community as a collective clientele needing heroic tricksters to combat (white) aggressors, on the one hand, but also expressing fear and condemnation of the internalized consequences of such activities when turned inward against the community itself:

> [B]ehavior involving trickery against a fellow sufferer, regardless of its motivations or nature, was not in the best interest of their community.... [Yet

their white] antagonists were tricksters of the worst sort—individuals who attempted to manipulate the force in nature to rob others of their very being. Therefore, enslaved Africans justified the conjurer's actions by virtue of their need to act as both protagonist and antagonist in order to protect the [community]. (p. 103)

On trickster elements in African traditional and African American cultures see Diedre L. Badejo, "The Yoruba and Afro-American Trickster: A Contextual Comparison," *Présence Africaine* 147:3 (1988): 3–17, and Robert D. Pelton, *The Trickster in West Africa: A Study of Mythic Irony and Sacred Delight* (Berkeley: University of California Press, 1980). I review Badejo's treatment of the Afro-American trickster, and also return to the issue of good and evil with reference to the folk traditional and black church censure of conjure as the 'devil's work,' in Chapter 5.

66. Zora Neale Hurston, *Jonah's Gourd Vine* (New York: Harper & Row, 1990), p. 176. Also in Hughes and Bontemps, *The Book of Negro Folklore*, p. 237.

67. Raboteau, *Slave Religion*, p. 33.

68. Hurston, *Moses Man of the Mountain*, pp. xxii, 73. A reference to this same magical book also appears in an ancient Egyptian text dating from the Demotic literature of the late period (7th century B.C.E.). This is not a place to pursue a philological correlation of Hurston's Book of Thoth with the Egyptian lore on which it depends, and to which it hearkens as a repository of 'Afrocentric' emblems. Nor can I review here the portrayal of the god Thoth or "Theuth" from another significant text in which he appears: Plato's dialogue *Phaedrus*. However, compare Hurston's depiction of the book with the opening passage of the Egyptian story of Setne Khamwas, known as Setne I and preserved on papyrus (No. 30646) in the Cairo Museum.

Prince Khamwas, son of King Ramses II and high priest of Ptah at Memphis, was a very learned scribe and magician who spent his time in the study of ancient monuments and books. One day he was told of the existence of a book of magic written by the god Thoth himself and kept in the tomb of a prince named Naneferkaptah (Na-nefer-ka-ptah), who had lived in the distant past and was buried somewhere in the vast necropolis of Memphis. After a long search, Prince Khamwas, accompanied by his foster brother Inaros, found the tomb of Naneferkaptah and entered it. He saw the magic book, which radiated a strong light, and tried to seize it. But the spirits of Naneferkaptah and of his wife Ahwere rose up to defend their cherished possession.
Miriam Lichtheim, *Ancient Egyptian Literature*, vol. 3 of 3 (Berkeley: University of California Press, 1980), p. 127.

The story continues and, along with a second story from the Greco-Roman period in Demotic literature (Setne II in Lichtheim, *Ancient Egyptian Literature*, vol. 3, pp. 138–51), provides material of great interest for an arcane tradition connecting ancient Egypt and Nubia to magical lore in African American culture.

69. Hurston, *Mules and Men*, p. 183. Levine, *Black Culture and Black Consciousness*, p. 50.

70. Indeed I claim for the formative nature of 'the conjure book' in black ⌐

folk religious culture what the literary theorist Henry Louis Gates, Jr., claims for the status of the "talking book" in black American letters: "The trope of the Talking Book is the ur-trope of the Anglo-African tradition." Gates, *The Signifying Monkey: A Theory of African-American Literary Criticism* (New York: Oxford University Press, 1988), p. 131.

2

Exodus

The tormented ones will want to discover (thus they have learned from Egypt) how they can powerfully conjure God to appear forthwith and help.

Martin Buber, *Kingship of God*[1]

In their reflections upon the meaning of the Emancipation Proclamation, Afro-American freed men and freed women and their folk theologians realized that a decisive event of biblical narrative had become an occurrence in their own historical experience. Lincoln's presidential order, following upon the cataclysm of civil war, demonstrated that the miraculous exodus of Hebrew slaves out of Egyptian bondage could become a mundane reality in contemporary terms. The implications of that lesson, reinforcing earlier apprehensions of divine providence and prophetic fulfillment, also promised future repetitions. The likelihood of ongoing recapitulations of biblical narrative became immediately accessible, even compelling, to the religious apprehension of thousands. Henceforth more than a minority of believers and converts would be convinced of the possibility that through prayer and expectation, through acts of obedience and righteousness, black folk could inherit divine promises of prosperity and freedom. Furthermore, an apparent precondition for such bestowals would appear to be their linkage to biblical models. That singular instance, the link between Lincoln's role in the emancipation and Moses' role in the Exodus, would distinguish itself as a kind of paradigm. In this manner a new development in the ancient tradition of biblical typology emerged in the collective psyche of a displaced people.

Typology is the hermeneutic (interpretive) tradition that links biblical types or figures to postbiblical persons, places, and events. Moreover, it is crucial for this hermeneutic that each particular dyad of a biblical "type" and its postbiblical "antitype" should be understood to represent a fulfillment of prophecy.[2] The origins of this tradition go back beyond the medieval period to the Christian Scriptures and the early church. Taken as a whole the two millennia have witnessed its formation into a uniquely Christian spirituality. Black religious figuralism also participates in that spiritual heritage. Nonetheless its adoption by Afro-Americans did not arise in a vacuum. Nor is it solely attributable to the pervasive influence of Puritan typology on all immigrant and relocated groups in United States cultural history. Puritan influence has been so compelling in that history that Werner Sollors has coined the term "typological ethnogenesis" to comprehend it. Sollors and other literary historians of Puritanism are convincing in the claim that disparate ethnic communities of whites, Jews, and peoples of color have inherited the Puritans' biblical figures for America. Just as they configured New England as "God's new Israel," as "Canaan" and as "Promised Land," in order to create their group identity as a New World "chosen people," so have other ethnic communities subsequently adopted and adapted that theological-political strategy.[3] Nonetheless a careful ethnographic investigation would reveal that each of them did so in distinctive ways, accommodating the Puritan hermeneutic to their antecedent traditions of spirituality.

For Afro-Americans the antecedent spirituality was distinctly extra-Christian. But one must do more than acknowledge the fact in passing. To take the West African sources of black American cultures seriously requires a structural or systematic incorporation of extra-Christian elements in our ethnographic investigations, and also (where relevant) in our theological formulations. In this chapter I attempt a structural treatment of black North American experience by stressing, on the one hand, the insistently ritual nature of black religion particularly as regards its appropriation of biblical typology. On the other hand I am careful to observe the bicultural character of black figural discourse. Maintaining a bicultural perspective ensures that the complexity of a culture and the density of its formations are displayed, rather than reducing black religion either to imputed African (ritual) elements alone or to conventionally Christian (theological) influences alone. In this connection one commentator has disclosed ritual elements operating in the figural theology of Puritan America[4] Indeed, on this basis I will indicate some of the same incantatory dynamics in Puritanism that I find in black religious data. On the other hand I do not here apply the term 'conjuring culture' to Puritan American spiritual transformations in the New World; I reserve the term for African American transformations because of their more clearly articulated pharmacopeic, ritual and magical orientations. The distinctively conjurational nature of black religion does not, I hasten to add, preclude a theological treatment of the tradition's Christian features as well.

The prominence of ritual in African and Afro-American cultures has been observed from a variety of disciplinary perspectives.[5] By ritual I mean social practices in which the creative impulse to pattern reality results in a repetitive structuring of human action. Ritual is a primal, and primary, mode of social interaction and performance. In operation it either precedes ideation, articulation of concepts, and symbolic expression (as in myth, poetry, or theatre) or subordinates these as means in its patterning of action. If we inquire to what end such processes aim, it appears that rituals are intended to be transformative more than representational. They are instrumental: focusing upon the "means to" rather than the "meanings of" transformed versions of reality.[6] So understood, rituals need not be limited to religious performances but can also include secular transformations. Neither need they be rigidly fixed by the invariant repetition of particular patterns, although continuity and conserving of similarities undergird their authoritative power to bond participants and induce their adherence. But in addition to invariant rituals, *homo ludens* (the human being as player) also improvises.[7] That is, we create supervisory modes or 'metapatterns' for altering the conventional patterning of our actions, while simultaneously observing the need for continuity and authority, in order to accommodate external contingencies and our own changing intentions in the structuring of reality. With respect to my subject in this chapter, a perennial source of emancipatory improvisations is the transformative process described by the anthropologist Victor Turner as *communitas*: the humanitarian bonding between persons in a community who intentionally dissolve class distinctions and conventional constraints in order to free each other for egalitarian relationships.[8]

Let us observe such improvisational play, with its communitarian intent, in the ritual strategies of black people in North America. Notable in this regard are ecstatic worship or spirit possession, on the one hand, and the folk tradition of conjure on the other. But by extension there are also aesthetic and political contexts in which ritual performances are featured. In the sections below I explore the ritual nature of black social and political figuration in the North American context, in a constructive effort that culminates in the presentation of an Afro-American repertory of biblically formed political configurations. In conclusion I also anticipate new developments in this black religious tradition of narrative reenactment— a tradition, I posit, which transforms the world through ritually improvised applications of biblical models.

God-Conjuring

In his literary theory Kenneth Burke also recognized ritual drama as the primal form of human action. Burke described ritual in metaphorical terms as the "hub" from which all other forms of distinctively human action radiate, like the spokes of a wheel. Human discourse, on this view, is best understood as "symbolic action" which retains genetic elements of its ritual origins.[9] It thereby conveys primal significations, as if continuing to function within a "ritual cosmos" (Zuesse) irrespective of changing contexts. Of course, some language users are more proficient in utilizing the power of discourse to recall and project ritual significations: poets and vocalists,

preachers and orators, writers and dramatists. Their varied inducements
to restore such a ritual cosmos are also an invitation to make the figurative
efficacious; to participate in symbols not only cognitively, but so as to
fulfill them on the scale of group actions and social dramas. "One must
learn to enact the world differently. One must see every *thing* as *symbol*.
It takes constant repetition and dramatization to achieve this ritual vision
of life."[10]

In his 1961 study of "the new African Culture," Jahnheinz Jahn described black religions in similar terms. Jahn spoke of African and African
American religions as traditions of "active worship" which "create" God,
and which "install the divine being as such." Intriguingly, Jahn's language
coincides with the conventional meaning of the word "conjure" as an act
of the imagination: to conjure up a picture, image, or idea. "Analogously
to the designation of an image," Jahn declares, "we may speak of the
designation of divinity."[11] This formulation suggests that in worship a
deity is conjured (summoned, evoked) in ways analogous to the imaging
or imagining of a designated object. In conjunction with Jahn's notion of
'making' or inducing the deity, I retrieve Kenneth Burke's elucidation of
incantation or the "incantatory": as a "device for inviting us to 'make
ourselves over in the image of the imagery.' "[12] Burke's formulation suggests a reciprocal reference: to designate a deity is also to image or re-
imagine the self. Moreover, such reciprocity allows for an extension of
incantatory dynamics beyond the primal context of ritual worship.

In its original context of ritual and worship, imaging or designating
of the deity may be said to occur integrally within the primal experience
of religious fascination, where the subject experiences terror or awe in the
presence of the holy (Otto). But under the exigencies of historical experience the designation of deity may acquire some intention other than
ritual or worship. From this perspective I propose the hypothesis that
black North American experience features a development from designating
or 'summoning' God in worship, to an intention to conjure God for
freedom. The best places to verify such a development are writings in
which religious expressions function rhetorically as ritual incantations—
incantations that reciprocally summon God and, in Burke's phrase, "invite
us to make ourselves over in the image of the imagery." The nineteenth
century "Ethiopian" texts of two black antislavery writers, David Walker
and Robert Alexander Young, are ideal for this purpose. These writings
are also remarkable in that they inaugurate black religious figuralism as a
literary tradition—the literary-religious tradition of Ethiopianism.

Black literary expression is called Ethiopian when it depends for its
rhetorical and prophetic force on Psalm 68.31 (King James Version):
"Princes shall come out of Egypt; Ethiopia shall soon stretch out her hands
to God." Albert Raboteau has described this text as "probably the most
widely quoted verse in Afro-American religious history."[13] We will return
to Ethiopia as a primary biblical figure in black religious tradition. More
immediately it is helpful to observe, in early Ethiopianist texts, a displace-

ment of conjuration from the sphere of ritual and worship to the existential sphere of human freedom. That displacement is evidently a *conditio sine qua non* for incantations like the following. First, in David Walker's *Appeal to the Coloured Citizens of the World* (1829) we read:

> Though our cruel oppressors and murderers, may (if possible) treat us more cruel, as Pharaoh did the Children of Israel, yet the God of the Ethiopians, has been pleased to hear our moans in consequence of oppression, and the day of our redemption from abject wretchedness draweth near, when we shall be enabled, in the most extended sense of the word, to stretch forth our hand to the Lord our God.[14]

This passage is notable for its *double* reference both to Exodus—"Pharaoh" and "the Children of Israel"—and to Ethiopia; indeed, Walker invokes "the God of the Ethiopians." This double use of biblical figures conveys as well the kind of incantatory reciprocity referred to in relation to Burke's expression "making ourselves over in the image of the imagery." For, reciprocally with the invocation of the God of the Ethiopians, the writer intends to remake Afro-Americans in the image of Hebrew slaves crying out under Egyptian bondage. Similar dynamics are evident in Robert Alexander Young's "Ethiopian Manifesto," which also appeared in 1829 and displays both the double reference to Exodus and Ethiopia and the reciprocal incantations of God and self.

> We tell you of a surety, the decree hath already passed the judgement seat of an undeviating God, wherein he hath said, "surely hath the cries of the black, a most persecuted people, ascended to my throne and craved my mercy; now, behold! I will stretch forth mine hand and gather them to the palm, that they become unto me a people, and I unto them their God."[15]

Here we may observe how a conjurational mode of spirituality is operative in the act of designating a deity. In the preceding passage a crucial element of such conjuration is that the "God of the Ethiopians" and an "undeviating God" are designated not only for the sake of worship—in order that "they become unto me a people, and I unto them their God"—but also for the sake of freedom, in that "the day of our redemption from abject wretchedness draweth near." The reference to freedom is crucial because it can be read not only as simple prophecy: that is, as a visionary prediction of the coming emancipation. Rather it is more profoundly understood as prophetic incantation: as religious expression intending to induce, summon, or conjure the divine for the realization of some emancipatory future. Again, such a "strategy of inducement" has been clearly described in Burke's literary theory:

> Neo-positivism has done much in revealing the secret commands and exhortations in words—as Edward M. Maisel, in *An Anatomy of Literature*, reveals in a quotation from Carnap, noting how the apparent historical creed: "There is only one race of superior men, say the race of Hottentots, and this race alone is worthy of ruling other races. Members of these other races are in-

ferior," should be analytically translated as: "Members of the race of Hottentots! Unite and battle to dominate the other races!" The "facts" of the historical assertion here are but a strategy of inducement (apparently describing the scene for the action of a drama, they are themselves a dramatic act prodding to a further dramatic act).[16]

Two differences between strategies of inducement as presented by Burke and that which we find in Ethiopianism should be noted. First, the Walker and Young texts disclose an explicitly religious strategy, a strategy in which divine providence is invoked. The second difference follows from the first. The "secret commands and exhortations" of Walker and Young are designed not only for human readers but preeminently for that divine reader who has been designated as "undeviating" and who has been conjured as the "God of the Ethiopians." But there is a more significant dimension of Ethiopianism in these texts.

The strategic use of language here consists not only in the rhetoric of inducement or incantation but also in that of biblical figuralism. Thus the figure in Psalm 68.31—Ethiopia stretching out her hands to God—becomes prophetic with respect to the situation and destiny of black people in early nineteenth century America: "when we shall be enabled, in the most extended sense of the word, to stretch forth our hand to the Lord our God" (Walker). Other variations are possible in the rhetorical use of this prophetic figure, as Young demonstrates by imagining or conjuring God seated on his throne and declaring: "Now, behold! I will stretch forth mine hand and gather them." In each instance it is the "God of the Ethiopians" who is designated, so that the biblical figure of Ethiopia is employed as an element (or prescriptively, as materia medica) within a strategy of inducement. Moreover, the figural identification of Ethiopia with 'all Africa' is juxtaposed with the God of the Bible in the phrase "God of the Ethiopians." Thus a symbol of Africa occurs in metonymic correlation with the deity of white Americans and signifies the conjoining and potential transformation of 'black' and 'white' religious traditions.

Finally, what is most consequential is that this God is designated for the purpose of human freedom. As the liberation theologian James Cone insists, it is the "God of the oppressed" who is invoked by the black freedom tradition.[17] Rather than move immediately to Cone's liberation theology, however, my task in this chapter requires that we remain at the pretheological level of religious ethnography. In this connection, the incantatory dimension in Ethiopianism can be construed as an effort to conjure-God-with-Scripture-for-freedom. By "stretching out their hands to God" in conversion, preaching, and ministry, the Ethiopianists imitate the Ethiopians of the prophecy in Psalm 68:31. This mimetic operation has its complement in the transformation of black American slaves who, by conversion, prayerful outcry, and longsuffering, become Mosaic slaves and thereby effectively induce God to grant them the same future of freedom as the Hebrew slaves were granted in the book of Exodus.

A final instance of black figural incantation in this brief review is the infamous "Address to the Slaves of the United States of America," delivered by Henry Highland Garnet in 1843. Garnet's speech explicitly incited slaves to revolt and violently overthrow their masters. It shared this feature with David Walker's *Appeal*, which explains why the text of Garnet's "Address" was appended to the 1848 publication of the *Appeal* by its promoters, and why both were banned in the South by their proslavery opponents. The Address employs only the Exodus figure, and the object of its incantatory inducements seems limited to the slaves as fellow human beings rather than also including the divine being. Despite such one-dimensionality Garnet's text employs other dynamics that claim our attention.

> You are not certain of heaven, because you suffer yourselves to remain in a state of slavery, where you cannot obey the commandments of the Sovereign of the universe. . . . The diabolical injustice by which your liberties are cloven down, *neither God, nor angels, or just men, command you to suffer for a single moment.* . . . You had better all die—DIE IMMEDIATELY, than live slaves and entail your wretchedness upon your posterity. If you would be free in this generation, here is you only hope. However much you and all of us may desire it, there is not much hope of redemption without the shedding of blood. If you must bleed, let it come all at once—rather DIE FREEMEN THAN LIVE TO BE SLAVES. It is impossible like the children of Israel, to make a grand exodus from the land of bondage. The Pharaohs are on both sides of the blood-red waters![18]

The strategy of inducement employed here involves the rather transparent tactic of arousing the slaves on the basis of their Christian piety and devotion. *If* you love God—the tactic asserts—whom you cannot fully obey and serve in your state of bondage, *then* you must murder your masters. Here we have, to say nothing more, a desperate inference. But more pertinent from an analytic perspective is Garnet's attempt to induce a higher degree of credibility and legitimacy for the inference by resorting to the figural elements of the biblical Exodus event: "It is impossible like the children of Israel to make a grand exodus from the land of bondage."

Two aspects of this reference to Exodus bear emphasizing. First, it is incantatory insofar as it engages its designated audience at the level of their spiritual and mimetic desire to be "like the children of Israel." In fact, we do well to read the passage as a 'counterspell'. It is Garnet's attempt to break or redirect the power which the Exodus figure wields over the slaves insofar as it induces them to wait for the same providential deliverence that attended the children of Israel. As the historian Timothy Smith has emphasized: "The Christian beliefs [the slaves] adopted enabled the African exiles to endure slavery precisely because these beliefs supported their moral revulsion toward it and promised eventual deliverance from it *without demanding that they risk their lives in immediate resistance.*"[19] But risking their lives for deliverance is precisely the disposition that Garnet, like Walker, sought to induce in his audience. In order to achieve that end he denied the literal correspondence between American slaves and

Hebraic slaves, for such correspondence implies a providential destiny that supersedes violent resistance. This counterconjurational requirement leads to a second point of emphasis.

Like a master conjuror, Garnet did not simply deny the prophetic power or the providential efficacy of the Exodus figure. That blundering tactic would subvert his own basis of spiritual influence on his audience! Rather, he denied only the similarity of situations between ancient Egyptian bondage and American bondage. He pointed out that the case of American slavery presented a greater difficulty than its biblical precursor: "The Pharaohs are on both sides of the blood-red waters!" By such language Garnet used the figure to reject the figure. But despite this counterfigural intention his rhetoric implicitly confirmed the vitality and utility of a conjurational tradition that uses biblical figures for incantatory effect.

Before leaving Garnet's strategy of inducement, however, we have the opportunity to observe certain resonances between it and the statement by the Jewish philosopher Martin Buber. In the passage that prefaces this chapter, Buber noted the magical worldview of ancient Israel in the context of the Exodus narrative. Indeed, coincident with Zora Neale Hurston, Harry Hyatt, and some of their folk sources reviewed previously, Buber too represented conjuration as originally an Egyptian magical tradition. Referring to the Hebrew slaves in Egyptian bondage, Buber observed that "the tormented ones will want to discover (thus they have learned from Egypt) how they can powerfully conjure God to appear forthwith and help." However, he proceeds to speculate, Hebrew slaves failed in their efforts to conjure God. Rather they encountered a God who declared, in Buber's words, "You do not need to conjure Me, but you cannot conjure Me either." This speculative reconstruction of an ancient collective subjectivity is Buber's way, as he explains, of accounting for the eventual atrophy of overt or explicit efforts to conjure God in Israel's religious history. "What is here reported is, considered in terms of the history of religion, the 'demagicizing' of faith."[20] Without engaging that claim at either its ethnographic or its theological level of assertion (is it an orthodox retrojection onto the ancient past of later standards of progress or purity?), I am struck by its congruence with Henry Highland Garnet's evident
> forbearance. Is his failure explicitly to 'conjure God with scripture for freedom,' in contrast to his Ethiopianist peers, evidence of his advancement of, or alienation from, a vigorous and compelling tradition?

Ethnogenesis

To designate a deity, I have suggested, means also to image a self. But this reciprocity need not refer only to individual selfhood; it may also include corporate identity. Thus, when a community or people designates its patron god or deities it simultaneously constitutes or reconstitutes its collective identity. In figural Ethiopianism, accordingly, we see the interplay between emancipatory aspirations on the one hand and corporate formation on the other—a conjunction with theological dimensions that I cannot address here.[21] The aspect of corporate formation, however, is also the focus of a literary and ethnohistorical work previously mentioned, Werner Sollors's *Beyond Ethnicity: Consent and Descent in American Culture* (1986). A study in literary history and criticism in the field of ethnic American literatures, the book is most useful here for its development of the concept of "typological ethnogenesis." By this term Sollors means simply "the sense of peoplehood that emerged among Puritans and ethnic groups [who] drew on typology."[22] He then proceeds to document the ubiquitous figuration of peoplehood that employs biblical typology in ethnic America. We need not repeat Sollor's convincing survey here. It is sufficient to note his conclusion: "In post-Puritan America, white and black, business and labor, Jew and Gentile, could follow typological patterns and *become biblical antitypes*."[23]

Sacred peoplehood through becoming a biblical antitype: that formula states the corporate interest of black America in biblical figuralism as manifested in Ethiopianism. The following three figures—Exodus, Ethiopia, and Egypt—are the major constituents of this typological ethnogenesis.

The Exodus figure

Slaves prayed for the future day of deliverance to come, and they kept hope alive by incorporating as part of their mythic past the Old Testament exodus of Israel out of slavery. . . . The Christian slaves applied the Exodus story, whose end they knew, to their own experience of slavery, which had not ended. . . . Exodus functioned as an archetypal event for the slaves. The sacred history of God's liberation of his people would be or was being repeated in the American South.

Albert J. Raboteau, *Slave Religion*[24]

In slave religion the figural vision of the emancipation as a type of biblical "Exodus" provides the paradigm for subsequent strategies and acts of political imagination. The Exodus figure has informed Afro-American political projects from the post-Reconstruction period to the recent civil rights movement, in which a premier instance is Martin Luther King, Jr.'s, leadership of the freedom movement. The figural dimension of King's leadership has been nicely articulated by the historian Lerone Bennett, Jr., as consisting of two foci: "his original choice of himself as a symbolic being . . . and the further fact that the movement was already based on the solid rock of Negro religious tradition."[25] That the tradition was biblical and typological explains both King's aptitude for, and success in, representing himself as a type of Moses. This representation had its climactic expression in his final and prophetic speech, "I've been to the mountaintop!" There King envisioned himself as Moses on Mount Pisgah who looks over into the Promised Land and affirms that, while he may not get there, someday his people will.[26]

A noteworthy aspect of Exodus figuration in slave religion is the element of inverting or improvising the conventional Puritan signification of the figure. On the one hand, "white Christians had identified the journey across the Atlantic to the New World as the exodus of a new Israel from the bondage of Europe into the promised land of milk and honey." On the other hand, "for the black Christian, as Vincent Harding has observed, the imagery was reversed: [the Middle Passage has brought his people to Egypt land, where they suffered bondage under Pharaoh. White Christians saw themselves as a new Israel; slaves identified themselves as the old."[27] The slaves' inverse identification evinces an improvisational propensity: a propensity not only for imitative appropriation but for variations and transformations of the available Euro-Christian materials and resources.

Here a theological view leads to further ethnographic insight. In religious perspective the figure of Exodus signifies God's liberation of Israel from Egyptian bondage by means of collective dependence on divine providence—that is, without the need for autonomous acts of violence. Thus the Mennonite theologian John Howard Yoder reminds us that in the Exodus event the people of the Lord do not take up arms. Rather, with an intensity that never recurs in subsequent biblical narrative, the community was able to claim (in place of their own military capabilities) that "the Lord is a Warrior!" (from "The Song of Moses," Exodus 15.3).

> The exodus experience is of a piece with the ancient Hebrew vision of Holy War. The wars of JHWH were certainly lethal, but they were not rationally planned and pragmatically executed military operations; they were miracles. In some of them (and the Red Sea is such a case), the Israelites, according to the record, did not even use arms. The combatant was not a liberation front or terrorist commando but JHWH himself. . . . The Red Sea event is for the whole Old Testament the symbol of the confession that the Israelites do not lift a hand to save themselves.[28]

In mimetic correlation with biblical narrative, the emancipation of Afro-American slaves occurred also without massive slave insurrections or armed struggle. The divine supercession of human (pharaonic) resistance, by means of providential developments vis-à-vis the victims, constitutes an essential element of correspondence between Exodus and the emancipation. This level of figural correspondence would be altogether missing in such cases as John Brown's insurrection or Nat Turner's revolt, the more so had they succeeded on a vastly larger scale.

But Exodus figuration is not the only phenomenon of its kind in black culture. Related forms of religious figuralism operate as well in other political projects of the nineteenth and twentieth centuries, including black nationalist projects of revitalization. As we will see, those projects were spiritual-political in that they sought to transcend the prevailing structures of social oppression by using religious discourse to render the New World and the Old a locus of communitas (Turner)—of freedom, wholeness, and cultural creativity for black peoples.

The Ethiopia figure

Princes shall come out of Egypt; Ethiopia shall soon stretch out her hands to God.

Psalm 68.31 (King James Version)

Probably eighteenth century missionary interpretation bequeathed to black believers the use of "Ethiopia" as a symbol for all Africa and for black people everywhere. White Christians may also have first interpreted the verse as a prophecy of the mass conversion of black people to the faith. Finally, it also seems to be a product of European chauvinism to claim that God's "providential design" ordained the slave trade in order to bring Christianity to Africa.[29] But gradually Afro-Americans cast the Christian fulfillment of the prophecy in their own perspective.

Generations of interpretation eventually developed into a literary-religious tradition called "Ethiopianism," which spanned the late eighteenth and early twentieth centuries, and engaged black communities on both the African and American continents. In Ethiopianism, Wilson Jeremiah Moses observes, Psalm 68.31 "came to be interpreted as a promise that Africa would 'soon' experience a dramatic political, industrial, and economic renaissance." This notion of African resurgence—resurgence against the background of the slave trade and Africa's nineteenth century humiliation by European colonialism—included an accompanying vision of the decline of the West. "The rise in the fortunes of Africa and all her scattered children would be accompanied by God's judgement upon Europeans." We will return to this emphasis in the discussion of apocalyptic themes in black religious experience. It suffices at this time to recognize, with Wilson Jeremiah Moses, that through their quasi-religious interpretations nineteenth and twentieth century black nationalists have effectively "conjured" with Psalm 68.31.[30]

The Egypt figure

It should be observed that Psalm 68.31 features both Egypt and Ethiopia as figures for prophetic fulfillment. The Ethiopia figure is explicitly situated in the conversion theme of the second half of the verse: "Ethiopia shall soon stretch out her hands to God." However, Ethiopianism as a social-political movement has also been energized by the first half of the prophecy regarding Egypt: "*Princes* shall come out of Egypt"—not slaves whether Hebrew or African! Here we must observe that the negative valuation of Egypt, which conventionally operates in Exodus figuration, is reversed in mature developments of Ethiopianism. For historically informed Ethio-

pianists ancient Egypt was understood to be a civilization of black rulers
instead of a prison house of black slaves. An affirmation and even celebration of Egypt occurred, as Gayraud Wilmore points out, among black
abolitionists and preachers of the nineteenth century and among black
intellectuals after the First World War.[31] Indeed we may recognize the
yearning for a Pan-African communitas in their bold, antistructural identification with Egypt and Ethiopia as the two great monarchies of ancient
Africa. In their countercultural figuration Egypt appears as a focus of ethnic
pride and reclaimed heritage instead of (conventionally) the land of bondage and oppression. The celebrated wealth and stature of Egypt as one of
the earliest human civilizations emerge in Ethiopianism through the retrieval of black people as leaders of culture and representatives of civilization.

In fact, in North America a resurgence of Egypt figuration is now
occuring which inherits from Ethiopianism a tradition of Pan-African
scholarship. This development promises to reverse the situation lamented
earlier in this century by the Pan-Africanist scholar W.E.B. Du Bois, who
described the historical amnesia afflicting black America in regard to the
"mighty Negro past" of Egypt and Ethiopia. Here we may speculate that
a new yearning for communitas is empowering the current figuration of
Egypt in black America. Reclaiming Egypt and Ethiopia as the two pillars
of a classical black heritage, today's Afro-American retrieves the historical
ground, once again, in Du Bois's expression, to "be a co-worker in the
kingdom of culture."[32] Rather than attempt to build a future black greatness in a historical vacuum, present-day Egypt figuralists struggle to expose
the "destruction of black civilization," and the "myth of the Negro past."
Moreover, they seek to reclaim their "stolen legacy" as the historical ground
for present participation in the next age of great black civilizations.[33]

Repertory

By configuration I mean a distinctive pattern of social-political arrangements that typify a particular period in a culture's historical development.
Of course, the term deliberately connotes the operative presence of an
imaged or 'iconic' figure embedded in each configural phase. In this study
the figures are predominantly biblical, of course, but it is evident that they
may be derived from diverse sources in history and culture. Note also that,
however providential the convergence of historical circumstance and biblical narrative, configurations are experientially and empirically grounded
in the intensive patterning of communities upon figural models. That is,
configurations are inseparable from the hermeneutic reflection on events
conducted by an interpretive community. In this section I show how the
operation of the Exodus figure in black American communities grounds
social-political patterns and arrangements that constitute an Exodus configuration, how the Ethiopia figure forms Ethiopia configurations, and so
on, with the Promised Land figure and other patterns.

Together these configurations constitute a catalogue or a repertory: a repertory of the Bible construed as a 'formulary' or 'conjure book'. Like a musical or dramatic repertory of selections, configurations constitute repeatable alternatives for social and political performances that are modeled on biblical narrative and figures. The following repertory consists of selected social dramas that may be occurring at any period in black historical experience. The potential for historical divergence from the Bible's narrative patterns is endless, of course, just as the Emancipation experience of the slaves differed in significant ways from the biblical Exodus. But it is the hermeneutic framework and its schematic congruence with biblical typology that make the notion of a repertory useful, and not a literal or even an allegorical fixation on exact correspondences between biblical history and black history. In any case, attempts to force exact and rigid correspondences remain uninformed by the ritual and improvisational dynamics of Afro-American spirituality. That spirituality is mimetic, not < in the sense of literalistically imitating its models but by creatively transforming them at the same time it thoroughly appropriates them.

The Exodus configuration

Whenever the Judeo-Christian tradition has been accessible to oppressed peoples, this scenario of election, captivity and liberation has captured the imagination.

Gayraud Wilmore, *Black Religion and Black Radicalism*[34]

In the Afro-American figural tradition it appears that all corporate liberation efforts can be configured, in the manner of ritual performances, as dramatic reenactments of Exodus, and their leaders envisioned as approximate types of Moses. We have already referred to the black freedom movement of the 1960s, and Dr. Martin Luther King's "Mosaic" leadership in it. But the freedom movement offers only the latest indication of the force of this figural principle. The earliest instance of the configuration antedates even the Emancipation-Exodus of the slaves in 1863. For < we must recall that there were prior emancipation events of various northern states, as well as the termination of the slave trade in Britain and its New World colonies in 1807. The New York State emancipation of 1827, in which the celebrated orator and activist Sojourner Truth was freed, provides a case in point. It should also be recalled that, after the Civil War era, there occurred the Exodus events of the post-Reconstruction period extending through the late 1870s and the 1880s. These were not emancipations but migrations: the beginning of the mass migrations of black Americans out of the South that culminated in the great urban and northward migrations of the early twentieth century.

Indeed, participants and contemporary observers alike used the term "exodus" to describe the migrations.[35] The historian Herbert Aptheker

tells us that already, during Reconstruction itself, black communities tended to move out of a state whenever radical legislators lost power. When Reconstruction in the South was finally terminated in 1877, numbers of black organizations and their leaders began planning both secretly and openly for mass relocation to the North. The first major move occurred
> in January 1879—the anniversary month of Lincoln's Emancipation Proclamation—starting from southern Louisiana. "Most of these men, women and children headed for Kansas—the land whence came John Brown."[36] Earlier in the 1870s Sojourner Truth had lobbied hard and long in Washington, and on lecture platforms around the country, for the United States government to set aside lands for black settlements in Kansas or "the West." She preferred Kansas to the missionary scheme of relocating blacks overseas to Liberia, and she expressed her views in the language of Exodus figuration:

> I have prayed so long that my people would go to Kansas, and that God would make straight the way before them. Yes, indeed, I think it is a good move for them. I believe as much in that move as I do in the moving of the children of Egypt [sic] going out of Canaan [sic]—just as much.[37]

In this manner Kansas was reconfigured as a specific locus of the American "Promised Land." This figural passion, which discovers or recreates Exodus typologies wherever possible, derives from both existential and religious elements in Afro-American social experience. As W.E.B. Du Bois acknowledged, freedom itself has acquired a sacred or numinous quality in black culture. "Few men ever worshipped Freedom with half such unquestioning faith as did the American Negro for two centuries," Du Bois wrote a generation or more after Lincoln's executive order that freed the slaves. "Emancipation was the key to a promised land of sweeter beauty than ever stretched before the eyes of wearied Israelites."[38] But in addition to this visionary dimension there were also social-political and economic aspects of the post-Reconstruction exodus. The former slave and abolitionist Frederick Douglass recognized this clearly; he described the migrations "as an assertion of power by a people hitherto held in bitter contempt; [and] as an emphatic and stinging protest against high-handed, greedy and shameless injustice to the weak and defenseless." Moreover, Douglass employed a pharmacopeic metaphor in elaborating the social-economic dimension of the migration of black laborers: "[by] wisely using the Exodus example, they can easily exact better terms for their labor than ever before. Exodus is medicine, not food."[39]

Douglass's remarkable metaphor combines biblical figuralism with the curative orientation of African American conjure traditions. (His personal encounters with conjure lore are related in his autobiography, *Narrative of the Life of Frederick Douglass, An American Slave*, 1845.) Moreover, the use of the Exodus figure in a political context featuring sustained domination and exploitation of black workers in the "Reconstructed" South shows that black Americans, in the generation after Emancipation, still had recourse to Exodus strategies as a remedy or cure for oppression.

Before concluding this configuration we should observe the remarkable representation by black women of the patriarchal figure of Moses. I refer not only to Sojourner Truth in her political leadership during Reconstruction. It was Harriet Tubman who was called, during her own lifetime, "General Moses" and "the Moses of her people." Tubman earned this distinction by her performance as a liberator of hundreds of slaves on the underground railroad. This instance shows how even the working-class black woman, otherwise stereotyped as "the mule of the world" (Hurston),[40] is configured as a type of Moses by the troping virtuosity of black religious figuration.

The Ethiopia configuration

Ethiopianism might have remained merely an escapist myth-system based upon Biblical proof-texts and confined to the circle of Negro church people had not a brilliant black scholar appeared on the scene in the 1870s and 1880's. . . . Their movement sprouted from the seed-bed of folk Ethiopianism. . . . His great contribution was toward the development of African "cultural nationalism."

St. Clair Drake, *The Redemption of Africa*[41]

In this passage St. Clair Drake has charted the transition from a specifically religious figuration of Psalm 68.31, "Princes shall come out of Egypt, Ethiopia shall soon stretch forth her hands to God" (King James Version), to its political impact in black nationalism. As we saw earlier, that Psalm was regarded as a prophecy of Africa 'stretching out her hands to God' through mass conversion to Christianity. Thus Ethiopianism had originally served to motivate black Christians in their own missionary efforts to "redeem" Africa by spreading the gospel and making converts there. That is, the tradition initially embodied a religious vision. Then arose a singular thinker who reconfigured Ethiopianism to promote, first, New World emigration back to Africa. Most radically however Edward Wilmot Blyden (1832–1912) extended Ethiopianism to create a post-Christian cultural nationalism. "When shorn of Christian beliefs about [Africa's] 'degeneration' and 'redemption' through conversion to Christ, Ethiopianist thinking leads to a belief that the forces are latent within Africa itself to 'redeem' it."[42]

In contrast to Ethiopia figuration, Edwin S. Redkey has described the emigrationist and black nationalist periods of Ethiopianism as a "Black Exodus." The Exodus figure applies as well as Ethiopia because of the back to Africa motivation of black nationalism; a motivation that operated as late as the Garvey movement of the 1920s (the largest mass movement in black history). But this variation in figural terminology serves only to highlight the improvisational aspects of a configural repertory. In this repertory, the Exodus and Ethiopia figures alternate as if to increase the amplitude—the imaginative or iconic power—of correspondences between black experience and biblical narrative. Thus the "flexibility in the Black story"[43] allows for black nationalism to be configured alternately by Exodus and Ethiopia figuration. The same crossing of the Atlantic Ocean by which Africans were transported to the Americas is reversed by emigration. In such a reversal, the configural imagination can envision the return to Africa as a crossing of 'the Red Sea'—*back to Egypt*! But, as we saw earlier with respect to Egypt figuration, rather than a place of bondage Egypt is transvaluated by black cultural nationalists as a positive representation of African greatness and potential destiny. By means of such improvisational transformation, a new Exodus is envisioned which leads out of captivity or exile in the New World.

Indeed, as we see in the conclusion, the New World as the site of exile or Diaspora is the most contemporary configuration of African American culture.

Americana

In this section I compare the African American uses of the Exodus figure with Puritan American antecedents in colonial New England. As already noted, biblical figuralism or typology is a Christian interpretive tradition that goes back to the early church and New Testament literature. It emerged as the authoritative way for Jewish Christians first, and Gentile Christians later, to read the Hebrew Scriptures prophetically: that is, by assigning Christian referents to sacred terms from the Jewish Scriptures. Christ as a second Adam, as a second Moses or David, or John the Baptist as another Elijah, are all examples of such prophetic or figural fulfillment. Typology or figuralism can be defined as a hermeneutic or interpretive tradition in which a person or place, object or event, is connected to a second entity in such a way that the first signifies the second and the second fulfills or encompasses the first. Moreover, as Erich Auerbach insisted, biblical figuralism is distinguished from other symbol systems (especially allegory) in that its figures always refer to actual, historical entities. So radical and all-encompassing is this realistic reference that the Bible promises to absorb *both* cultural and individual histories by its figural frame of reference:

> All other scenes, issues, and ordinances . . . the history of all mankind, will be
> given their due place within its frame, will be subordinated to it. The Scripture
> stories . . . seek to subject us, and if we refuse to be subjected we are rebels
> . . . we are to fit our own life into its world, *feel ourselves to be elements in its
> structure* of universal history.[44]

The formation of individual believers by means of the same imitation of biblical types has been the traditional subject of the "lives of saints" or hagiographies. Such narratives focus on heroes of mimetic proficiency: those whose lives so conform to the models authorized in sacred texts that they themselves have become new exemplars in continuity with those texts.

The theologian James Wm. McClendon, Jr., has extended the hagiographic interest in exemplary lives with his innovative project in narrative studies, "biography as theology." The focal point for this project is the exemplary character of persons (not only canonically recognized saints but < also ordinary persons) whose lives embody and display the theology of their community of reference. As models or exemplars they are human expressions or embodiments of their community's religious convictions. The term "exemplar" means a preeminent example, a model or pattern most worthy of imitation. The exemplary lives selected by McClendon for theological reflection include those of Dag Hammarskjøld and Martin Luther King, Jr. Speaking of the dominant images in these lives as "the key" to their biographies, McClendon illustrates how "the convergence of such images in a particular person helps to form his characteristic vision or outlook."[45] As we might expect, when discussing the "characteristic vision" of King it is the Exodus and Moses "images" that predominate.

> King understands his work under the image of the Exodus; he is leading his people on a new crossing of the Red Sea; he is a Moses who goes to the mountaintop, but who is not privileged to enter with his people into the promised land. These are major images, and for [King] there are others as well.[46]

This figural portrait of King is one example of McClendon's "biography as theology," which programmatically treats individual lives and their characteristic vision as informed by biblical images. By contrast, I treat collective rather than individual character. (Elsewhere I have referred to this distinction in terms of "ethnography as theology" in contrast to "biography as theology."[47])

The prominence of the Exodus figure in the life of King is not a coincidence; we have already seen that Exodus is a compelling figure for Afro-American culture at large. The black theologian James Cone, for instance, claims that "almost all blacks in America—past and present—have identified Egypt with America, Pharoah and the Egyptians with white slaveholders and subsequent racists, and blacks with the Israelite slaves."[48] In this connection we may posit here a kind of continuum between biography and ethnography, similar to that which Sacvan Bercovitch observes between biography and history in the social discourse of Puritan New England. As Bercovitch remarks, the linking term in this continuum is *exemplum* or "the exemplary."

> For the seventeenth-century Puritan, *exemplum fidei* denoted a type of Christ; and what he meant by type pertained equally to biography and to history. ... Biographically, the New Englander and the Israelites were correlative types of Christ; historically, the struggles of the New England saints at that time,

in this place . . . [were] the deeds Christ was now performing through them in America.[49]

Because of their conviction that an individual life could constitute an instance or "example for faith," Puritan histories of the colonies "read like spiritual biographies of an elect land."[50] Since the groundbreaking work
> of Perry Miller in early American studies there has come into being a substantial literature concerning the figural transformation of New England into a type of biblical Israel. The present section briefly highlights this literature in order to locate black religious figuralism in a larger context of North American figural discourse. But of course the primary focus is an ethnography of black religious figuralism. In rendering connections between these two figural traditions one need not insist on (nor alternatively deny) direct Puritan influences on black traditions of figural discourse. At any rate there is sufficient evidence to suggest the pervasive influence of typological thinking throughout North American history and culture. Sacvan Bercovitch's *Puritan Origins of the American Self* (1975), alongside Werner Sollors's examination of "typological ethnogenesis" in United States cultures, indicate a nearly universal influence of Puritan typology. Moreover Wilson Jeremiah Moses's work in black literary history corroborates specific instances of typological influences on Afro-American religion and culture. The "black jeremiad," for example, is a literary genre cognate with the so-called American jeremiad of Puritan tradition. Here
> however I only suggest a family resemblance between the African American and the mainstream Puritan American figural traditions, as a way of better understanding the mimetic aspects of Afro-American conjurational spirituality.

It is remarkable that the Exodus figure should be common to such dissimilar communities. The implications of the resemblance between the two traditions are more significant for this study than the question of historical sources and documented influence—an influence that awaits adequate investigation by historians. The lack of research, however, cannot be due to the marginality of black religious figuralism. On the contrary, the preliminary treatment given previously has already substantiated the centrality of biblical figures in black religious expression. Of course, as many observers have commented, the identification of Afro-American slaves with Israel's sufferings in the 'house of bondage' possessed more literal force than the same identification of white Americans. Again, Vincent Harding's comment about reversed (or inverse) images—America as "Promised Land" for the latter group, and America as "Egypt" for the former—illustrates one way to express the differing aspects of realism involved.[51]

> Despite such contrasts in their use of biblical figures, by the mid-nineteenth century black Christians in North America had more in common with their white antagonists than with religious cultures anywhere else in the world. For, as reciprocally alienated but nonetheless proximate

religious communities they shared the hermeneutics of figuralism generally, and the centrality of the Exodus figure in particular.⌐ On the basis of
this hermeneutical and theological connection I judge, along with the
historian David W. Wills, that "black religious leaders were also more
involved than has been realized in appropriating, preserving, and defending the broadly Puritan tradition from which so much in American evan-
gelism is derived." Accordingly I proceed to display what Wills calls the
"shared framework" between African American and Puritan American
traditions, or what Paul Holmer called a common "grammar of faith."[52]
I am specifically interested in a common *figural* grammar, but also in the
differences between Puritan typology as one "mode of spiritual
perception"[53] and Afro-American modes of spiritual perception.

In pursuing this matter of similarity and difference it is useful to hypoth-
esize a conjurational or incantatory element operating in each Christian
community.[54] Bercovitch provides a point of departure in his reference to
the Puritans' extrapolations from the "spiritual sense" of Scripture—its
sensus spiritualis. Going beyond orthodox exegeses of Scripture, the Puritans rendered their own experience as a coequal source of theological
reflection. Their ostensible biblicism, Bercovitch argues, functioned as a
mask which veiled or disguised their "inversion" of the traditional relationship between Scripture and experience.

> Early New England rhetoric is a titanic effort to secularize traditional images
> without abandoning the claims of exegesis. The clergy compensated for their
> extreme subjectivism in substance by an extreme orthodoxy in approach....
> [They] instituted what might be called, for want of a better term, a rhetoric
> of inversion.... Discarding the difference between plural and singular, moving ⌐
> freely between historiography and spiritual biography, they inverted the no-
> tion of *exemplum fidei,* who stands for the elect community, into the notion
> of a church-state that is an elect Christian, *in imitatio Americae.*[55]

We may thus ascribe to Puritan typology an incantatory impulse, comparable in effect (however unconscious) to magical performances in the
operation of spells or solemn invocations. The incantatory factor consists
in the designation or 'invocation' of an entire people as a sacred image or
icon. To refer again to Burke's useful expression, incantatory rhetoric
functions "as a device for inviting us to 'make ourselves over in the image
of the imagery.'" I propose that the figure of an "American Israel" was ⌐
intended, defensively and iconographically, to reverse the European representation of America as a "howling wilderness" and of its colonists as
the offscouring of the Old World among the native peoples or 'barbarians'
of the New World.[56] Offensively, the iconic intent was to eclipse the
representation of Old World cultures as preferential sites of divine activity
in history, and to claim the New World instead as the contemporary focal
point of God's redemptive activities.

On this view we may understand the Puritans' mimesis, in which New

England appears as the biblical Israel, as a form of incantational self-representation. Indeed, one likely connection between Puritan and African American uses of the Israel figure resides in the incantatory element found in their respective preaching traditions. As the historian Donald G. Mathews observed, "the incantational preaching which New Light Baptists brought from New England was very much like the interaction between leader and worshipers in African religion."[57] Coincidentally, the incantatory style of such Baptist preachers may have inured that denominational tradition to generations of black converts during the course of successive encounters with Christianity, from the period of the Great Awakening in the eighteenth century to the great mass of conversions following the emancipation of 1865.

However, there is a toxic aspect of the Puritan typological tradition. As Bercovitch argues, the figure of "America" as "God's New Israel" (Cherry) has become a distorted and pernicious mythic symbol. The figure has come to symbolize a destiny that is familiarly invoked and, in the conventional sense 'conjured up,' by such mythic-symbolic constructs as "American dream," manifest destiny, "redeemer nation," and chosen peoplehood. "Only in the United States," Bercovitch claims, "has nationalism carried with it the Christian meaning of the sacred. Only America, of all national designations, has assumed the combined force of eschatology and chauvinism."[58] As Bercovitch also makes clear, this usurpation of sacred symbols did not occur accidentally. Rather, it was an extension of the Puritans' initial inversion of Scripture and experience. We may attribute America's figural excesses to the Yankee descendents of the Puritans, for whom "typology took on the hazy significance of metaphor, image, and symbol; what passed for the divine plan lost its strict grounding in Scripture; 'providence' itself was shaken loose from its religious framework to become part of the belief in human progress . . . [and] sacred history [translated] into a metaphor for limitless secular improvement."[59]

Bercovitch maintains, however, that these developments were not transformations of the typological heritage but essentially extensions and adaptations. For it was the Puritans themselves who had already transformed New England into an antitype of the biblical figures "Canaan" and "Israel." How could they have foreseen the future distortion of their labored exegeses? Had the 'genetic' or embryonic elements of their own distortion—their own errors of inversion—been made clear to them, could they have found the way to a more authentic figuralism? Bercovitch suggests that the Puritan fathers were incapable of heeding a number of contemporary efforts to warn them. One of the most vigorous alarms was sounded by Roger Williams (1603–1683). Writing at length to John Cotton from England, Williams castigated the colonists for their theocratic claims to the title of "New Israel." Insofar as such use was legitimate (that is, warranted by Scripture), he argued, the title applied to all "saints" or believers. But as colonists of America, or as citizens either of England or of New England, no one could claim separate distinction under the title.

Furthermore, with regard to their conviction that a special providence was involved in establishing the preeminence of New England, and in vouchsafing there a continual fulfillment of prophecies, Williams remonstrated: "If this be not to pull *God* and *Christt* and *Spirit* out of Heaven, and subject ∠ them unto *naturall*, sinful, inconstant men, . . . let Heaven and Earth judge."[60]

With great acuity Williams's remarks expose the incantatory or magical element in Puritan typology: its impulse to compel "God and Christ and Spirit" out of Heaven and thus harness sacred energies for human enterprises. From the perspective of the present study the error of this attempt resides not in its imputedly manipulative character—which in any case can be otherwise construed (for example, as a synergism in which human agency participates or cooperates with divine providence).[61] Rather the problem resides in the repeated (if not inevitable) failure of civil society to configure itself genuinely as an antitype of a biblical figure. (Recall here Reinhold Niebuhr's thesis of Christian realism in *Moral Man and Immoral Society*.[62]) It may be existentially impossible for a collective to become an∠ authentic biblical exemplar, an unambiguous example for faith. That was certainly the view of Roger Williams, who declared without qualification and categorically: "*America* (as *Europe* and all nations) lyes dead in sin." To Williams's assertion of orthodoxy Bercovitch adds the clarification that as individual, anonymous, or generic Christians, "the Saints in New England, like all saints, belonged to the figural Israel. But *as New Englanders* they could no more arrogate that identity to themselves than could their neighbors, the Indians (Williams called *them* Americans), or the inhabitants of the Old World they had abandoned."[63]

Behind this typological insistence lies a theological distinction with consequences for the ethnographic comparison of Puritan American and African American typological traditions. It is the classical Western theological distinction between natural and supernatural existence. Natural existence in the classical view is inherently ambiguous; only the transcendence of temporal and material conditions will permit the manifestation of unambiguous life. This principle is found in various forms in the Christian gospels, but its most graphic articulation is found in St. Augustine's *The City of God* (413–426). Augustine insisted that the earthly church or ∠ people of God is inevitably mixed during its temporal existence. No matter how mightily a community strives for righteousness, it remains composed of two groups: people of the city of God, *civitas dei,* and people of the city of the devil, *civitas diaboli.* Never is it possible under the conditions of existence to separate the two, and so be able to declare that a living community fully constitutes an exemplar of the faith. In Roger Williams's more graphic terms: "Nature knows no difference between [people] . . . in blood, birth, bodies. . . . [Any land is] a ship at sea with many hundred souls . . . papists and protestants, Jews and Turks . . . whose weal and woe is common."[64]

Yet this view of reality, which recognizes the union of opposites in

concrete experience, is also consistent with the conjurational worldview
of African American folk tradition. That worldview has already been introduced with the concept of the world as a pharmacosm in which tonic
and toxic aspects coinhere in the same phenomena. I will return to this
issue as part of a philosophical treatment of wisdom perspectives in black
culture and in the analysis of the biblical figure of Apocalypse. Here it is
sufficient to posit a difference in worldview between the two traditions:
a Puritan American tradition that projects its figural identity and destiny
monolithically, as an unambiguous representation of biblical exemplars,
and a African American tradition that is cognitively predisposed to view
its reality (its tortuous past, blighted present, and foreseeable future) as
simultaneously tonic and toxic.

Notes

1. Martin Buber, *Kingship of God*, 3rd. ed., trans. Richard Scheimann (New
York: Harper & Row, 1967), pp. 105–6.
2. Puritan theology in early America understood the type-antitype dyad to
be prophetically or divinely constituted, "the God who predetermines history
having lent it this power." As quoted by Ursula Brumm, *American Thought and
Religious Typology* (New Brunswick, N.J.: Rutgers University Press, 1970), p. 23.
On dyads and "figural prophecy" see the definitive study by Erich Auerbach,
"Figura," in *Scenes from the Drama of European Literature*, ed. Wald Godzich and
Jochen Schulte-Sasse (Minneapolis: University of Minnesota Press, 1984), pp. 29,
56, 72.
3. See Sacvan Bercovitch, *The Puritan Origins of the American Self* (New
Haven: Yale University Press, 1975), and Conrad Cherry, ed., *God's New Israel:
Religious Interpretations of American Destiny* (Englewood Cliffs, N.J.: Prentice-Hall,
1971). On "typological ethnogenesis" see Werner Sollors, *Beyond Ethnicity: Consent
and Descent in American Culture* (New York: Oxford University Press, 1986),
pp. 50, 57.
4. Sacvan Bercovitch, *The American Jeremiad* (Madison: The University of
Wisconsin Press, 1978), pp. 132–76.
5. For example, see Evan M. Zuesse, *Ritual Cosmos: The Sanctification of
Life in African Religions* (Athens: Ohio University Press, 1979). On the primacy
of ritual as a "continuity" (Raboteau) between African and Afro-American cultures
see Dona Richards, "The Implications of African-American Spirituality," in *African
Culture: The Rhythms of Unity*, ed. Molefi Asante and Kariamu Asante (Westport,
Conn.: Greenwood Press, 1985), p. 213.
6. These are the terms of John Middleton's distinction between the "magico-
religious," and the mythological or cosmological, in *Magic, Witchcraft, and Curing*,
ed. John Middleton (Garden City: N.Y.: The Natural History Press, 1967), p.
ix.
7. See Johan Huizinga, *Homo Ludens: A Study of the Play Element in Culture*,
trans. R.F.C. Hull (New York: Roy Publishing, 1950).
8. Victor Turner, *Dramas, Fields, and Metaphors: Symbolic Action in Human
Society* (Ithaca, N.Y.: Cornell University Press, 1974), p. 274. Cf. also the treatment

of *communitas* in his *The Ritual Process: Structure and Anti-Structure* (Ithaca, N.Y.: Cornell University Press, 1969), pp. 96–113.

9. Kenneth Burke, *The Philosophy of Literary Form: Studies in Symbolic Action* (New York: Vintage Books, 1957), p. 87.

10. Zuesse, *Ritual Cosmos*, p. 7.

11. Jahnheinz Jahn, *Muntu: The New African Culture* (New York: Grove Press, 1961), p. 219.

12. This is a formulation that Burke borrowed from John Crowe Ransom. Kenneth Burke, *The Philosophy of Literary Form: Studies in Symbolic Action* (New York: Vintage Books, 1957), p. 100.

13. Albert J. Raboteau, "The Black Experience in American Evangelicalism: The Meaning of Slavery," in *The Evangelical Tradition in America*, ed. Leonard Sweet (Macon, Ga.: Mercer University Press, 1984), p. 197.

14. David Walker, *David Walker's Appeal* (1829), ed. Charles M. Wiltse (New York: Hill & Wang, 1965), p. xiv.

15. Robert Alexander Young, "The Ethiopian Manifesto," in *A Documentary History of the Negro People of the United States*, ed. Herbert Aptheker (Secaucus, N.J.: Citadel Press, 1974), p. 91.

16. Burke, *The Philosophy of Literary Form*, p. 6.

17. James H. Cone, *God of the Oppressed* (New York: Seabury Press, 1975). This is perhaps the author's most definitive statement of black theology as a liberation theology that draws upon the indigenous sources of black religious experience and cultural productions—sermons, songs, and stories—as distinguished from academic or political formulations.

18. Henry Highland Garnet, "An Address to the Slaves of the United States of America," in *Black Nationalism in America*, ed. John H. Bracey, August Meier, and Elliott Rudwick (Indianapolis: Bobbs-Merrill, 1970), p. 71ff. Alongside David Walker's "Appeal" this is the most provocative and incendiary antislavery statement in American history that features as well a powerful use of biblical (specifically Exodus) figuralism.

19. Timothy L. Smith, "Slavery and Theology: The Emergence of Black Christian Consciousness in 19th Century America," *Church History* 41 (1972):498.

20. Buber, *Kingship of God*, p. 106.

21. In this connection the black theologian and pastor Olin P. Moyd has defined redemption in dual terms: not only liberation but also corporate formation. Within black American religious tradition, Moyd claims, "redemption has not been limited to liberation from oppression; it has also meant the divine process of being formed into a community where love and justice would prevail." By this dual emphasis Moyd intends to restore the full sense of the term as explicative of Afro-Christian faith. That full meaning has been reduced, on his view, in conventional theologies that limit redemption to salvation from sin and guilt. On the other hand the corporate or "confederation aspect" of the term has been neglected even by contemporary black theologians. Engaging in his own project of hermeneutic retrieval, therefore, Moyd wants to restore the "salvation-from-oppression dimension of redemption which has been neglected by Euro-American theologians" while reclaiming the element of corporate redemption found in "folk religious expressions which has been overlooked by Black theologians." Finally, Moyd suggests that this more adequate and wholistic sense of redemption has provided the theological rationale for moral action and ethical reflection in black religion and culture. Olin

P. Moyd, *Redemption in Black Theology* (Valley Forge, Pa.: Judson Press, 1979), pp. 16, 33, 28.

22. Werner Sollors, *Beyond Ethnicity: Consent and Descent in American Culture* (New York: Oxford University Press, 1986), p. 42.

23. Ibid., p. 50; emphasis mine.

24. Albert J. Raboteau, *Slave Religion: The "Invisible Institution" in the Antebellum South* (New York: Oxford University Press, 1978), p. 311.

25. Bennett continues intriguingly: "What King did now—and it was a high achievement—was to turn the Negro's rooted faith in the church to social and political account by melding the image of Gandhi and the image of the Negro preacher and by overlaying all with Negro songs and symbols that bypassed cerebral centers and exploded in the well of the Negro psyche." Lerone Bennett, Jr., *What Manner of Man: A Biography of Martin Luther King* (Chicago: Johnson Publishing Co., 1964), p. 72.

26. Martin Luther King, Jr., "I See the Promised Land," in *A Testament of Hope: The Essential Writings of Martin Luther King, Jr.*, ed. James Melvin Washington (New York: HarperCollins Publishers, 1991), p. 286.

27. Raboteau, *Slave Religion*, p. 251. Raboteau cites the Harding comment as follows: Vincent Harding, "The Uses of Afro-American Past," In *The Religious Situation*, ed. Donald R. Cutter (Boston: Beacon Press, 1969), pp. 829–40. On Afro-American "symbolic reversals" see Lucius T. Outlaw, "Language and Consciousness: Towards a Hermeneutic of Black Culture," *Cultural Hermeneutics* 1 (1974): 403f.

28. John Howard Yoder, "Exodus and Exile: The Two Faces of Liberation." *Cross Currents* (Fall 1975): 299f. Yoder's own reflections on corporate formation follow: "Peoplehood is not the product of liberation; peoplehood with a history and a trust in the God who has led the fathers is prior to liberation. This has implications for contemporary thinking...." Ibid., p. 303.

29. St. Clair Drake, *The Redemption of Africa and Black Religion* (Chicago: Third World Press, 1970), p. 41.

30. Wilson Jeremiah Moses, "The Poetics of Ethiopianism: W.E.B. DuBois and Literary Black Nationalism," *American Literature: A Journal of Literary History, Criticism, and Bibliography* XLVII:3 (November 1975): 412, 414. Moses, *Black Messiahs and Uncle Toms: Social and Literary Manipulations of a Religious Myth* (University Park: Pennsylvania State University Press, 1982), p. 160.

31. Gayraud S. Wilmore, *Black Religion and Black Radicalism: An Interpretation of the Religious History of Afro-American People* (Maryknoll, N.Y.: Orbis Books, 1983), p. 121.

32. Significantly linking black creative "powers" to the two major black monarchies of antiquity, Du Bois lamented that "these powers of body and mind have in the past been strangely wasted, dispersed, or forgotten. The shadow of a mighty Negro past flits through the tale of Ethiopia the Shadowy and Egypt the Sphinx. Throughout history, the powers of single black men flash here and there like falling stars, and die...." W.E.B. Du Bois, "Of Our Spiritual Strivings," *The Souls of Black Folk* (New York: New American Library, 1969), p. 46.

33. Cf. Chancellor Williams, *The Destruction of Black Civilization: Great Issues of a Race from 4500 B.C to 2000 A.D.* (Chicago: Third World Press, 1974), and Melville Herskovits, *The Myth of the Negro Past* (Boston: Beacon Press, 1958), and George G. M. James, *Stolen Legacy* (San Francisco: Julian Richardson & Associates, 1976).

34. Wilmore, *Black Religion and Black Radicalism*, p. 37.

35. Herbert Aptheker, ed., *A Documentary History of the Negro People in the United States*, vol. 2, *From the Reconstruction Era to 1910*, 5th ed. (New York: Citadel Press, 1951), p. 713.

36. Ibid.

37. Arthur H. Fauset, *Sojourner Truth: God's Faithful Pilgrim* (1938; reprint, New York: Russel Co., 1971), p. 175.

38. W.E.B. Du Bois, *The Souls of Black Folk* (New York: New American Library, 1969), p. 47.

39. Aptheker, *A Documentary History*, pp. 725–26. Aptheker's citation locates the Douglass excerpt in an address delivered before the American Social Science Association, September 12, 1879, and published in Boston in the *Journal of Social Science* XI (May 1880): 1–35.

40. The full passage reads, in folk dialect: "Honey, de white man is de ruler of everything as fur as Ah been able tuh find out. Maybe it's some place way off in de ocean where de black man is in power, but we don't know nothin' but what we see. So de white man throw down de load and tell de nigger man tuh pick it up. He pick it up because he have to, but he don't tote it. He hand it to his womenfolks. *De nigger woman is de mule uh de world* so fur as Ah can see." Zora Neale Hurston, *Their Eyes Were Watching God* (1937; reprint, New York: Harper & Row, 1990), p. 14. Emphasis mine. On the black woman as the (related) biblical figure of "Hagar-in-the-wilderness," see Delores S. Williams, *Sisters in the Wilderness: The Challenge of Womanist God-Talk* (Maryknoll, N.Y.: Orbis Books, 1993), pp. 22–29, 117–19, 196–98.

41. Drake, *The Redemption of Africa*, pp. 54, 61f.

42. Ibid., p. 71.

43. Harold Dean Trulear, "The Lord Will Make a Way Somehow: Black Worship and the Afro-American Story," *The Journal of the Interdenominational Theological Center* XIII:1 (Fall 1985): 101. Cf. also James Cone, "The Story Context of Black Theology," *Theology Today* XXXII: 2 (July 1975): 145.

44. Erich Auerbach, *Mimesis: The Representation of Reality in Western Literature*, trans. William Trask (Garden City, N.Y.: Anchor Press/Doubleday, 1952), p. 14f. Cf. Auerbach, "Figura," pp. 11–76.

45. James Wm. McClendon, Jr., *Biography as Theology: How Life Stories Can Remake Today's Theology* (Nashville, Abingdon Press, 1974), p. 90.

46. Ibid., p. 93.

47. See my treatment of "ethnography as a source of theology" in Theophus H. Smith, "The Biblical Shape of Black Experience: An Essay in Philosophical Theology," pp. 27–35. Ph.D. dissertation, Graduate Theological Union, 1987.

48. James H. Cone, *For My People: Black Theology and the Black Church* (Maryknoll, N.Y.: Orbis Books, 1984), p. 63.

49. Bercovitch, *The Puritan Origins*, p. 35f.

50. Ibid., p. 114.

51. Harding, "The Uses of the Afro-American Past," pp. 829–40.

52. In contrast to the range of cross-cultural studies in anthropology and ethnography, Wills attests that "there has been relatively little attention, within the context of American religious history, to the encounter of blacks and whites. This encounter has indeed often not been thought of as a central theme in our nation's *religious* history but it surely is. . . . Puritanism is studied so carefully largely because of its apparent impact on our collective sense of purpose and value. Studies

of pluralism recurrently acknowledge that our pluralism has not characteristically involved sheer diversity but variety within a shared framework. It is precisely in the context of the encounter of blacks and whites, however, that our efforts to forge a common culture have been most severely tested." David W. Wills, *Black Apostles at Home and Abroad: Afro-Americans and the Christian Mission from the Revolution to Reconstruction*, ed. David W. Wills and Richard Newman (Boston: G. K. Hall, 1982), pp. ix, xi–xii. Paul L. Holmer, *The Grammar of Faith* (New York: Harper & Row, 1978).

53. This is Stephen Manning's expression in his essay "Scriptural Exegesis and the Literary Critic," in *Typology and Early American Literature*, ed. Sacvan Bercovitch (Amherst: University of Massachusetts Press, 1972), p. 58.

54. Cf. Jon Butler, *Awash in a Sea of Faith* (Cambridge, Mass.: Harvard University Press, 1990) on the prevalence of spiritualism in nineteenth century American life. Although the theoretical basis for this hypothesis cannot be elaborated here, it is evident that it presupposes a rethinking of Puritan typology, and indeed the formation of American culture at large, from an Africanist or an African Americanist perspective.

55. Bercovitch, *The American Jeremiad*, p. 114f.

56. Michael Wigglesworth, "God's Controversy with New England," in Cherry, *God's New Israel*, p. 44. I examine Wigglesworth's expression in more detail when treating the figure of Wilderness in Chapter 3.

57. Donald G. Mathews, *Religion in the Old South*, in Chicago History of American Religion Series, ed. Martin E. Marty (Chicago: Chicago University Press, 1977), p. 191.

58. Bercovitch, *American Jeremiad*, p. 176. Cf. Bercovitch, *Puritan Origins*, p. 108, and Sollors, *Beyond Ethnicity*, pp. 42–50.

59. Bercovitch, *American Jeremiad*, p. 93f.

60. Bercovitch, *Puritan Origins*, p. 110. Commenting on the intensity of Williams's theological censure, Bercovitch adds that he "debated with the baffled outrage of a man who just could not fathom his opponents' obstinacy. To some degree, we sense the same tone in the arguments of many others at home and abroad."

61. This perspective abandons the discredited view of magic as manipulative versus religion as petitionary or supplicative. As I indicate in the Introduction, one person's magic is another person's religion: that is, one's own manipulation can always be construed as supplicative and someone else's petition as manipulative. Hence the distinction collapses into ambiguity.

62. Reinhold Niebuhr, *Moral Man and Immoral Society* (New York: Scribner's, 1932).

63. Bercovitch, *Puritan Origins*, p. 110.

64. Ibid.

3

Law

[My servant] will not fail or be discouraged
till he has established justice in the earth;
and the coastlands wait for his law.

Isaiah 42.4

In this chapter I complement the emancipatory focus of the preceding chapter by attending to that other narrative pole in the book of Exodus: law. The advent of law in a people's history—the historical moment when a culture first establishes its identifying tradition of law—often coincides with a "wilderness" experience: a transition from relative anarchy to relative order. Thus the figure of Wilderness arises as a prominent feature in the biblical narrative of ancient Israel. Indeed the law-wilderness connection can become a pronounced feature for postbiblical cultures that identify with ancient Israel. Before turning to such instances I will state the case first for biblical narrative. Recall that the book of Exodus features not only the miraculous liberation from Egypt but also the promulgation of the first Mosaic law code—the ten commandments delivered by the prophet at Mount Sinai. Moses-the-lawgiver, and the divine origination of his law at Sinai, are narrative elements just as central to the book of Exodus as the character of Moses-the-liberator and his miraculous deliverance of the people at the Red Sea.

Recall too that the more extensive law codes, the deuteronomic and levitical, are narratively framed as extensions of the Exodus event. Inventories of laws are embedded in the chronicle of events found in the books of Leviticus, Numbers, and Deuteronomy. The events themselves are depicted as occurring within the context of Israel's postexodus experience in the wilderness. (Significantly, the original Hebrew title, "In the Wilderness," is changed to the Book of "Numbers", because it includes a census of the people.) Indeed, in textual terms the three books are literary ex-

tensions or sequels of the book of Exodus. Traditionally known as the
third, fourth, and fifth books of Moses, they are suitably construed as law
books framed by historical narration. They span foundational events in
the life of the first generation (as narrated) out from Egypt, and also
catalogue the juridical developments attending its formation as a society
> under law. These developments are represented as occurring between the
liberation from Egypt, in Exodus 15, and the entry into the promised land
of Canaan (with the crossing of the Jordan River and the conquest of the
city of Jericho,) in Joshua 1–6.

The figural relations between the book of Exodus and the books of
law also corroborate Northrop Frye's typological schema in *The Great
Code*. Just as Exodus represents a type of Genesis—specifically an "ethnogenesis" (Sollors)—so here the books of law are types of Exodus narration. The significance of these figural relations for the present study will
be evident in the sections "Judges" and "Wilderness." There we will see
how similar codes—ranging from Exodus to Wilderness and Law—are
found operating in both Puritan and African American historical experience and figural imagination. Such recapitulations may be regarded alternatively, of course, as providential events or as natural occurrences arising
inevitably in the experience of any first generation undergoing social-
political transformation.

At any rate, in this chapter I focus on the figural and conjurational
aspects of law in black American experience, limiting the scope of the
discussion to two seminal eras: the beginnings of black people's engagement with American law in the early colonial and revolutionary periods,
and their post-Reconstruction engagement with law after the failure of
democratic reforms in the South. (I do not maintain a strict chronological
sequence: my structure is determined by the logic of the biblical figures
treated, not by historical narration. Accordingly, I alternate between the
colonial and the postslavery periods as elucidation of the figures may
require.) In addition to the emphasis in preceding sections on figural
discourse, I will show how black Americans have attempted to render
American law as a social curative or *pharmakon* (Greek: medicine, poison)
for transforming the destructive reality of slavery. After illustrating such
conjurational uses of the law, I turn to the racist abuse of embodied persons
as *pharmakoi* (Greek: scapegoats, ritual victims). The analysis of the Wilderness figure encompasses these issues and closes Part I.

Judges

In the year 1700 the celebrated Puritan chief justice, Samuel Sewall, published the earliest antislavery tract to appear in New England. *The Selling
of Joseph: A Memorial* took the biblical figure of Joseph, the youngest son
of Jacob, who was sold into bondage by his own brothers (Genesis 37),
and applied it to the buying and selling of Africans in the Atlantic slave
trade. Just three years earlier the same Justice Sewall had publicly repented

of his role in the now infamous Salem witch trials. "On Fast Day, January 1697, Sewall stood up in church as the minister read his declaration of guilt: he was eager, he said, 'to take the Blame and shame' of his grievous mistake."[1] In this manner Sewall publicly acknowledged his error as one of the judges who condemned innocent persons to death in the Salem hysteria. In his antislavery "Memorial" of 1700 we find him confessing again, this time in published form, the injustice and culpability of his fellow countrymen involved in the slave trade. What sort of memorializing, remembering, or anamnesis was the old Puritan attempting to render in his later years? What kind of anamnesis was he offering to his compatriots, this "Judge of the old Theocracy / Whom even his errors glorified . . . the Puritan / Who the halting step of his age outran" (Whittier)?[2]

In 1700 Sewall not only published his Memorial he also began legal proceedings for the manumission of a particular slave publicly known as "Black Adam." This Adam was the slave of another judge, John Saffin, who in 1694 while in Virginia hired out Adam for seven years of indentured servitude to his tenant farmer, a Mr. Shepard. Shepard, apparently against the practices of his landlord, treated Adam liberally: not only allowing him to eat at his own table with his wife and children, but also providing him with fertile land for his own use and even promising him his freedom (on Saffin's authority presumably) at the end of the seven year period in 1701. Nevertheless, according to Saffin's narration of the events, such benevolence did not succeed in accommodating the slave to the conditions of his servitude. Evidently Adam, so far from showing gratitude and a diligent commitment to the interests of his masters, exhibited instead the natural characteristics of a man deprived of his liberty: offended dignity, surliness and insubordination, intimidating behavior, and other attitudes so threatening that Shepard grew concerned for his own safety and that of his family. He finally urged Saffin to take Adam back. The slaveholder did so, and attempted to get other means of profit out of Adam by renting him to one man and then another until taking him to Boston.

But in Boston the slave's recalcitrance continued. In Saffin's words, "he had nothing to do but work in the Garden, make Fires and the like, was kindly used, [and] did eat of the same as the English Servants did." Thus, for misguided reasons Saffin expected Adam's gratitude, presumably for according him comparable treatment as shown to "the English Servants" while nonetheless holding him in bondage. But it seems that 'equal' treatment as an indentured servant rendered Adam all the more impatient for true freedom. After all the presumed 'privileges' extended him he became, Saffin later reported, "so quarrelsome and contentious, calling the Maids vile names, and threatening them (as they said) that they were sometimes afraid to be in the Room with him."[3] Indeed, in his Memorial Sewall himself acknowledged such behavior and described it as the manifestation of a more general temper among many slaves experiencing "Uneasiness . . . under their Slavery." In Sewall's view their "continual aspiring after their forbidden Liberty, renders them Unwilling Servants."[4] Eventually outraged by the intractability, the lack of docility and pliability, in his slave, Saffin retaliated during the spring of 1700 by threatening to deny Adam his anticipated freedom in the following year. At this point a Brother Belknap[5] came to Adam's aid by petitioning the General Court to intervene on his behalf. Belknap, calling for the manumission of both Adam and his wife, successfully persuaded Justice Sewall to consider the merits of the case. Finally in November 1703, after three years of litigation, a jury of Sewall and three other judges of the Superior Court declared Adam free. "Saffin, relentless, again petitioned the Assembly, but to no avail. In 1710 Saffin died. A few years later the name of Adam, a 'Free Negro,' appears in the Boston town records."[6]

But we are not allowed simply to enjoy this happy ending. In the confrontation between the two colonial judges, Judge Saffin left behind for posterity his own publication to answer Justice Sewall's Memorial. In 1701, while preparing for a court hearing to retain possession of Adam, Saffin published "A Brief and Candid Answer to a late Printed Sheet, Entituled [sic] The Selling of Joseph." The work has been described as the "only forthright defense of slavery in the Continental colonies until the time of the Revolution." At the end of his "Candid Answer" to Justice Sewall, Saffin appended this doggerel:

> "The Negroes Character"
> Cowardly and cruel are those
>   *Blacks* Innate,
> Prone to Revenge, Imp of
>   inveterate hate,
> He that exasperates them,
>   soon espies
> Mischief and Murder in their
>   very eyes.
> Libidinous, Deceitful, False
>   and Rude,
> The Spume Issue of Ingratitude.
> The Premises consider'd, all
>   may tell,
> How near good *Joseph* they
>   are parallel.[7]

It is clear that we must read Saffin's invective in the light of his difficulty in holding on to Black Adam. But its most significant feature emerges from the background of biblical interpretation common to New Englanders of that day. Saffin, albeit negatively, took seriously Sewall's typically Puritan figuration of the biblical Joseph as applied to black slaves. Indeed, himself a judge, Saffin attempted to counter the propriety of that biblical reference with his parody of a legal brief: "The Premises consider'd, all / may tell, / How near good Joseph [Blacks] / are parallel." However far this parody falls from the strictures of legal argumentation, it shares with

Sewall's Memorial the strategy of persuasion by means of literary figures and tropes. Indeed, neither author employed rigorous argument. Saffin's text was a sixteen-page attack on Sewall, which, to follow commentator Sidney Kaplan, is "too devious (and noxious) to detail here." In terms of its rhetoric, however, Kaplan describes the pamphlet as "a tangle of casuistry—the first in a long American line, prototype of the Biblical defense of slavery in the era of Calhoun, in which scriptural snippits are manipulated to fortify a theology of white superiority and black bondage."[8] I return to Saffin's rhetoric momentarily. First it is useful to examine as well the rhetorical strategy of Saffin's judicial opponent, Samuel Sewall.

It was not by rational argument but by means of biblical figures, skillfully applied to New England's slave trade, that Sewall sought to persuade his compatriots against the trade. In that literary performance we can see a characteristically Puritan "strategy of inducement" operating with "incantatory" effects, to borrow again Kenneth Burke's terms of analysis. Rhetorical effects are strategies of inducement when, "apparently describing the scene for the action of a drama, they are themselves a dramatic act prodding to a further dramatic act."[9] Such devices abound in Sewall's work. His rhetoric constituted a dramatic act by representing his slave trading peers iconographically as biblical types—in this case, Joseph's brothers who sold him into bondage. On the one hand Sewall expounded the biblical text: "*Joseph* was rightfully no more a slave to his Brethren, than they were to him: and they had no more Authority to *Sell* him, than they had to *Slay* him." On the other hand he applied that text directly to the situation of his fellow countrymen, in their "taking Negroes out of *Africa*, and Selling of them here."[10] By exposing in that manner their implicit, figural identity with Joseph's brothers, Sewall's strategy was to induce in his peers an alternative persona: namely, his own model of the emancipator who facilitates the manumission of slaves and the termination of the trade.

It is a matter of historical debate how far *The Selling of Joseph* refers directly to the circumstances of Black Adam's manumission. Kaplan argues that "black Adam was very much on Sewall's mind as he opened his Bible and took up his quill," but other historians dispute the connection.[11] In any event Sewall's extensive reference to the biblical Adam of the garden of Eden resonates coincidentally, if not deliberately, with Black Adam of the garden in Boston. What is certain is that Sewall employed the Adamic figure inclusively, as a representation of universal human equality and the natural right to liberty. "It is most certain," he wrote in the Memorial, "that all Men, as they are the sons of *Adam*, are Coheirs." Elsewhere with incantatory effect he capitalized the triad, "GOD LIBERTY, and ADAM." Accordingly not one, but two characters from the book of Genesis were central to Sewall's strategy: by thus juxtaposing the Joseph and the Adam figures he sought to induce his readers to replace their figural identity with Joseph's brothers and instead become Adamic brothers, or "Coheirs" of African peoples. Here we find yet a third figure in Sewall's rhetorical strategy. His text also features one of the earliest American representations of "Ethiopia" as a figure for all Africa and all Africans. Indeed, by the time the pamphlet concludes the author has left Joseph behind and relies exclusively on Adamic and Ethiopian figuralism. "These *Ethiopians*, as black as they are; seeing they are the Sons and Daughters of the first *Adam*, the Brethren and Sisters of the Last ADAM [Christ], and the Offspring of GOD; They ought to be treated with a Respect agreeable."[12] It is evident in this passage how vigorously Sewall employed rhetorical incantation in order to evoke more fraternal brothers for the "Ethiopians" among his slave trading readership.

Elsewhere in the Memorial, however, Sewall found it necessary like Saffin to engage in his own counterfiguration. He knowledgeably anticipated a typological schema that has traditionally reinforced slavery. According to this schema, just as all human beings are descendents of Adam and are 'cursed' by perpetual expulsion from a paradisal state (the Garden), so also all Africans are descendents of a biblical progenitor whose offspring was cursed to be "a bondman of bondmen to his brothers"—evidently in perpetuity. "*These Blackamores*," as Sewall represented the tradition, "*are of the Posterity* of Cham [*sic*], *and therefore are* under *the Curse of Slavery.* Gen.9.25, 26, 27.*" That sophistic and now notorious defense for enslaving Africans is referred to as the "curse" or "myth of Ham" (Cham). It has a long and ignoble history in American biblical interpretation, extending into the midtwentieth century.[13] Indeed, Sewall's response was among the first refutations published in America. "*Canaan*," he observed in the biblical account, "is the Person Cursed three times over without the mentioning of *Cham*.... Whereas the Blackamores are not descended of *Canaan*, but of *Cush*." Moreover Sewall proceeds immediately to quote the Scripture passage, previously described as the single most significant verse in Afro-American biblical hermeneutics, Psalm 68.31:

> *Princes shall come out of Egypt* [Mizraim] *Ethiopia [Cush] shall soon stretch out her hands unto God.* Under which Names, all Africa may be comprehended; and their Promised Conversion ought to be prayed for.

Remarkably we can see in Sewall's displacement of the figure of Canaan by the figure of Ethiopia an eighteenth century expression of the nineteenth and twentieth century tradition of Ethiopianism. But this proto-Ethiopianism is conveyed by a Puritan rather than an African American writer. Perhaps the Memorial provided a literary model for later developments in black American religious discourse. In any case, Sewall's antecedent text includes two major features characteristic of later Ethiopianism: the use of the name "Ethiopia" in such a manner that "all Africa may be comprehended;" and a prophetic interpretation of Psalm 68.31 with the missionizing implication that the "Promised Conversion [of African peoples] ought to be prayed for." The persistence of these features in subsequent black literary expression provides documentary support for regarding Puritan typology in New England as one of the sources of Ethiopianism in North America.

It is illuminating to return to Judge Saffin's counterfigural treatment in his poem, "The Negroes Character." Apart from our evaluation of his poetry, of his characterizations of Black Adam, or of his ethnic stereotypes of black people generally, Saffin did in fact raise an important issue. What are the criteria for rendering black Americans as antitypes of biblical figures? What would make "the Negroes character" genuinely "parallel" or "near good Joseph"? When we actually attend to the hermeneutic requirements of biblical figuralism, Sewall's use of the Joseph figure to apply to all enslaved blacks must be reappraised. I have emphasized one of the most important criteria for such applications in the discussion of Americana with reference to Puritan figuration of New England as New Israel. Puritan figuration in that instance was heterodox (as Bercovitch observed) because Christian typology requires that the postbiblical antitype be a true exemplar of its biblical type—that is, a model worthy of imitation within the community of the faithful. On the contrary, New Englanders (just like historical Israel) were unable as a collective to achieve the moral and spiritual stature of an individual exemplar. Unfortunately, when we measure Sewall's Joseph figuration with this criterion we discover the same Puritan theological error indicated in Chapter 2. In each case the figure is applied indiscriminately: that is, without reference to (or effort to confirm) the truly exemplary nature of the object to which it is applied. In the present instance, in place of exemplary referents for Black Adam we are confronted by the slavemaster's portrayal of a man "proud and surlie"[14] and chafing under the conditions of his servitude. However biased and distorted that portrayal, there is no effort in our sources to affirm or deny that the moral character of Black Adam was in any way 'above average' or extranormal. Normality or mediocrity in this regard cannot justify enslavement, of course. But neither is it praiseworthy or exemplary in the theological sense: to be an example for faith; to be worthy of reverence and imitation.

Sewall himself unwittingly provided data for the application of such a criterion in his Memorial. In arguing against slavery he declared: "I am sure, if some Gentlemen go down to the *Brewsters* to take the Air, and Fish: And a stronger party from *Hull* should Surprise them, and Sell them for slaves to a Ship outward bound: they would think themselves unjustly dealt with; both by Sellers and Buyers."[15] Such an argument intends to underscore the natural human outrage merited by the injustice of enslavement. By succeeding in its intention, however, the argument also underscores the entirely ordinary and expected character of such outrage—the justified resentment, counterintimidation, and vengefulness such as Saffin charged against Black Adam. In theological terms: to the degree that Black Adam's responses to slavery were in fact Adamic (nonexemplary) in the sense of natural and ordinarily human, precisely to that degree do they link him to the first Adam in contradistinction to the last Adam—that is, Christ—as a true exemplar. More to Saffin's point, however, mere attributions of naturalness or normality prevent Black Adam or his peers from appearing as true biblical exemplars.

Here we are in a position to perceive a crucial distinction between the incantatory strategies of Puritan American figuration and the conjurational performances of African American figuration. However praiseworthy we find Justice Sewall's antislavery efforts, we must also judge his figural strategy to be merely rhetorical or literary. It lacked the performative and embodied nature that characterizes conjurational forms of black religious figuralism, which was emphasized in the discussion in Chapter One. The embodied and materialist character of conjurational performance, when correlated with biblical figuration, requires that postbiblical models or antitypes concretely substantiate or manifest their biblical prototypes. Mere verbal or literary ascription is insufficient to achieve the performative depth or the existential concreteness of material transformations of reality. Such transformations are the genuine telos or intent of conjure performances, and not merely a verbal or conceptual change. Sewall's incantatory strategy—ascribing to African slaves the narrative role of the biblical Joseph—may have effected a conceptual change in early America's public discourse. The Memorial was certainly a catalyst in the antislavery debate. Its figural and incantatory strategy was nonetheless ineffectual in transforming reality insofar as it did not, and perhaps could not within the limits of Puritan culture, reconstitute Africans in colonial America as embodied exemplars of biblical figures. In the most radical analysis this lack of efficacy in Puritan figuration signifies a lack of power, both spiritual and material. The power to transform concretely was precisely the problematic issue in black religious projects of conjuring culture, as we shall see.

The task of reconstituting African Americans as embodied biblical exemplars was properly conceived and executed by black religionists themselves. I have examined the most familiar context for such achievements in the Exodus and Ethiopian figuration of black Americans. Although I will continue to display features of embodied figuration as constitutive in black religious projects of conjuring culture, I anticipate later analysis by recalling a recent instance of embodied figuration that manifests the Joseph figure in a remarkable manner. With almost prophetic attention to the terms of Saffin's comparison of "the Negroes character" and "good Joseph," a contemporary black American lifestory exemplifies at least two Josephite virtues as portrayed in biblical narrative: Joseph the dreamer, both the diviner of dreams and the facilitator of their reality, and Joseph the reconciler of estranged brothers. Taking the latter feature first, recall that the narrative climax (Gen. 45) of the Joseph story emphasizes his transformed character as reconciler. Here at the end of the story he does what he could not or would not do at its beginning: He makes himself intimate with his brothers, he entreats them to be at peace with themselves and himself, and he restores open communication where before "they could not speak peaceably to him."[16] Joseph's new conciliatory character is given a final touch when the brothers, now departing for a return trip to their father, Jacob, receive Joseph's informed admonition: "Do not quarrel on the way" (Gen. 45.24).

On this showing the Genesis narrative portrays the 'goodness' (Saffin) or exemplary character of Joseph as the virtue of reconciliation. Of course the fact that Black Adam or any slave was not conciliatory to his masters, in a manner "parallel" to Joseph's embracing of his brothers, does not justify a censure of "the Negroes character." (Again, not to be exemplary is not a legitimate ground for condemnation, abuse, or enslavement.) An appropriate search for correspondence between Black Adam and the biblical Joseph cannot focus simply on the abstract virtue of reconciliation itself. Legitimate correspondence must also take into account the existential or narrative context within which such a virtue is practiced. The narrative context within which Joseph effects his reconciliation to his brothers contains, foremost, the intervention of divine providence in Israel's history. Integral to that intervention is the dramatic reversal in Joseph's political rise to power in Egypt. The power relations at the end of the Joseph story, where the abusive brothers are now supplicants standing before their elevated former victim, are clearly incommensurate with the historical and political context of colonial slaves. A more appropriate context for comparison is a contemporary one, in which the power relations between black and white Americans are closer to parity (albeit only relatively). I refer briefly to the life story of that practitioner of reconciliation and ethnic 'brotherhood' Martin Luther King, Jr.

The biographical contours and the social-historical context of King's life render him a more authentic antitype of the biblical Joseph than Saffin's Black Adam. Like the Egypt of the Genesis story, King's society acknowledged and rewarded his prophetic gifts—his utopian dreaming and 'scheming'—as socially desirable assets. His Nobel Peace Prize was one of many occurrences in which this descendent of slaves earned the acclaim of his compatriots. Among other parallels is the overarching narrative framework: a dearth (or 'famine') of social justice and ethnic unity had come upon the land, and a man of King's gifts and character was sought by national leaders to assist the restoration of civil order and the integrity of law. Finally, we may recall King's dream of ethnic harmony and his labor toward a "beloved community"—that vision of great reversals and that praxis of reconciliation that rendered his life notably 'parallel to good Joseph'. Indeed, the parallel has been strikingly acknowledged by the memorial placed at the Lorraine Hotel in Memphis, Tennessee, where King was assassinated in 1968.

The Memphis memorial hearkens back to Sewall's Memorial with remarkable resonance. The eighteenth century text, with its antislavery message and its Adamic figuration, calls implicitly for the reconciliation of all peoples as "brothers" (not the reprehensible brothers of Joseph but the cocreated brothers of Adam). With a different but convergent emphasis the Memphis memorial paraphrases and inscribes Genesis 37.19–20. The

inscription, which reads, "Come, let us slay the dreamer," constitutes a
recent instance of figural expression in public discourse.[17] In more ways
than one, the inscription from Genesis marks a fateful recapitulation of
the Joseph story in an Afro-American life. For not only was King acknowledged as "the dreamer" of his society. His "I Have a Dream" address,
delivered at the Lincoln Memorial in Washington, D.C., on August 28,
1963, invoked explicitly the dream of a future Adamic "brotherhood."

> I have a dream that one day on the red hills of Georgia, sons of former
> slaves and sons of former slave-owners will be able to sit down together at
> the table of brotherhood. . . .
>   I have a dream that one day, down in Alabama, with its vicious racists,
> with its governer having his lips dripping with the words of interposition and
> nullification, that one day, right there in Alabama, little black boys and black
> girls will be able to join hands with little white boys and white girls as sisters
> and brothers. I have a dream today![18]

The social-historical distance between the context of King's powerful
speech that day and the context of powerlessness among colonial slaves is
vast. The crucial factor in creating that distance is not merely time, but
the intervening change in the power differential between black and white
Americans. In closing the gap between black and white access to power
in the New World, religious vision and political struggle were twin engines
of social transformation. A crucial element in such transformation was the
transformation of American law. My hypothesis is that the earliest encounters of black people with American law disclose conjurational aspects.
Some of the evidence for this hypothesis can be found in the incantatory
strategies which African Americans share with Puritan Americans. But, as
we have seen, the Puritan invoking of biblical figures as applied to postbiblical experience could too easily lack material reality and historical sub-
stantiation. Not only the theological and ethical demands of Christian
orthodoxy, but also the performative and materialist orientation of folk
practices, require more concrete forms of figural fulfillment for African
Americans.

Pharmakon

pharmakon: *I. a drug, medicine (Homer, etc.); 2. in bad sense, an
enchanted potion, philtre, so a charm, spell, enchantment; also a drug,
poison (Sophocles, Euripides). II. a remedy, cure (Hesiod); 2. also, a
means of producing (Plato).*

H. G. Liddell, *An Intermediate
Greek-English Lexicon*[19]

African American facility with the civil-religious discourse of the republic
first emerged in the context of revolutionary hope and egalitarian expectations. It was during the independence movement of the 1760s and 1770s

that, for the first time since lifelong bondage had been legislated for blacks a century earlier, general emancipation seemed to be in reach. Optimism in this regard derived not only from political and military events. A nascent class of aspiring religious leaders were also moved by the biblical and theologically based debates of the period. Before the revolutionary events, during the Great Awakening or religious revivals of the 1730s and 1740s, significant numbers of blacks had been exposed to the religious enthusiasm of the Methodist and Baptist traditions. They had also encountered elements of theological-political radicalism as conveyed by those post-Reformation communities.

In its renewal of Reformation zeal and piety, the Great Awakening introduced black converts to the political possibilities of debate and dissent between the various church bodies. It also introduced their communities to the theological requirements of such debate and dissent. Black leaders realized for the first time, as the historian Peter H. Wood has claimed, that "established authority could be challenged through Christian doctrine itself. This was a lesson which attentive black Americans learned quickly, and which subsequent generations of leaders would not soon forget."[20] On the eve of the American Revolution black leaders were even hoping for their own freedom and equality as spoils of war. Their hopes had been raised high by the potent combination of Christian revivalism and the Jeffersonian rhetoric of the Declaration of Independence. The political vision of the philosopher-statesman and chief architect of that document, Thomas Jefferson, extended to human equality within the European tradition of the "rights of man." But the founder's vision was moderated in the Declaration. Jefferson himself had sought to abolish slavery as a consequence of independence from England. Such hopes were dashed of course by concessions made to establish and maintain confederation with the southern colonies during the constitutional conventions of 1786 and 1787.

Subsequently, even the religious liberty to worship as fellow believers with white Americans was denied black converts. "By the time the Constitution was approved many blacks and whites could foresee an eventual day of racially integrated Christian worship, even in the South. Such a vision was not to be."[21] Such visionaries could not know that their freedom to worship in American churches would be systematically curtailed, decade after decade, by the major denominations. Yet already by the 1790s that fate was sufficiently threatening that some black religious leaders felt impelled to begin movements for separate worship. In this context they launched the independent black church movement. The phenomenal growth of black churches, of course, did not occur until the advent of massive black conversions to Christianity following full emancipation in 1865. But a permanent foundation of black preachers, proselytizers, and "folk theologians"[22] had already emerged between the Great Awakening of the 1730s and the revolution of 1776. Most significantly, the subsequent literature of independent and missionary black church communities indicates that these founders were grounded in biblical texts and Christian doctrines generally. In particular, they understood the hermeneutic requirements of biblical typology or figuralism.

According to Exodus figuration, a formulaic remedy is required to secure providential acts of divine liberation: that is, to conjure-God-for-freedom. Leaders of an oppressed community must represent themselves as types of Mosaic slaves, repeatedly beseeching their rulers in the name of God "to let my people go" (Ex. 7.16, 8.20, 9.1). Two formal petitions from the revolutionary period, conforming precisely to such figural and conjurational requirements, are useful for illustration here. The first petition was delivered "for the Representative of the town of Thompson" to authorities in Boston by a committee of slaves in April 1773. It displays several rhetorical devices that employ imitative or mimetic strategies. One such device features a pharmacopolitical remedy for American slaves that was simultaneously (homeopathically) toxic for the American slavocracy. "Sir," the petition began,

> the efforts made by the legislative of this province in their last sessions to free themselves from slavery, gave us, who are in that deplorable state . . . [to] expect great things from men who have made such a noble stand against the designs of their *fellow-men* to enslave them.[23]

By imitating the independence efforts of the colonists against their respective oppressors, the British, these slave petitioners gave those efforts a double valence. American independence became tonic/toxic by being rendered a remedy for slave petitioners and, at the same time, an indictment against American slaveholders. By means of a (conjurational) trick of discourse the petitioners identified their own civil liberties with the newly achieved liberties of the young republic: "We cannot but wish and hope Sir, that you will have the same grand object, we mean civil and religious liberty, in view in our next session." It is instructive that the petition cites "religious" in addition to civil liberty. A religious prescription appears with incantatory effect subsequently, at a central place in the petition. Its tone of abject humility is reductively (mis)understood if only taken as evidence of the docility of the petitioners. The spiritual intention here is conjurational rather than self-abasement for its own sake, in that the rhetoric conforms to the Exodus formula of conjuring-God-for-freedom.

> Since the wise and righteous governor of the universe, has permitted our fellow men to make us slaves, we bow in submission to him, and determine to behave in such a manner as that we may have reason to expect the divine approbation of, and assistance in, our peaceable and lawful attempts to gain our freedom.[24]

Could humility as a 'strategy of inducement' (Burke) be any more transparent here: "we bow in submission to [God]" *in order* that we may receive "divine approbation of, and assistance in . . . lawful attempts to gain our freedom"?

A second petition featuring less polished discourse was delivered to the Massachusetts House of Representatives in May 1774. The opening lines echo the preceding petition: the petitioners begin by professing themselves to be "a Grate Number of Blacks of this Province who by divine permission are held in a state of Slavery within the bowels of a free and Christian Country—Humbly Shewing." But this humility is also strategic and incantatory, as the provocative reference to "a free and Christian country" hints. A 'trick' of persuasive rhetoric is in operation here, as evinced by subsequent clauses that invoke both religious and democratic principles precisely in order to challenge the officials claiming to represent those principles:

> ... Humbly Shewing
> That your Petitioners apprehind we have in common with all other men a naturel right to our freedoms ... [that] there is a great number of us sencear ... members of the Church of Christ how can the master and the slave be said to fulfil that command Live in love let Brotherly Love contuner and abound ... [that] nither can we reap an equal benefet from the laws of the Land which doth not justifi but condemns Slavery.

But despite the foregoing claim that slavery is unsanctioned by "the laws of the Land," a subsequent clause (nearly heartbreaking in its familial pathos) inadvertently admits the existence of a now century-old legal precedent: the state sponsored and lifelong imposition of trans-generational bondage.

> or if there had bin aney Law to hold us in Bondage we are Humbely of the Opinion ther never was aney to inslave our children for life when Born in a free Countrey. We therfor Bage your Excellency and Honours ... [that] our children be set at lebety at the yeare of twenty one for whoues sekes more petequeley your Petitioners is in Duty ever to pray.[25]

The duty to pray for their children's liberation hints at the petitioners' resistance to oppression. But the forcefulness of that resistance is concealed by a rhetorical trick: it is located at the end of the petition where one is typically in the posture of begging or 'praying' the legislators for judicial redress.

Eventually black American letters developed more overt and provocative modes of expression in the "black jeremiad." This term has been coined by the Afro-American literary historian Wilson Jeremiah Moses. Moses describes the black jeremiad as a literary genre that arose as early as the revolutionary period, in mimetic relationship to the Jeffersonian rhetoric of the revolution. Indeed, in this connection he quotes Jefferson's jeremiadic fragment which stated his reflections on American slavery—"I tremble for my country when I reflect that God is just,"—and observes that "black writers in the early republic were to repeat the themes of the Jeffersonian jeremiad with enthusiasm for the next several decades." Moses highlights other mimetic aspects of the genre as well: its Afro-American nationalism in resonance with "European and Anglo-American nationalism," including the themes of chosen peoplehood and a divine providence watching over historical developments. But in contrast to the ostensible submission to that providence that we see in the petitionary genre, Afro-American jeremiads were more free to disclose harsher sentiments.

> For obvious reasons most black writers did not wish to be identified as inflammatory pamphleteers. But this did not mean that [they] consistently eschewed the rhetoric of violence, confrontation, or racial chauvinism.
>
> Early black writers in America often referred to themselves as a chosen people.... [They] came to make thinly veiled references to slave rebellions as means whereby God might effect his justice.[26]

It is intriguing to compare Moses's comments on the Afro-American jeremiad with those of his counterpart, Sacvan Bercovitch, on the Puritan American jeremiad. In particular, both literary historians address the ostensive character of their respective genres—their conspicuously caustic tone and calamitous import as jeremiads. Each scholar insists that these features contrast sharply with the genre's trajectory as persuasive discourse, and with the evident intentions of the authors. Bercovitch's treatment argues that the Puritan jeremiad employed the rhetoric of criticism and castigation in order more thoroughly to enlist its audience in upholding the zealous idealism of its critique. That is, the real objective was not condemnation or repudiation but precisely the opposite: ideological affirmation and reenlistment in an ideal cause. In a similar manner, Moses argues that among black writers "the rhetorical threat of violence was not based on any real desire for a racial Armageddon. Black jeremiads were warnings of evils to be avoided, not prescriptions for revolution. When Ethiopia stretched forth her hands, it would not be to take up a sword but to embrace her erstwhile enemies."[27] In choosing *not* to 'prescribe' (borrowing Moses's salutary term) violent revolution, writers of black jeremiads chose instead biblical and democratic prescriptions for their legal and civil rights grievances. I regard this prescriptive aspect of the jeremiad as evidence of the 'pharmaconic' propensity or inclination in African American rhetorical traditions. That pharmaconic intent is conveyed by incantatory devices that invoke civil, religious, and moral laws alongside the direful consequences to be incurred for violating such laws. Using his own terms, Wilson Jeremiah Moses provides an illustration of this jeremiadic strategy in the following case.

Writing to Thomas Jefferson in 1791, the Afro-American author and architect Benjamin Banneker created a moment of convergence between the two traditions, Puritan and African American. Banneker affirmed what Puritan writers had declared for decades, that America had prospered amid the nations of the world because God had favored its people with the best of laws, natural and divine (compare biblical Israel's reception of divine law in the wilderness). Having evoked this common figural heritage he then declared the natural rights of slaves to freedom under God, and the punishment to ensue if those rights continued to be thwarted. Against this background Moses describes the following jeremiadic element in Banneker's rhetoric.

> Banneker's letter was thus more than an appeal to Jefferson's sincerity and sense of justice; it was almost a legalistic argument, attempting to manipulate fears of violating the covenant. Banneker was appealing to a very real belief in America's divine mission and providential destiny. He was also appealing to the Enlightenment view that a glorious future could be expected only if the nation conformed to universal natural law.[28]

By conjuring with the same priniciples of law, divine and natural, that Jefferson had invoked in the founding of the young republic, Banneker's jeremiadic strategy was to render them tonic to his own cause and toxic to the nation's antidemocratic and self-contradictory policies. In Moses's terms, Banneker's letter affected a legalistic tone by arguing that "Americans must extend the provisions of their natural rights covenant to include the blacks, or deny the covenant itself." This prescriptive formulation clearly displays the tonic/toxic dichotomy employed in the Afro-American jeremiad: either include black Americans as citizens under the covenental, civil, and juridical traditions of the republic (tonic), or else forfeit (toxic) those covenental benefits and privileges that have been secured by the republic through its foregoing allegiance to divine and moral law. By implication, in this formulation law itself is the crucial element—a tonic or toxic *pharmakon* depending on its use or abuse by the republic. We will see how black Americans discovered that law alone was insufficient as a social curative in this regard: Another pharmaconic agent was needed to counter the toxicity of American law itself, as it became increasingly the medium of institutionalized racism and of routinized, systemic violence against black people.

Pharmakoi

> *The (rite of the) pharmakos was a purification of this sort. . . . If a*
> *calamity overtook the city by the wrath of God, whether it were famine*
> *or pestilence or any other mischief, they led forth as though to a sacrifice*
> *the most unsightly of them all as a purification and a remedy to the*
> *suffering city. They set the sacrifice in the appointed place, and . . . burnt*
> *him with fire . . . and scattered the ashes.*
>
> Tzetzes, *Thousand Histories*[28]

The post-Reconstruction era is perhaps most distinguished for terminating a brief democratic episode in the South, during which black legislators were elected and held public office. Other effects derived from, and then accelerated, the unconstitutional measures taken to end such participation in political powersharing: the disenfranchisement of Afro-Americans as voters, the advent of segregation and "Jim Crow" structures (legalized discrimination), and the beginning of lynching as an American social practice. On the one hand, the new political and economic structures functioned to ensure the political powerlessness and the poverty-class status of black people, just as securely as if they were a subject people living under dictatorship. In the midst of a democratic republic, the new laws constituted a massive zone of de facto if not de jure domination. On the other hand, the near epidemic levels of lynching established a pattern of victimization so profound and enduring that it reconfigured black social identity with yet another biblical figure: the 'victimary' figure of the scapegoat.

We know the word "scapegoat" from the ancient Hebraic practice recorded in Leviticus 16.21–22: "Aaron shall lay both his hands on the head of the living goat, and confess over it all the iniquities of the children of Israel . . . that the goat may bear upon him all their iniquities to a land apart." But there is another ancient tradition of scapegoating, and a tradition with more significance for African American experience and culture: the ancient Greek scapegoat, a human being called a *pharmakos*. As James G. Frazer reported, "the Athenians regularly maintained a number of degraded and useless beings at the public expense; and when any calamity, such as plague, drought, or famine, befell the city, they sacrificed two of these outcasts as scapegoats."[30] These *pharmakoi* were reserved for times of plague or (read) civil discord, when the evils infecting the community were ritually ascribed or transferred to them. The continued presence of such ritually infected persons thereby became toxic to the community, and they were driven out with rites that coincidentally if not deliberately resulted in the victims' death.

A similar scapegoating of Afro-Americans, as a class of *pharmakoi* in a culture of domination, is especially evident in the American ceremonial practice of lynching. The routine occurrence of the practice postdated the abolition of slavery. More precisely, lynching began in the 1880s as a backlash or reaction to the voting enfranchisement and legislative empowerment of black southerners during Reconstruction. The execrable practice did not end until the 1930s.[31] Thus the phenomenon endured a half-century, and has indelibly impacted Afro-American social existence with the marks of victimization. Prior to emancipation, the slavery period featured the victimization of black people as part of the institution of slavery itself. Its most blatant forms included murder, maiming, beatings, rape, and the hangings or other forms of execution that followed attempted escapes and organized revolts. But lynching in southern communities after Reconstruction was qualitatively different in its ritual or ceremonial character, on the one hand, and in its vengefulness on the other. I do not refer simply to vicious elements like live body burnings or dismembering of genitals. (Coincidentally, comparable features of ancient Greek scapegoating rituals included both immolation of victims and beating of victims' genitals. Various forms of dismemberment or *sparagmos* yielded cathartic effect for participants in such sacrificial rites.[32]) More particularly I am interested in the fact that such procedures often occurred alongside distorted Christian affirmations and sanctions. (Consider the ritual of cross-burning and other religious manifestations by the Ku Klux Klan and similar terrorist and hate organizations.)

What accounts for the convergence of religion and violence in such overt forms of victimization? What can account for lynchings as quasi-religious ceremonies, particularly in the context of a tradition like Christianity which so highly emphasizes its love ethic? Ironic convergences like these are cognate, in my view, with the political convergence in American history of democratic traditions, on the one hand, and ethnic domination on the other. I posit a homology in American culture, between the paradoxical convergence of Christianity and violence and the paradoxical convergence of democratic traditions and institutional racism. In order to illuminate such cognate phenomena and their role in configuring black American experience, I have borrowed the terms of analysis available in a recent controversial theory. It is a contemporary theory of religion and (its constitutive role in) violence, which includes in its purview a theory of law and (its contitutive role in) violence. In applying this theory to black American experience, I am most interested in understanding the homology between religion and law in relation to violence. Because these are my overriding interests, I am of course open to the possibility that other theories may possess greater explanatory power. At the same time, however, the following analysis appears more conducive to this study than other relevant theories.

The most provocative treatment of ritual scapegoating has been provided by the literary theorist René Girard. Girard is a French American scholar whose theory of ritual sacrifice and scapegoating has extended his work into a wide variety of disciplines. Among religion scholars he is best known for his claim that "generative violence," or violence as the "primitive sacred," best explains the deep structures of religion and culture from human prehistory to the present. Girard began his scholarly career as a literary critic and theorist focusing on the phenomena of imitation, desire, and "mimetic rivalry" in novels. What he discovered in literature, however, led him to extend his critique to the fields of psychology (human desire), anthropology (cultural formation), and the history of religions (violence and the sacred). Most provocatively for Western readers, Girard now insists that specifically Christian traditions of scapegoating and ritual violence (against Jews, to take a notable example) are consistent with the sacralization of violence typical in all religious cultures throughout history. He highlights this identification or equation between violence and 'the sacred' (itself a disputed category among religion scholars) and calls it the "primitive sacred" of the human species. Gods, revered ancestors, and spirits are personifications of this primitive sacred, so that 'in the name of God' individuals and societies have felt themselves religiously empowered to execute the most ferocious forms of human violence. However, for Girard the ritual structuring of violence emerges in human prehistory as a preventive or prophylaxis against unstructured forms of violence: the virulent, contagious, reciprocal violence in which every group in a community is turned against every other group without means of resolution.

To prevent the ultimate catastrophe of an entire community completely destroyed by its internal rivalries and hostilities, we have learned (so the theory claims) how to interpose sacrificial strategies that cathartically and routinely release conflictual passions and reestablish communal harmony. Using structured violence or conflict to forestall greater violence and conflict, sacrificial processes 'fight fire with fire' and thereby appear to be the best expedient available for solving social conflicts. Historical observation can readily survey a wide spectrum of such expedients, from the most overt to more attenuated or diluted forms; from murder and mass murders to expulsion or ostracism, persecution or harassment. According to Girard even modern judicial systems, which are designed to transcend scapegoating and the lawlessness of a society ruled by revenge (as in blood feuds), nonetheless operate as covert systems of sacrifice.

> If our own system seems more rational, it is because it conforms more strictly to the principle of vengeance. Its insistence on the punishment of the guilty party underlines this fact ... [it] *rationalizes* revenge and succeeds in limiting and isolating its effects in accordance with social demands.... In the final analysis, then, the judicial system and the institution of sacrifice share the same function, but the judicial system is infinitely more effective. However, it can only exist in conjunction with a firmly established political power. And like all modern technological advances, it is a two-edged sword, which can be used to oppress as well as to liberate.[33]

With regard to law, sacrificial and scapegoating processes operate in the following configurations of culture: (1) where cultures initially lack the institutions of law and order that are capable of arbitrating conflicts and that can transcend cycles of aggression and revenge (primitive or precivilized societies); (2) where institutions of law and order have deteriorated or, for whatever reason, are no longer effective (anarchic, transitional, or revolutionary societies); (3) where institutions of law and order mediate the domination of a minority elite over the rest of society (dictatorship or totalitarianism); (4) where established institutions mediate the domination of a majority over a minority population (semidemocratic or republican societies). In the latter case, most relevant for this study, the transcendence of arbitrary violence and vengeance that legal structures are intended to provide functions selectively. That is, under domination such transcendence does not achieve full rationalization, to use Girard's term. It is not extended universally or absolutely, but is contingent upon the society's need for unanimity, or need for a redirecting of social hostilities away from a less expendable person or class, for example. Thus when contingencies require it, the legal or judicial process itself can function as a medium of scapegoating. And it can do so with all the religious sanction formerly harnessed by primitive societies in the form of ritual sacrifice.

In the same way that sacrificial victims must in principle meet the approval of the divinity before being offered as a sacrifice, the judicial system appeals to a theology as a guarantee of justice. Even when this theology disappears, as has happened in our culture, the transcendental quality of the system remains intact. Centuries can pass before men realize that there is no real difference between their principle of justice and the concept of revenge.[34]

This last formulation is consistent with the cumulative experience of Afro-Americans as *pharmakoi*. Even since their former victimization as property black Americans have remained in the position of prospective victims. As dictated by the economic and political needs of the majority culture, their legal status as citizens has been violated with little compunction by nearly unanimous decisions in both national and local communities. Mob 'justice' operating with religious sanction in the form of lynchings (notably in the case of the Ku Klux Klan) have been replaced by criminal justice systems that selectively target ethnic minorities, the poor and uneducated, and consolidate public sentiment and political power on that basis. Here we may apply Girard's theory by using the cognate terms developed in this chapter: social conflict and violence in the dominant culture function as a poison or toxic *pharmakon* and are purged or rendered benign by means of a scapegoat or *pharmakos* class. In formulaic expression: the pharmakos (scapegoat) is a tonic pharmakon (antidote, cure) for purging the toxic pharmakon (poison, disease) of the community's internal conflicts. For Girard this anthropological complex of relations constitutes the fundamental basis of his scapegoat theory. But for the purposes of this study, the semantic relations of the pertinent terms display a deeper dimension in Afro-American practices of conjuring culture. Going beyond Girardian theory and analytic interests, our subject matter directs us to consider the evidence for conjurational reversals of scapegoat-as-antitoxin phenomena.

Let us grant the hypothesis that human beings are rendered as scapegoats through sacrifical social processes—extralegal and legal, religious and nonreligious. Therein they convey tonic effects for a society that exploits them as antidotes to its internal conflicts and rivalries. Within the terms of such a hypothesis, a conjurational culture that is skilled in the operation of tonic/toxic dichotomies may readily conceive a pharmacopeic strategy for reversing its own social configuration as a class of scapegoats. Such reversals are typical of conjure tricks: one conjuror's spells and charms can be turned against another's by superior craft or more potent materials. Here I shift the perspective from the nation at large and consider the case of black Americans intraspecifically. On this view, malignant forces (both material and immaterial) have rendered their social existence toxic to themselves. As perennial victims or scapegoats *pro durante vita* (for the duration of life, in the legal language of seventeenth century colonial legislatures), their mode of being is continually in danger of becoming annihilatory, contingent upon the operation of social forces over which they have been intolerably powerless. To exert or reclaim power, some way must be con-

structed for countering this state of affairs either directly, by gaining social-
political power that terminates victimization altogether, or by a pharmacopeic antidote: by the 'trick' of turning victimization against the victimizers and to the benefit of the victims.

In the absence of decisive political power and economic resources for
directly terminating their victimization, Afro-Americans have had recourse
to a variety of indirect means. In subsequent discussions of wisdom and
praxis some of these strategies are examined more closely. They derive in
part from the slavery period through such trickster exploits, and proverbial
wisdom formulas, as "puttin' on ole massa"[35] and 'dismantling the slavemaster's house with the slavemaster's tools.' But the culture's trickster
and wisdom practices are also evident in its conjurational efforts to address
victimization. Most pertinent to the present analysis, these efforts have
taken the form of imitating or representing Afro-American victims as
biblical victims, in order to dispel (by counterspell) their continued victimization. I reviewed a precursor of such efforts in an earlier section
(Judges) of this chapter, where Samuel Sewall attempted a Puritan figuration of African Americans as a type of the biblical victim Joseph. In a
subsequent analysis of gospel figurations, I likewise review Harriet Beecher
Stowe's effort to render black Americans as biblical victims with her Uncle
Tom figure. But in each instance these white authors, one writing in the
eighteenth century and one in the nineteenth, failed to represent their
subjects in ways that did not invite sustained or even increased victimization.

Indeed, this is the biblical-figural conjure trick whose performance
long eluded Afro-American themselves, as practitioners of religious and
cultural transformation, up to the midtwentieth century. How does one
reconfigure the embodied person herself or himself, who is configured by
the culture as a victim or scapegoat, using the biblical forms and democratic
traditions revered by the society and its institutions? How has black America solved the problem of rendering American law and religion as cures
for violence? How have black freedom movements and civil rights strat-
egists learned to correlate the (pharmaco)toxic nature of violence, on the
one hand, and the 'pharmaconic' operations of conjuring culture on the
other? How has the culture deployed the focal point of such a correlation,
the embodied person herself or himself, for such purposes? We have seen
that the intended or surrogate victim of violence can be rendered as a
pharmakos who is also a *pharmakon*—a scapegoat who is thereby a cure.
The quest for a means to terminate the former, by manipulating curative
performances of the latter type, must be addressed again in the theological
chapters of Part III when we turn to gospel formations of black America.

Wilderness

*Thou shalt remember all the way which the Lord thy God led thee these
forty years in the wilderness, to humble thee, and to [test] thee, to know*

what was in thy heart, whether thou would keep his commandments or not.

Deuteronomy 8.2

We are acting over the like sins with the children of Israel in the wilderness.

Nicholas Street, "The American States Acting over the Part
of the Children of Israel in the Wilderness and Thereby
Impeding Their Entrance into Canaan's Rest" (1777)[36]

In the writings of W.E.B. Du Bois, that Afro-American heir to New England culture and Puritan sensibilities, we find the biblical Wilderness figure embedded in a jeremiadic fragment. At the turn of the twentieth century, two centuries after its figural inception in Puritan discourse, Du Bois invoked the forty-years-of-wilderness figure found in the book of Exodus. In the Afro-American figural universe Wilderness can signify either the post-Reconstruction period of the late nineteenth century, or the early colonial period (to which I turn in this section). With subtlety (almost undetectably), Du Bois applied the Wilderness figure to the failed pursuit of a democratic "promised land" in the South after the emancipation of the slaves in 1865.

> Years have passed away since then,—ten, twenty, forty; forty years of national life, forty years of renewal and development, and yet the swarthy spectre sits in its accustomed seat at the Nation's feast... the Nation has not yet found peace from its sins; the freedman has not yet found in freedom his promised land.[37]

In this passage Du Bois depicts with pathos the aging freedmen and freedwomen's slow realization—decade after decade from the 1860s to 1900—that their ancestral vision of Promised Land was "a dream deferred" (Langston Hughes). Instead of a biblical Promised Land the fate of that generation was reconfigured in the pattern of ancient Israel's wilderness experience. Moreover, true to the Bible's Wilderness figure their experience included the advent of new legal and juridical traditions. Yet the new laws lacked any of the benign forms of socialization or communal formation to be found in Mosaic prohibitions and statutes. They were distinguished instead for their uniformly alienating, oppressive, and toxic effects. In fact, they were devoid of the egalitarianism or communitas (Turner) found in either biblical or American civil law—a communitas long desired by black folk in America since the colonial period.

Du Bois's use of the biblical figure of post-Exodus Israel, wandering in the desert or wilderness for forty years before entering the promised land of Canaan, has been invoked by earlier generations of Americans. In fact "wilderness" has been a readily available metaphor of American experience ever since Michael Wigglesworth described the continent as "a waste and howling wilderness" in his 1662 poetic jeremiad or lament

"God's Controversy with New England." But Wigglesworth's solely negative figure of wilderness, "Where none inhabited / But hellish fiends, and
brutish men / That Devils worshiped," was soon tempered by a missionary
use of the figure. In 1670, Samuel Danforth's famous election day address
revised the image by enshrining the Puritan misssion in New England as
an "errand into the wilderness."[38] Danforth's address assessed the nature
and the failure of this "errand" and, as Sacvan Bercovitch points out, was
scheduled signficantly for the fortieth anniversary of the Puritan landing.
In Bercovitch's view the address is typical of the genre of the American
jeremiad as both public lamentation and exhortation. The term "jeremiad"
initially characterized the public ministry (627–580 BCE) of that biblical
prophet, Jeremiah, whose gloomy criticisms and doomsday predictions
were notorious in his lifetime and remain so today. Bercovitch, however,
qualifies this use of the term (and thereby departs from the magisterial
scholarship of his predecessor in Puritan literary history, Perry Miller).
He insists that, precisely by confessing the community's failings, an American jeremiad signifies the "persistence" and binding power of the founders'
dream.

In Puritan America the founders dreamed of New England as a promised land of Christian piety, moral rectitude, and civic virtue under the
laws of church and state. Forty years before Danforth assessed the Puritan
errand in the American wilderness, John Winthrop had exhorted his fellow
passengers arriving from Southampton, England, on board the *Arbella*:
"consider that wee shall be as a Citty upon a Hill, the eies of all people
are uppon us."[39] That number, forty years, was portentous for the Puritan
imagination because of its biblical significance, specifically as found in the
passage quoted from Deuteronomy: "remember all the way which thy
God led thee these forty years in the wilderness." Significantly for the
present study, the number recurs with prophetic force in the figural context
of the earliest African American experience of a biblical Wilderness: the
wilderness test of colonial America.

It is commonly known that the first group of African slaves in North
America arrived in Jamestown, Virginia, in 1619. Less familiar is the fact
> that for the first forty years there existed a period of relative parity between
black and white indentured persons. In figural iconography this corresponds to the biblical forty years of testing in the wilderness. In related
terms the historian Lerone Bennett describes those forty years as a fateful
moment of communitas (Turner) as compared to the next three centuries
of interethnic strife. Bennett claims that, even in a context where chattel
slavery and indentured servitude were normative, it was a time of egalitarian actualities and possibilities among ethnic groups in the American
colonies. "Slavery or modified slavery was a distinct possibility for all
disadvantaged people—Indians as well as black and white immigrants. So
was an open society. [But] socioeconomic forces . . . tilted the structure in
the direction of Negro slavery."[40] Historians may differ on the inevitability

of that tilt, but it is clear that the colonies were motivated by economic considerations to distinguish between black and white forms of servitude. From the 1660s onward, each colonial legislature in succession extended < the legal condition of captive Africans from indentured servitude to *pro durante vita*: "for the duration of life."[41] But whereas an economic concern was primary in determining subsequent developments, those developments were activated by a religious and legal crisis.

As African slaves became increasingly familiar with British custom and law, numbers of them began claiming the right of manumission or freedom on the basis of their Christian baptism. They reasoned that having submitted to the rite of baptism, they simultaneously acquired a legal right to manumission as newly Christianized and (therefore) 'patriated' subjects of the English king. A growing incidence of these demands led to anxiety and protests from slaveowners, not indeed on religious grounds but with a view to the financial losses they would incur if this inconvenient tradition were enforced. Documents from the colony of Maryland, for example, explicitly refer to the loss of property and labor that would ensue unless legislation were enacted. Accordingly, legislation was enacted. Laws were promulgated throughout the colonies "obliging negroes to serve *durante vita* . . . for the prevencion of the dammage Masters of such Slaves must susteyne by such Slaves pretending to be Christ[e]ned[;] And soe pleade the lawe of England."[42] Six colonial legislatures in the early eighteenth century thereby denied the relationship between Christian baptism—freedom "in Christ"—on the one hand, and social freedoms on the other. There was, incidentally, some attempt at a religious rationale for the new legislation. Church officials and missionaries considered the laws crucial for encouraging planters to permit religious instruction of their slaves without the fear of losing their source of labor. They reckoned that "slaveholders refused them permission to catechize their slaves because baptism made it necessary to free them." By enacting the new statutes, Virginia like other colonies espoused the pious view that "diverse masters, freed from this doubt, may more carefully endeavor the propagation of christianity."[43]

The legal basis of the planters' fears derived from the role of baptism as a rite of passage to citizenship in 'Christendom' since the medieval period. As typical in theocratic states, English law also linked religious identity and citizenship. But economic profit based on domination required colonial governments to contravene any such linkage in this case. On reflection it becomes clear that the consequences of this policy, however unrealized at the time, were disastrous for the theological integrity of American Christianity. The new laws implicitly separated spiritual and social-political existence with a virulence that has characterized and bedeviled American churches from the inception of the republic. Among the slave converts themselves, however, some initially refused to accept this overturning of legal tradition. They tenaciously held on to the emancipatory consequences of their new religious status. So tenacious were the convictions of many that authorities in Virginia had to resort to violence and bloodshed in order to extirpate a longstanding relationship between Christian identity and emancipatory expectation. As they realized that those expectations were being subverted, some slaves apparently considered more overt means of liberation.

> Rev. James Blair, official representative of the bishop of London in Virginia [in 1731] . . . announced that his suspicions had been confirmed: "notwithstanding all the precautions which the ministers took to assure them that baptism did not alter their servitude, the negroes fed themselves with a secret fancy that it did, and that the King designed that all Christians should be made free. And when they saw that baptism did not change their status they grew angry and saucy, and met in the night-time in great numbers and talked of rising."

Blair did not doubt that some of the slaves were "sincere Converts," but thought that the great majority of them sought baptism "only . . . in hopes that they will meet with so much the more respect, and that some time or other Christianity will help them to their freedom." Whether they were sincere believers or merely desperate and hopeful human beings, the Virginia conspirators' passion for freedom was sufficient to raise the spectre of revolt. The threat of rebellion was suppressed by whippings, organized patrols, and the executions of four leaders.[44]

From an idealistic perspective it is ironic that a bishop of the English church was the indirect agent of this eradication of human hope and Christian possibility. Already by 1667, it is true, Virginia had declared that "the conferring of baptisme doth not alter the condition of the person as to his bondage or freedom." But subsequently in 1727, sixty years later, the bishop of London himself (Edmund Gibson) found it necessary to issue an official letter of assurance to American planters. With stunning theological and ethical myopia the Bishop wrote, "Christianity, and the embracing of the Gospel, does not make the least Alteration in Civil Relations."[45] This declaration constitutes an abysmal act of self-nullification and voluntary impotence among ecclesiastical authority. More to be marveled at, however, was the persistence of emancipatory expectations among slave converts a generation and more after Virginia first mandated the civic impotence of baptism. Indeed, such expectations endured out of all proportion to the relatively small numbers of manumissions actually granted on the basis of baptism.[46] What explains this perseverance of expectation in the potency of a civil-religious rite of passage? What accounts for the slaves' tenacious belief in the ritual and legal efficacy of emancipatory baptism?

Here we should recall that, in addition to the legal precedence in English polity for correlating baptism with the rights of citizenship, a Christian tradition of 'emancipatory baptism' extends back to the time of the Emperor Constantine and his establishment of Christianity as the state religion of the empire (323 C.E.). The emancipatory implication of baptism was also a signal element in the left-wing movements of the Protestant Reformation in sixteenth century Germany and central Europe. In each one of these historical antecedents, the ritual use of rebaptism or "anabaptism" provided a form of resistance to the domination of ecclesiastical authorities in alliance with established state power.[47] There is no warrant or necessity to posit any direct influence of these precedents on the developing notions of emancipatory baptism among slaves in colonial America. However, an indirect influence is probable in terms of the egalitarian nature of early Baptist revivals, missions, and church life among slave communities (in contrast to Anglican attitudes). The historian Mechal Sobel, in this connection, reminds us that "early Baptists (along with Quakers and Methodists) widely shared these strong antislavery feelings. The Baptist appeal to and acceptance of blacks, both as equal members and as preachers, is the best evidence of the depth of their feelings about the spiritual equality of the races."[48]

Of course Baptist forms of interethnic communitas deteriorated as the churches became an acceptable southern denomination in the early nineteenth century. But the earlier egalitarianism of Baptist traditions, in combination with an openness to ecstatic forms of worship and religious expression, suffices to explain their prodigious influence in the development of black North American Christianity. Indeed, Sobel goes so far as to postulate a convergence between the white "Baptist worldview" and an "African/American sacred cosmos." We need not subscribe wholly to the details of her historical reconstruction in order to entertain its explanatory possibilities as a hypothesis.

> The Baptist invitation came at a supremely opportune moment, when enough blacks were sufficiently comfortable with spoken English to appreciate the Baptist message and uncomfortable enough in their bifurcated Sacred Cosmos to hunger for a renewed one.... Blacks accomplished an almost impossible task, one that was not much appreciated then and has not been widely understood since ... they became Baptist Americans at the same time as they made a quasi-African Sacred Cosmos at home in the New World.[49]

The tenacity of slave belief in the emancipatory efficacy of baptism, and in defiance of the prevailing legal and political realities, provides data for the hypothesis of an emergent African American sacred cosmos, in which one insistently conjures freedom by means of ritual performances like Christian baptism.

Now I return to the African American worldview of the "pharmacosm" proposed in the discussion of Genesis. There I speculated that continuities between African traditional religions and black American culture have resulted in a pharmacopeic worldview, in which the cosmos is constituted and reconstituted by healing and harming processes and practices. Here I reemphasize in passing the specific feature of Bakongo sacred medicines and ritual performance, by means of which a ritual expert attempts to reestablish the ancestors' vision of the cosmos in some idealized form of balance, perfection, and vitality. Furthermore I suggest that a surrogate source for the ancestors' cosmic vision in black American culture has been the Bible. That is, I propose that the Bible has functioned as an alternative source of formulas and materials (metaphorical and metonymic) for African people(s) deprived of their indigenous means for 'curing,' revising, or revisioning culture. After the discussion in this chapter, however, it is appropriate to posit that not one, but two, sources of Euro-American religion and culture were comprehended early in the matrix of transformative strategies devised by African Americans to conjure American culture.

Not only was the Bible regarded as an efficacious source of curative transformations. Also available and useful were the <u>democratic ideals and republican political traditions</u> that distinguished the process of nation-building in colonial North America. Those two sources of United States culture and consciousness have been identified as the biblical and republican elements by Robert Bellah, sociologist of American religion and culture. Whereas I have attempted to locate the operation of 'biblical republicanism' as a factor beginning in the colonial period of black American experience, Bellah and his colleagues emphasize the contemporary emergence of biblical and republican elements in black social and political strategies. They are quite explicit, indeed, in noting how these combined "strands" of American national life have operated with transformative effect. For example, biblical republicanism was a distinctive factor in the successful oratory of the black civil rights activist Martin Luther King, Jr.

> King's articulation of the biblical and republican strands of our national history enabled a large number of Americans, black and white, to recognize their real relatedness across difference. King characterized legal disenfranchisement, poverty, and unemployment as . . . glaring failures of collective national responsibility . . . [and] reawakened recognition by many Americans that their own sense of self was rooted in companionship with others . . . [others] whose appeals to justice and solidarity made powerful claims on their loyalty.[50]

But this ability to make powerful appeals of justice and solidarity by combining the biblical and republican strands of American culture, and to challenge and reform its legal and social-economic structures, did not spring de novo from the oratory of King and his generation. It is embryonic, as we have seen, in the earliest engagement of black religious and political leaders with the legal institutions of the republic. Moreover it bears some of the defining features of the conjurational spirituality identified in preceding chapters, and explored in further detail later.

Notes

1. Samuel Sewall, *The Selling of Joseph: A Memorial*, ed. Sidney Kaplan (Amherst: University of Massachusetts Press, 1969), p. 28.

2. Ibid.

3. Ibid., p. 37.

4. Ibid., pp. 7, 10.

5. This Belknap is not further identified in Kaplan's editorial notes or in the *Dictionary of American Biography*.

6. Ibid., p. 44f.

7. Ibid., pp. 42, 43.

8. Ibid.

9. Kenneth Burke, *The Philosophy of Literary Form: Studies in Symbolic Action* (New York: Vintage Books, 1957), pp. 6, 100.

10. Ibid., p. 11.

11. Ibid., pp. 35, 38.

12. Ibid., pp. 38–39.

13. Latta R. Thomas, *Biblical Faith and the Black American* (Valley Forge, Pa.: Judson Press, 1976), pp. 27–82. Cf. Gayraud S. Wilmore, *Black Religion and Black Radicalism: An Interpretation of the Religious History of Afro-American People*, 2nd ed. (Maryknoll, N.Y.: Orbis Books, 1983), p. 119.

14. Sewall, *Selling of Joseph*, p. 36.

15. Ibid., p. 15.

16. The great dramatic reversals of the Joseph story portray this cumulative transformation: the darling of his father sold to Gentiles by his own brothers (ch. 37.28); his rise to favor in Potiphar's household in Egypt and late imprisonment for alleged adultery (ch. 39); his liberation from prison and accession to the court of Pharaoh based on his gift of interpreting dreams (chs. 40–41); the denouement in which Joseph, as prime minister of Egypt, provides for his father and brothers during famine, receives their homage (they do indeed bow "themselves before him with their faces to the ground") (ch. 42.6), and embraces them rather than avenging his old grievance (chs. 42–45). Surely the concatenation of these reversals serves to effect and to make intelligible an inner *character* reversal. It is true that Joseph retains his characteristic pride in his powers of dream divination and in his mastery of situations (ch. 44.15). It is true that Joseph the prime minister tests his brothers in order to see whether, as before with him, they will allow the youngest boy (this time, Benjamin) to go into slavery while they return to Jacob and report the loss of yet another favorite son (ch. 44.14–34). Nonetheless, he has also become reconciler to his estranged brothers: "So Joseph said to his brothers, 'Come near to me, I pray you. . . . I am your brother, Joseph, whom you sold into Egypt. And now do not be distressed, or angry with yourselves, because you sold me here; for God sent me before you to preserve life.' And he kissed all his brothers and wept upon them; and after that his brothers talked with him" (Gen. 45.4:15).

17. Cf. William Bradford Huie, *He Slew the Dreamer* (New York: Delacorte Press, 1968).

18. Martin Luther King, Jr., "I See the Promised Land," in *A Testament of Hope: The Essential Writings of Martin Luther King, Jr.*, ed. James Melvin Washington (New York: HarperCollins Publishers, 1991), p. 219.

19. *An Intermediate Greek-English Lexicon*, ed. H. G. Liddell (Oxford: Oxford University Press, 1983), p. 855.

20. Peter H. Wood, " 'Jesus Christ Has Got Thee at Last': Afro-American Conversion as a Forgotten Chapter in Eighteenth-Century Southern Intellectual History," *The Bulletin of the Center for the Study of Southern Culture and Religion* 3:3 (November 1979): 5.

21. Ibid., p. 7.

22. St. Clair Drake, *The Redemption of Africa and Black Religion* (Chicago: Third World Press, 1970), p. 48.

23. Joanne Grant, ed., *Black Protest: History, Documents, and Analyses, 1619 to the Present* (New York: Fawcett World Library, 1968), p. 28.

24. Ibid., pp. 28–29.

25. Ibid., p. 30.

26. Wilson Jeremiah Moses, *Black Messiahs and Uncle Toms: Social and Literary Manipulations of a Religious Myth* (University Park: Pennsylvania University Press, 1982), p. 32.

27. Ibid., pp. 32–33.

28. Ibid., p. 34.

29. Quoted in Jacques Derrida, *Dissemination*, trans. Barbara Johnson (Chicago: University of Chicago Press, 1981), p. 133.

30. J. G. Frazer, *The Golden Bough* (New York: S. G. Phillips, 1954/59), pp. 540–41, cited in Derrida, *Dissemination*, p. 133. Cf. J. E. Harrison, *Prolegomena to the Study of Greek Religion* (New York: Meridian, 1903), p. 102.

31. For figures for lynchings before 1885 and up to 1933, see William Z. Foster, *The Negro People in American History* (New York: International Publishers, 1954), pp. 361, 392, 420–22, 456, 480.

32. Frazer, *Golden Bough*, pp. 541, as cited in Derrida, *Dissemination*, p. 132. Cf. René Girard, *Violence and the Sacred*, trans. Patrick Gregory (Baltimore: The Johns Hopkins University Press, 1977), pp. 98, 100, 131–32, 199, 247.

33. Girard, *Violence and the Sacred*, p. 23.

34. Ibid., p. 24. Cf. the chapter on law in *Curing Violence: Religion and the Thought of René Girard*, ed. Theophus H. Smith and Mark I. Wallace (Sonoma, Calif.: Polebridge Press, 1994).

35. See the following collection of slave narratives of trickery and deception: Gilbert Osofsky, ed. *Puttin' on Ole Massa* (New York: Harper & Row, 1969). See also the treatment of the Afro-American trickster character in Lawrence W. Levine, *Black Culture and Black Consciousness: Afro-American Folk Thought from Slavery to Freedom* (Oxford: Oxford University Press, 1977), and in John W. Roberts, *From Trickster to Badman: The Black Hero in Slavery and Freedom* (Philadelphia: University of Pennsylvania Press, 1989).

36. Cherry, *God's New Israel*, p. 70.

37. W.E.B. DuBois, *The Souls of Black Folk*, (New York: New American Library, 1969), p. 47.

38. Ibid., p. 44. Bercovitch, *American Jeremiad*, p. 11.

39. Cherry, *God's New Israel*, p. 43.

40. Lerone Bennett, Jr., *Confrontation: Black and White* (Baltimore, Md.: Penguin Books, 1965), pp. 21–22.

41. Coincidentally, one can trace to this period the earliest propagation, specific to North America, of negative stereotypes regarding black peoples. This perspective is consistent with the theoretical view that such stereotypes are ideological justifications of the economic causes of domination after the fact; they are not causes preceding domination but disguised rationalizations pretending to explain it. Cf. Bennett, *Confrontation*, pp. 24ff.

42. Raboteau, *Slave Religion*, p. 99.

43. Ibid., pp. 98–99.

44. Ibid., p. 123.

45. Lester B. Scherer, *Slavery and the Churches in Early America 1619–1819* (Grand Rapids, Mich.: Eerdmans, 1975), p. 90.

46. Ibid.

47. The heterodox practice of rebaptizing believers was first instituted by the martyred bishop and church father Cyprian in the third century CE and subsequently ratified by the Council of Carthage in 256. But the practice is most memorably associated with the Donatist church of Roman North Africa in the fourth century. After the Great Persecution under the Roman Emperor Diocletian (303–5) it was discovered that "too many people, including highly placed clergy, had sacrificed [to the Emperor], and feeling against these *traditores* ('surrenderers') was intense." W.H.C. Frend, *The Donatist Church: A Movement of Protest in Roman North Africa* (Oxford: Clarendon Press, 1952), p. 10. With the Edict of Milan in 313, the emperor Constantine ended Roman persecution of Christians, restored church property, granted special privileges to the clergy, and, coincidentally, promoted to election as bishop of Carthage a priest (Caecilian) who represented official church pardons for the compromised Christians. The rival candidate, Donatus, failed with his followers to reverse Constantine's advocacy and Donatism subsequently permeated all of Roman Africa as the popular though unorthodox church. In defiance of imperial authority, on the one hand, and the orthodox Catholic bishops on the other, the Donatists focused their protest by rebaptising their own church members, who had previously been baptized in the imperially privileged or state church. Frend has claimed that "the Donatist Church was the first and only Church in the ancient world which was prepared to challenge the established social and economic order in the name of social justice." W.H.C. Frend, "Heresy and Schism as Social and National Movements," in *Religion Popular and Unpopular in the Early Christian Centuries*, vol. XXV (London: Variorum Reprints, 1976), p. 14. On the emancipatory aspects of Donatist protest see the counter-argument of A.H.M. Jones, *Were Ancient Heresies Disguised Social Movements?* (Philadelphia: Fortress Press, 1966).

The second major instance of rebaptizing believers is associated with the Anabaptists or *Wiedertaufer* (again-baptizers) of the sixteenth century Protestant Reformation in central Europe. More radical even than Luther and the mainstream Protestants (who still proceeded to establish national state churches despite their break with Roman Catholic transnational authority), the Anabaptists protested by rebaptizing new members irrespective of their civil-religious baptism as citizens of their nation-states. They "wanted a church free from the guardianship of the state, accepting such members as joined out of an uncoerced decision.... This was the first goal of Anabaptism ... a fellowship of those who ... desired earnestly to be Christian ... [in] a new kind of church.... Infant baptism was at that time not only a churchly practice, but also a civil obligation. The collision ... [with] the city-state of Zurich ... was immediate." Fritz Blanke, "Anabaptism and the Reformation," in *The Recovery of the Anabaptist Vision*, ed. Guy F. Hershberger (Scottsdale, Pa.: Herald Press, 1957) pp. 60–61.

48. Mechal Sobel, *Trabelin' On: The Slave Journey to an Afro-Baptist Faith* (Princeton: Princeton University Press, 1988), p. 87.

49. Ibid, p. 101.

50. Robert N. Bellah, Richard Madsen, William M. Sullivan, Ann Swidler, Steven M. Tipton, *Habits of the Heart: Individualism and Commitment in American Life* (New York: Harper & Row, 1985), p. 252. See pp. 28–31 for a discussion of the "biblical and republican strands" with reference to the democratic ideals of freedom and equality and slavery (especially p. 31).

II

THEORETICAL PERSPECTIVES

There is always more in myths and symbols than in all of our philosophy.

Paul Ricoeur, *Freud and Philosophy*[1]

Two decades ago, in the early 1970s, the Afro-American philosopher Lucius Outlaw announced the need for an interpretive philosophy of black North American culture. Outlaw's early work called specifically for "a hermeneutic of black cultural productions." To hermeneutics, the Western theory and discipline of interpretation, were assigned the following projects in the study of African American culture: "a study of black religious language and symbolism in sermons and spirituals; a study of the language of the blues; and a study of black political writings and speeches, particularly protest language; a study of poetic and literary languages and symbolisms."[2] In the preceding chapters I have responded to Outlaw's still outstanding call from the perspective of my own religious studies interest in the ethnographic data of black culture in North America. More specifically, I have examined ethnographic, ethnohistorical, and literary-historical sources for insight into the phenomena of conjuration and biblical figuralism in black North American culture.

I began by describing the cultural intentions operating in conjurational phenomena found in black America's use of biblical figures like Moses; then I examined conjurational processes operating in folk strategies for inducing freedom; finally I addressed 'pharmaconic' strategies for curing violence by means of legal procedures and institutions. I will continue to treat materials drawn from ethnographic and literary-historical sources. But I will also turn from primarily ethnographic interests to more theoretical and deliberately interpretive endeavors. My purpose in these theoretical chapters is to explore the significance, for a theory of conjuring

culture, of the religious forms and other cultural materials that Outlaw
catalogued. In this regard my interests converge with Outlaw's quest for
a "Black Philosophy."[3] Insofar as that quest envisioned a hermeneutic of
black culture it proceeded implicitly in the line of development of Western
hermeneutics, from its beginnings in medieval Bible exegesis to its modern
theoretical concerns with all cultural productions.[4]

> Two psychosocial benefits of such a philosophy can be usefully high-
lighted here. African Americans need a philosophy that is able to address
what Outlaw has called "the existential crisis of divided self-consciousness"
and "the struggle for cultural integrity."[5] Throughout the following discussion we will see various manifestations of the existential condition of
a people experiencing divided self-consciousness. Indeed, Outlaw understands black cultural integrity to consist in a resolution of the double or
divided consciousness that has been described, in terms now classical in
black studies, by W.E.B. Du Bois: "It is a peculiar sensation, this double-
consciousness . . . One ever feels his twoness,—an American, a Negro; two
souls, two thoughts, two unreconciled strivings; two warring ideals in one
dark body, whose dogged strength alone keeps it from being torn asun-
der."[6] A merging of the black and white cultural selves, and thus a resolution of the problem of psychic and social disintegration, has been and
continues to be the precondition for black cultural integrity. In Outlaw's
view such resolutions are achieved "via a transformation of consciousness"
leading to authentic existence by means of what he calls "symbolic reversal." Such transformations consist of

> endeavors in *symbolic reversal* (reversal of symbolism) whereby one moves on
> the level of symbolic meaning . . . from imposed determination of one's (a
> people's) existence to those generated by oneself (by the people themselves)
> in the process of living . . . that existence in its authenticity.[7]

In the succeeding analysis I will explore aspects of Afro-American
spirituality that constitute traditional avenues for achieving transformations of consciousness and authenticity of existence through reversals of
symbolism. Such reversals are grounded in the dualistic condition of "double consciousness" as depicted by Du Bois. Dualistic existence is more
broadly and commonly denoted by the term "biculturality." Indeed, the
reality of belonging to two distinct cultures conditions black experience
in a wide variety of areas. I will be most concerned with the areas of
aesthetic production, specifically in the analysis of spiritual songs, and with
cognition, orality, and literary production in the discussion of wisdom
traditions. Black aesthetic and oral performances in particular may be
described in terms of dualistic patterning involving reversals of meaning
(Outlaw) or "signifying" (Gates), and also "style-switching" or the alternation of culture-specific codes (Marks). Such reversals and alternations
are clearly evident in black America's biblical figuration (for example,
America as Egyptian bondage and/or Promised Land). At the most generalized level of cultural performances (whether oral or musical, literary,

dramatic, or ritual), cultural forms identified as Afro-American or "black" have been variously correlated with polarized forms identified as Euro-American or "white."

In concluding Part II the focus will shift from black America's spiritual and aesthetic quest for transformations of consciousness, to prophetic aspects of conjurational phenomena. Indeed, my examination of prophetic elements in the lives of Nat Turner and Sojourner Truth will issue in new criteria for evaluating prophecy in the Afro-Christian tradition, and therein prepare the way for subsequent theological discussions in Part III.

Notes

1. Paul Ricoeur, *Freud and Philosophy: An Essay on Interpretation* (New Haven: Yale University Press, 1970), p. 527. Cf. Shakespeare, *Hamlet*, act 1, sc. 5, lines 166–167, "There are more things in heaven and earth, Horatio, than are dreamt of in your philosophy."
2. Lucius T. Outlaw, "Language and Consciousness: Towards a Hermeneutic of Black Culture," *Cultural Hermeneutics* 1 (1974): 411.
3. Lucius T. Outlaw, "Black Folk and the Struggle in 'Philosophy,' " *Radical Philosophers' Newsjournal* (April 1976): 21–30.
4. The term "hermeneutics" itself is derived from a number of related words in ancient Greek: *hermeneuein*, the verb form meaning "to interpret," and *hermeneia*, the noun form for "interpretation." The form *hermeios* designated the priest of the oracle at Delphi, and together all the forms proceeded from—or led to—the name of the Greek messenger god and sacred trickster, *Hermes*. Thus the primal meaning of the word includes the task of communicating messages from the gods. More generally, hermeneutics designates the translation of what is obscure for the benefit of human understanding. In its oldest, medieval use the term denoted the interpretation of one particular domain of obscure messages and meanings: biblical texts. Thus medieval hermeneutics, as the system or theory of biblical exegesis and interpretation, retained directly the etymological reference of the word to its original religious and oracular context. However, the discipline was subsequently secularized during the European Enlightenment, as the Bible came to be perceived as merely one ancient text among others. Hermeneutics became the methodology for the discipline of philology in addition to biblical studies. That development was confirmed in the nineteenth century by the German theologian Friedrich Schleiermacher (1768–1834), who established hermeneutics as the science of understanding any classical text or literary artifact whatever.

But Lucius Outlaw's use of the term, to apply to philosophical reflection on black cultural productions generally, reflects a much larger extension of hermeneutic fields. To include as objects of interpretation a culture's oral and literary expression (religious and secular), its musical and political discourse, and so forth, requires a vast augmentation of the discipline. In the intervening developments between the work of Schleiermacher and that of Outlaw, hermeneutics has proceeded from the interpretation of sacred texts and classical literature exclusively, to its current inclusion of any expression of human life. Those developments owe their success to the philosophical labors of Wilhelm Dilthey (1833–1911). Following upon the work of Schleiermacher, Dilthey established hermeneutics as the methodological basis for all the human sciences or cultural studies (*Geisteswissen-schaften*, as distinguished from the natural sciences or *Naturwissenschaften*). What the various cultural studies or humanities hold in common is the attempt to understand the distinctive aspects of human existence in terms of its personal or individual, cultural and social-historical manifestations.

5. Outlaw, "Language and Consciousness," p. 403.

6. W.E.B. DuBois, *The Souls of Black Folk* (New York: New American Library, 1969), p. 45.

7. Outlaw, "Language and Consciousness," pp. 403–4.

4

Spirituals

These were spiritual things, not physical artifacts. The orishas were in those early churches. The gods were in our voices. . . . When those Black men began to handle those [musical] instruments, they were directly in contact with their gods.

Jimmy Stewart, "Introduction to Black Aesthetics in Music"[1]

In this chapter I explore conjure as a defining spirituality among black North Americans that bears both covert and transparent aspects. I focus first on the covert aspects that derive from traditional practices beginning in the slavery period. Second, I consider certain iconographic modes that allow for increased openness and transparency. In the first case, conjuring culture by means of biblical figures operates covertly in black America because conjure, in its traditional form as folk magic, has been censured and marginalized since the earliest days of missionary supervision of black Christianity. In reaction to such censure, the use of biblical figures to conjure culture involved a collective strategy to cloak, mask, or disguise conjurational operations by employing approved religious content—namely biblical and Christian content. In this connection the historian Albert Raboteau has reprised the much debated thesis of the anthropologist Melville Herskovits: that "some elements of African religion survived in the United States not as separate enclaves free of white influence but as aspects hidden under or blended with similar European forms."[2] Such legitimation strategies are rarely explicit and conscious, of course, and may even be forgotten or repressed if ever rendered transparent. My task in this book is to retrieve, from either disguise or repression, such conjurational strategies and intentions and thus render them as fully transparent as possible. With this task in mind I am responding in particular to Lucius Outlaw's concerns for "the mediation of our people's traditions; and, most importantly, the achievement of increased self-transparency."[3]

A useful area for exposing and displaying the masking devices that

conceal conjurational phenomena is that of music. I begin by exploring
the concealment of incantatory effects in black music, as an analogue to
the covert quality of conjuring with biblical figures. That is to say, I treat
this occlusive feature in aesthetic terms first. We will see that the occluding
of incantatory intent in black music, religious and secular, is due to the
codes it employs as a principal feature of the aesthetic and religious expression of black peoples in the Americas. Raboteau too indicates that the
culture area of music, in addition to folklore and language, evinces a
masking or 'blending' effect with reference to African and European musical sources. In addition he highlights two areas of "commonality" between African traditional religions and European religious or folk belief:
"ecstatic behavior" or (more narrowly) spirit possession, and "magical folk-
belief," particularly conjure.

Finally, the two highlighted areas of possession and conjure are corroborated in Gayraud Wilmore's observations of 'continuities' between
African and Christian spirituality in slave religion.

> Whether the missionaries desired it or not, Christianity had to provide *some*
> aspects of spirituality that were continuous with the African experience. . . .
> the missionary could not, in good conscience, depreciate the presence and
> mysterious work of the Holy Spirit in the life of the believer. This work could
> readily be interpreted by the slaves as identical with conjuration and the
> Orisha-possession of his ancestral religion.[4]

The concealment of such African continuities in African *American* spirituality is illumined by certain performance codes of black musical expression
as discussed in the analysis of ecstatics. In the discussion of iconics, I will
shift from musical to oral and literary expression in an effort to display
the kind of transparency that is possible when black religious expressions
and representations are more readily decoded or decipherable. Decipherment of religious figures and performances reveals what is concealed by
coded forms of signifying and brings to view an iconographic dimension
that continues to fascinate, sustain, and empower African Americans in
their contemporary religious and political transformations.

Aesthetics

> *It is only in his music . . . that the Negro in America has been able to tell*
> *his story. It is a story which otherwise [is told] . . . compulsively, in symbols*
> *and signs, in hieroglyhpics; it is revealed in Negro speech and in that of*
> *the white majority and in their different frames of reference.*
>
> James Baldwin, "Many Thousands Gone"[5]

I quote the novelist and essayist James Baldwin not to advance the veracity
of his claim of uniqueness for black musical expression. Rather, I am more
interested in the rhetorical intention and the structure of the claim: the
combined use Baldwin made of an aesthetic referent (music) alongside a

narrative referent (story) to evoke simultaneously the opacity[6] and the transparency of black American experience. Baldwin's corollary claim is compelling as well: that the significance of black experience is *concealed* within codes—in "symbols and signs, in hieroglyphics"—on the one hand. On the other the significance of cultural experience is *revealed* in the differences and engagement of black and white discourse. In the following discussion I consider the opaque or coded nature of black America's musical and ecstatic expression; later I take up the more transparent or revelatory dimensions of its bicultural discourse. I begin by noticing the irony of Baldwin's claim, the irony of claiming that exemplary communication occurs in the relatively opaque medium of musical expression (opaque relative to the clarity of meaning that is normally ascribed to discursive modes of communication). As if deliberately to heighten that irony the historian and essayist Lerone Bennett has elaborated Baldwin's claim as follows:

> No writer, so far, has fingered pain and dread with the exquisite agony of a Billie Holiday. No Negro poet, no philosopher or pundit has said as much about the human condition as Charlie Parker or Bessie Smith. . . . Negro philosophy still awaits its Sartre; Negro literature still awaits its Kafka. . . . James Baldwin said once that there has not yet arrived a sensibility sufficiently tough to make the Negro experience articulate.[7]

Again, without attending to the substance of such claims I am interested in the two authors' rhetorical intentions. Implicit in the Baldwin and Bennett assertions is the notion of a black aesthetic—a distinctive ethos of artistic sensibility and cultural creativity that is crucial for communicating Afro-American experience, and also for criticizing the depth and authenticity of such communication.

The concept of a distinctive black aesthetic in America derives from an older era and an earlier generation of black writers in the twentieth century. It was the Harlem Renaissance artists of the 1920s (such as Zora Neale Hurston and James Weldon Johnson), and its chief theoretician of *The New Negro* (1925), Alain Locke, who first proclaimed a distinctively black American artistic tradition originating from folk and popular culture. Locke and his collaborators successfully demonstrated that black culture harbored vast and untapped resources of original (not merely imitative) creative expression. Their claims were further advanced by the so-called Black Arts movement of the 1960s and 1970s, which inherited the unfinished agenda of the Harlem Renaissance. The movement successfully articulated and popularized the concept of a black aesthetic tradition that is distinguishable from Euro-American or Western values and artistic conventions. In the words of the literary theorist Houston Baker, Jr.:

> The guiding assumption of the Black Arts Movement was that if a literary-critical investigator looked to the characteristic musical and verbal forms of the masses, he could discover unique aspects of Afro-American creative expression—aspects of both form and performance—that lay closest to the verifiable

emotional referents and experiential categories of Afro-American culture. The result of such critical investigations...would be the discovery of a "Black Aesthetic"—a distinctive code for the creation and evaluation of black art.[8]

Baker's "distinctive code," or codes, for creating and evaluating black art has received a measure of empirical specificity and verification in the ethnomusicological research of Morton Marks. In his well-received essay "Uncovering Ritual Structures in Afro-American Music," Marks presents evidence of a ritual practice spanning both the United States and the Caribbean that he calls, borrowing a term from sociolinguistics, "style-switching." The essential feature of African American style-switching is the alternating of patterned expression from the forms of one culturally identified style system to those of another. Marks claims further that this alternation functions to induce trance behavior during sacred as well as secular ritual events. " 'Africanization' is expressed on the musical plane [for example] by the addition of a complex rhythmic accompaniment to a staid Protestant hymn tune, whose melody line is altered in the direction of a percussive singing style. But...there is more going on than what is conventionally called expressive behavior." *The change in style is generating a ritual event, namely spirit-possession.*"[9]

The key to inducing ritual spirit possession or trance phenomena is the abrupt shift from one element of the dualism to the other. "What is crucial for the discussion is that switching is always from a 'white' style to a 'black' style, from a European to an African one."[10] The polarization of styles between two cultural systems, one identified as African and the other as European or American, constitutes the duality. But the phenomenon as a whole is not only musical or aesthetic; it is also spiritual and ecstatic. That is, Marks's research demonstrates how this one realm of black aesthetic production contains codes or cues for inducing the phenomenon of spirit possession. Style-switching signals the opportunity for trance states based on cultural expectations deriving from folk religion and its spiritual traditions. Such expectations are now culturally 'at large,' however. That is, they are not limited to religious contexts but are also conveyed through such secular venues as the entertainment industry. The inducement of trance behavior through style-switching can occur in commercial radio broadcasts of "soul music," for example.[11] In this manner style-switching may be regarded as a by-product of culture contact throughout the Americas. Moreover the phenomenon can also be derived from antecedent, West African performance rules, yet without the double meaning systems and significations that characterize its operation among New World cultures.

Reisman observes that the oscillation between "noise" and "order" is probably a widely distributed West African speaking convention. However, in Africa it occurs without its attendant New World meaning as an expression of cultural duality and the symbolic movement between the categories of different cultures.... The "Africanization" of behavior...is an underlying performance rule throughout much of the New World, and [is everywhere] connected with a break from "order," usually represented by European-derived forms, into seemingly "disorderly" group behavior.

Marks further comments that observers who are uninitiated in such performance rules tend to misrepresent or ignore complex and group forms of black expression as mere noise or disorder. Because they are ignorant of the "underlying organizing principles" of the performances, they also miss the level of meaning that is generated by the 'break' from order into apparent disorder. (I will return to the claim that style-switching is generative of meaning.) Marks notes that this dual symbolization provides a kind of record or deposit of culture contact between Africans and Europeans, on the one hand, and a record of mutual acculturation between them on the other.[12] Moreover, the concept of style-switching adds to previous formulations of African American biculturality an empirical and structural basis for demonstrating duality in a specific domain of black culture: music. But the duality of cultural expressions operates in black discourse also. Indeed, Marks himself emphasizes the ubiquity in the African diaspora of "double systems in virtually every sphere of human activity. In whatever aspect of social or cultural organization, in the linguistic, religious, economic, legal, medical and aesthetic-expressive domains, there exists a pairing or opposition of an official institution or form and its vernacular or Afro-American counterpart."[13]

In a similar vein the black literary critic Ralph Ellison has observed double systems in African American culture. In a markedly lucid passage Ellison notes how "forms of transcendence" were dualistically cultivated in the Afro-American musical tradition from the early to the middle twentieth century. That period featured the creative prominence of musicians who felt impelled to master both American jazz and European classical music.

> Culturally everything was mixed, you see, and beyond all question of conscious choices there was a level where you were claimed by emotion and movement and moods which you couldn't put into words. Often we wanted to share both: the classics and jazz, the Charleston and the Irish reel, spirituals and the blues, the sacred and the profane. . . . [T]here were certain emotions, certain needs for other forms of transcendence and identification which I could only associate with classical music. I heard it in Beethoven, I heard it in Schumann.[14]

Like Marks, Ellison shows how black musical traditions employ cultural dualism between forms identified as black or African, and those identified as white or European. But where Marks observes trance behavior associated with spirit possession, Ellison indicates the phenomenon of sociocultural transcendence: "You got glimpses, very vague glimpses, of a far different world than that assigned by segregation laws, and I was taken very early with a passion to link together all I loved within the Negro community and all those things I felt in the world which lay beyond."

Ellison's "transcendence" of ethnic segregation and cultural isolation (Du Bois)[15] through biculturality provides a broader, more extensive instance of style-switching. What Marks and Ellison share, however, is a recognition that the phenomena of black aesthetic duality conceal (and hence can reveal through decipherment) deeper dimensions of cultural existence—deeper than the superficial treatment of style as imitation or ornamentation. In the next section I explore those more profound levels of meaning and existence in relation to the ecstatic aspects of black music and black experience.

Ecstatics

The blues are sung, not because one finds oneself in a particular mood, but because one wants to put oneself into a certain mood. . . . [They] do not arise from a mood, but produce one. . . . The spiritual produces God, the secularized blues produce a mood.

Jahnheinz Jahn, *Muntu: The New African Culture*[16]

In the figural schema of this book the present chapter holds the place that is represented by the Hebrew Psalms among biblical books. The Psalms are musical and literary vehicles of inspiration and religious exaltation, on the one hand, and of lamentation and spiritual consolation on the other. In a convergent vein, Alain Locke has suggested that the religious music of the American slave—the "spirituals"—functions in black North American culture in a manner analogous to the role of the book of Psalms in ancient Hebrew and Jewish cultures. As black America's indigenous Psalter, the spirituals display "an epic intensity and a tragic profundity of emotional experience, for which the only historical analogy is the spiritual experience of the Jews and the only analogue, the Psalms."[17] Another observer has extended Locke's analogy, indeed, to encompass the entire Bible in a manner most congenial to the biblical organization of this study. The ethnomusicologist Harold Courlander, considering the spirituals together with other "Negro religious songs," has observed that

if these songs are arranged in a somewhat chronological order, they are equivalent to an oral version of the Bible. Each song presents in a capsulized or dramatic form a significant Biblical moment. . . . [Thus] it would be possible to put a large body of Negro religious songs together in a certain sequence to produce an oral counterpart of the Bible; if printed, they would make a volume fully as thick as the Bible itself.[18]

The longstanding name for this psalmlike genre, "spirituals," is appropriate not merely because they express religious attitudes or convictions. After all, W.E.B. Du Bois, as if to render their religious orientation secondary, called the spirituals the "sorrow songs." Others, like Zora Neale Hurston, have refused that melancholic designation and have insisted that the spirituals convey cathartic qualities that reminds one of the ironic,

double-edged vitality of the blues. (Consider the designation of the spirituals not only as <u>sorrow songs</u> but also as 'jubilee' <u>songs</u>—a designation
made popular by the name of the Fisk Jubilee Singers, a group of Fisk
University black students who traveled the United States and Europe in
the late nineteenth century singing these songs before mass audiences.[19])
In this connection the black theologian James Cone, in his illuminating
study *The Spirituals and the Blues* (1972), has represented the genre of
blues music as a "secular spiritual."[20] The implication of this ironic ap-
pellation is clear: that the spirituality of "the spirituals" can coinhere in
"the blues" because their common spiritual quality is not limited to religious expression, but exceeds such expression and derives from a more
comprehensive source. What is that common source?

First it is important to acknowledge, with Cone and other observers,
that conventional religious opinion denigrates the blues. Even black church
communities have typically censured these songs as 'the devil's music'; as
merely indecent or vulgar musical expression. Blues music is often reduced
in import to some of its (admittedly) earthy expressions of hedonistic,
erotic, or libidinous desire. Alternately the blues are considered the quin-
tessential expression of fatalism and self-pity. "For many people, a blues
song is about sex or a lonely woman longing for her rambling man," Cone
acknowledges. "However, the blues are more than that. To be sure, the
blues involve sex and what that means for human bodily expression, but
on a much deeper level."[21] We categorically preclude a deeper level of
meaning by limiting the blues to expressions of self-pity or the desire for
physical consolations, just as we limit the spirituals to expressions of lamentation and the desire for religious consolations. On the contrary, the
conventional view misses the presence of a spiritual impulse in the blues.
"Black church people ... did not understand them rightly. If the blues are
viewed in proper perspective, it is clear that their mood is very similar to
the ethos of the spirituals. Indeed, I contend that the blues and the spirituals flow from the same bedrock of experience, and neither is an adequate
interpretation of black life without the commentary of the other."[22]

With notable corroboration Lerone Bennett too has remarked that
black folk tradition, in its earliest manifestations, does not differentiate
reality into sacred and secular "strains." On the contrary,

> the tradition and the Negro who bears that tradition cannot be understood
> without holding these two contradictory and yet complementary strains—
> sacred/secular—together in one's mind. This is, I think, the essential genius
> of the Negro tradition which did not and does not recognize the Platonic-
> Puritan dichotomies of good-bad, white-black, God-devil, body-mind. ... The
> Negro tradition, read right, recognizes no such dichotomy. The blues are the
> spirituals, good is bad, God is the devil and every day is Saturday. The essence
> of the tradition is the extraordinary tension between the poles of pain and
> joy, agony and ecstasy, good and bad, Sunday and Saturday. One can, for
> convenience, separate the tradition into Saturdays (blues) and Sundays (spir-
> ituals). But it is necessary to remember that the blues and the spirituals are

not two different things. They are two sides of the same coin, two banks, as it were, defining the same stream.[23]

Despite Du Bois's designation of the spirituals as "sorrow songs," and despite the conventional use of the term "blues" to mean melancholy, the plaintive tone evident in each genre is preliminary to their full trajectory. Each genre aims, in its own way, to perform a spiritual operation. In the researcher Jahnheinz Jahn's formulation, "the spiritual produces God, the secularized blues produce a mood."

Jahn's assertion follows from his own research into the ways that black religion and culture in North America are continuous with ancestral West African cultures. In particular, as we saw in the discussion of "God-conjuring," Jahn claims that worshippers in African traditional religions are active or creative in inducing the manifestation of deities—in 'producing' the gods. Despite the admitted and highly articulated differences between such religions and Christianity, in which the worship relationship is "unequivocally determined by God alone," Jahn judges that black North American Christianity also exhibits this inductive or evocative feature. "With the designation of a Christian God Christian standards penetrate the cult, above all the sharp separation of good and evil; but the nature of worship, the *service* of God, remains to a great extent African. For God is not only served but invoked, called up and embodied by the faithful."[24] Here I would qualify Jahn's assertion to refer instead to the intersubjective experience of the worshippers, whether African or African American. Rather than producing the God whom they 'invoke, call up and embody,' I maintain that they collectively create the phenomenal conditions conducive to their subjective apprehension of the divine. This modified view is consistent with the anthropological treatment of possession in Sheila Walker's study *Ceremonial Spirit Possession in Africa and Afro-America*, which highlights the role of "cultural determinism" in such phenomena.[25]

The spirituals and the blues are correlative, on this view, because they produce or generate altered states in the subjective and intersubjective experience of their performers and their audiences. Both genres are spiritual, to vary the terms, because they effect a transformation of spirit from melancholy moods and psychic depression to elation and subjective exaltation. The term most commonly used for designating this transformed state is "ecstasy," which is understood here in its etymological sense: the condition of "standing out" from oneself—from one's normal psychic state or centered self—so as to experience states that exceed normal conditions. Let us take the case of the spirituals and the blues in juxtaposition. Alain Locke has presented the case for the transformative nature of the spirituals in precisely the terms just reviewed—as sorrow songs in search of transcendence and exaltation.

Indeed [the spirituals] transcend emotionally even the very experience of sorrow out of which they were born; their mood is that of religious exaltation, a degree of ecstasy indeed that makes them in spite of the crude vehicle a classical expression of the religious emotion.[26]

In similar terms Ralph Ellison has stated the case for the transformative nature of the blues.

> The blues is an impulse to keep the painful details and episodes of a brutal experience alive in one's aching consciousness, to finger its jagged edge, and to transcend it, not by the consolation of philosophy, but by squeezing from it a near-tragic, near-comic lyricism.[27]

Blues songs serve to transcend negative experiences and feelings by recapitulating or reprising the negative, but precisely in such a way (ecstatically) that they transform the negative. Such transformation is ironic, of course. (A term more pertinent to this study is "homeopathic," that is, the use of a dis-ease to cure disease. Consider, as one among hundreds of examples, the ironic self-affirmation that operates in this lyrical declaration of unrelenting negation: "If de blues was whiskey, / I'd stay drunk all de time. / If de blues was money, / I'd be a millioneer."[28] In these lyrics the negative experience of unrelieved misery is reconfigured (through the images of "drunk" and "millioneer") by a kind of boasting bravado. James Baldwin observed similar forms of irony and transcendence not only in the blues but also, by extension, in gospel music and jazz—two contemporary analogues to the older sacred-secular correlation of the spirituals and the blues. "One hears in some gospel songs, for example, and in jazz ... especially in the blues ... something tart and ironic, authoritative and double-edged. [By contrast] white Americans seem to feel that happy songs are *happy* and sad songs are *sad*."[29] Instead of sad songs producing sadness, for those persons initiated in the ritual structures of the culture these sad songs induce precisely the opposite of sadness. The result is "ironic" and "double-edged" by virtue of its power to escape the negative. That power is achieved precisely by articulating the negative, but the negative is imbued or impressed with the performer's ecstatic proficiency and virtuosity. To return to my earlier emphasis on concealment, we may say that the blues conceal their ecstatic import to everyone but the initiated listener. Only such a listener can appreciate that the ostensibly sorrowful tone is only "camouflage," as the commentator Jahnheinz Jahn declares.

> The melancholy is a camouflage. ... If we read the text of the blues songs without prejudice and notice the double meaning, which all authors emphasize, we find them mocking, sarcastic, tragi-comic, tragic, dramatic and accusing, often crudely humorous ... [but] only exceptionally ... melancholy."[30]

Jahn's emphasis on the "double meaning" of the blues is consistent with Morton Marks's broader acknowledgment of "double systems in virtually every sphere of human activity" throughout the African diaspora. But Marks's research lets us emphasize the oscillation between the bipolar elements of those diverse double systems: oscillation between the poles of existential negation and secular consolation (the blues); between lyric lament and religious consolation (the spirituals); between black-identified and white-identified performance styles (Marks); and so on. Here it is appropriate to pose the question, Does such oscillation itself generate ecstatic states? Is there a logic of ecstatic inducement—inducing the subject to 'stand-outside-oneself'—that can account for the efficacy and appeal of these double systems? An adequate response to this query cannot be attempted here. However, the oscillation between poles may indicate that the goal of such performances is not performance or ecstasy for its own sake. Rather, I propose that the goal is something else and that ecstasy is a familiar, culturally induced and anticipated by-product (but only a by-product or epiphenomenon) of this more fundamental trajectory. Ecstasy is epiphenomenal, on this view, to transformations of reality that more precisely constitute the compelling goal or telos of such performances.

This hypothesis finds corroboration in the religious experience of enslaved Africans in North America. According to the mystical philosopher Howard Thurman, the spirituals are a "monument" to the slaves' experience of spiritual transformation: most notably the transformation of their dehumanizing existence by religious experiences of transcendent care and affirmation. In this regard Thurman found it "amazing," and still unfathomed today, how black Christian converts were able to counter the most toxic aspects of their newly acquired faith tradition. They were highly at risk, he recognized, as Christian initiates who were also legal chattel. One of the earliest problems of theological formation under such conditions was how to avoid construing the Christian God—a providential, suprahistorical deity according to conventional doctrine—as the primary agent of their lawful bondage and dehumanization. "When the master gave the slave his God, for a long time it meant that it was difficult to disentangle religious experience from slavery sanction." (Compare the so-called myth of Ham, the view that God through Noah cursed and ordained all black peoples to perpetual servitude under white peoples—a nineteenth century white Christian perspective that still finds adherents today.) On the contrary, the slaves' religious songs testify to an altogether different apprehension of the Christian God and of their own humanity.

> The existence of these songs is in itself a monument to one of the most striking instances on record in which a people forged a weapon of offense and defense out of a psychological shackle. By some amazing but vastly creative spiritual insight the slave undertook the redemption of a religion that the master had profaned in his midst.[31]

The "redemption" of Christianity, as a vehicle for redeeming the slaves' own abused humanity, is the transformation that the spirituals evinced and, to some degree, effected. Parallel to such religious transformations are secular transformations that also revalue or transvaluate black existence and cultural identity as positive and desirable. I propose that the meaning produced in diverse sectors of black American culture—particularly the meaning of its aesthetic, ecstatic, and ritual performances—is best understood in terms of such identity transvaluations. In this regard I agree with the conclusive statements of Morton Marks on ritual structures in black music: we must not "view Afro-American ritual as mere rote behavior which contains zero information." Identity affirmation is the hermeneutic content that Marks too indicates for style-switching as a feature of black aesthetics and ritual performance. "What emerges from this analysis is the conclusion that social meaning is generated through the alternation and interplay among a set of codes . . . and that one social meaning generated by these rituals consists of *statements of cultural identity*. Thus, these rituals may be called individual and/or societal statements about such an identity." But ritually articulated identity statements can be derived not only from performances that feature style-switching. Identity transvaluations and declarations are also a feature, I submit, of ecstatic and conjurational performances that similarly display one's proficiency in deploying the constitutive elements of New World cultures. In diverse modes of expression, on this view, the practitioner or performer is "stating his or her personal religious and cultural commitments by fulfilling them on the spot."[32]

Identity affirmations may also involve the element of covert communication. In particular, conjurational performances may both reveal and conceal the nature of the practices as a feature that, in an earlier historical period, ensured the practitioner's personal survival or vocational integrity. Subsequently, concealment may no longer be required by circumstances; it may nonetheless continue to function in the form of conventional codes that are transmitted to succeeding generations who now employ them without awareness of their masking intent. It is evident in this regard that I am extrapolating from other religious and cultural phenomena, in order to support my claim for the pervasiveness of conjurational phenomena that also conceal their operation. I have been doing so throughout this chapter, of course, by reference to black musical expression. As a final extrapolation I cite the comment of another observer, who has also noted the prevalence of linguistic codes in distinctively black American styles of communication. The linguist Grace Sims Holt explicitly acknowledges that such codes no longer serve a survival strategy. They nonetheless continue to function and, moreover, constitute modes of identity affirmation while concealing their import from uninitiated observers.

> What developed as a necessary mode of communication has become an integral part of the language system of blacks, though the necessity is not as great as it was in the beginning. . . . [M]ore prevalent in the nonreligious society of today's blacks, [such] communication is specifically designed to baffle any white within hearing distance of the conversation. These codes are also used by blacks to say to one another that "I'm *really* a brother."[33]

Concluding this passage Holt further underscores the deceptive aspect of coded communication in the encounter between black and white cultures. Moreover, she highlights the performer's gratification derived from the proficient use of such deception. "Much satisfaction is gained from the fact that by using this device white society may be 'put down' in the presence of whites, without their having the faintest notion that they are the objects of ridicule." I now turn to conjurational performances that more readily *reveal* than conceal their operation, and that gratify their practitioners by virtue of their transparency over against their opacity.

Iconics

The Afro-American artist is . . . a conjuror who works JuJu upon his oppressors; a witch doctor who frees his fellow victims from the psychic attack launched by demons of the outer and inner world.

Ishmael Reed, *19 Necromancers from Now*

but there's only one search: Wandering this world is wandering that, both inside one transparent sky.

Jelaluddin Rumi, *Open Secret*[34]

> Probably the most familiar example of coded communication in black music comes from the spirituals, not from the blues. I refer to the thesis, popularized by the music historian Mark Miles Fisher, that some slave spirituals conveyed secret messages among conspirators engaged in underground meetings and in plots to escape from their plantations. Probably the most persuasive evidence of that claim can be found in the songs "Go Down, Moses" and "Steal Away to Jesus." In these two songs particularly it is possible to construe religious exhortations in terms of double meanings. In "Go Down, Moses" a divine or semidivine liberator, a "Moses" figure, is summoned (conjured) to "go down" and tell Pharoah (massa) to "let my people go." And "Steal Away to Jesus" may be construed as a communal summons or inducement to individual slaves to 'steal' themselves from bondage to freedom. Such exhortatory summonings or inducements bear obvious incantatory or conjurational aspects, among which is a conspiratorial element. Recall here an archaic meaning of the word, con-jure: to 'conspire,' in the sense of swearing together, as does a jury.[35]

However, other commentators warn against a reductive approach that treats the spirituals indiscriminately as coded speech or covert communication. "If songs of the type of 'Steal Away to Jesus' and 'Go Down Moses' are to be considered conscious disguises for political, temporal meanings, a large part of the religious repertoire must be placed in the same category. Every reference to crossing the Jordan could be interpreted to mean escape to the North; every battle of the Israelites might be read to mean the battle for Negro freeedom . . . and every trumpet blast interpreted as Emancipation Day."[36] This representation of the Fisher thesis is a caricature composed of excesses, of course. Nonetheless it provides a useful, cautionary note for my concluding emphases. Comparable excesses would attend a unilateral or monolithic effort to reduce African American spirituality solely to its ecstatic aspects. On the contrary, we can also observe in black religious traditions an alternate mode of spirituality that is contemplative or revelational more than ecstatic or emotivist (more Apollonian than Dionysian, to employ the familiar Western mythopoeic categories).

This alternate spirituality incorporates, while displacing the priority, of covert communication as examined in the preceding sections. Because the intention of this alternative tradition is increased transparency—as if one were looking through an open window rather than deciphering cryptic codes or, more precisely, as if the codes were an 'open secret'—I describe this alternate tradition as the iconographic or 'iconic' dimension of African American spirituality. The term is most familiar, of course, with reference to the pictorial Christian art of the Eastern Orthodox traditions, in which an icon mediates the believer's apprehension of spiritual realities by means of earthly representations. Through the icon, that is, both spiritual and everyday worlds dwell "inside one transparent sky" (Rumi). Before discussing the spiritual dimensions of African American iconography, however, let me propose a likely historical framework for the emergence of this alternate mode of religious apprehension and manifestation among black peoples in the New World.

I conjecture that a postemancipatory shift occurred in black religious experience—the shift from a covert use of conjurational strategies to a more transparent mode of spirituality—without however diminishing the covert tradition. My hypothesis is informed by a related hypothesis concerning a shift in the use of deceptive strategies or coded devices as featured in black storytelling following the Civil War. Lawrence Levine has observed that, after emancipation, the popularity among black Americans of trickster tales (featuring, for example, the animal tricksters Brer Rabbit and the Signifying Monkey) gave rise to a new storytelling tradition. By contrast the new tradition celebrated black heroes who confronted their antagonists and white oppressors directly (for example, the transitional figure High John the Conqueror, or the fantastic railroad laborer John Henry): that is, without resorting to guile and deception.[37]

> Direct confrontation was not a quality totally unknown to the heroes of antebellum slaves. The Biblical figures that played so important a role in slave mythology where anything but tricksters: David confronted Goliath, Moses the Pharoah, Samson the Philistines, Jesus the religious and secular authorities of his time. These figures retained importance in the twentieth century, but the weakening of the sacred universe rendered them less immediate and potent. The crucial change marking black folklore after emancipation was the development of a group of heroes who confronted power and authority directly, without guile and tricks, and who functioned on a secular level.[38]

On this view, after emancipation black Americans could now risk (dare, aspire to) secular tales and models of direct confrontation with the formerly near-absolute power of their white antagonists.

We need not accept Levine's analysis as factual. It may be the case, for example, that what he observes as a shift in the historical data is actually a shift in our perception of that data. It could be that incidences and representations of direct confrontation by black Americans remained more or less constant before and after emancipation, but that our sources for documenting such themes after emancipation are more copious (for whatever reason). A similar situation may obtain with reference to my parallel claim, a claim that bears a 'family resemblance' to Levine's analysis. It may be that obscure, deceptively coded, and covert forms of conjurational performance, as contrasted with transparent practices, have remained more or less constant in black American history. It could be that the sources for documenting transparent performances are more copious (for whatever reason) in more contemporary periods. In any event such transparent modes of conjurational performance can now be explored alongside the more traditional representations of conjure as a system of occult, esoteric, or covert practices. To that exploration I turn next.

> *The slaves were not permitted the Bible, but every chronicler reports their love for it and on the way the book itself became a symbol of liberation.*
> *... Of course, the content of the Bible, its message of hope and liberation, meant much to people denied the book as object or literacy as access. ...*
> *[And] they regarded the book numinously as did their white brothers and sisters.*
>
> Martin Marty, "America's Iconic Book"[39]

Black religions in the New World are not uniformly ecstatic in character. Although religions of 'the spirit' (ecstatic spirit possession) appear to predominate, we can also find more contemplative instances of black 're-ligions of the book.' Such a counterindication is provided by Rastafari-anism, a messianic tradition indigenous to Jamaica but which also has its North American communities of dispersion (diaspora). The Rastafarian movement emerged immediately after the crowning in 1930 of Ras Tafari ('Prince' Tafari), as Ethiopia's Emperor Haile Selassie: as the "King of Kings, Lord of Lords, and the conquering Lion of the Tribe of Judah."[40] Selassie was embraced by expectant Jamaicans as the messianic black king who had been prophesied by the celebrated 'race leader,' Marcus Garvey, in the late 1920s. Garvey prophesied a savior who would deliver his people from their exile in the New World and restore them to Africa. In that context of prophetic expectation, Haile Selassie's Ethiopian titles and prophetic designations converged with the nascent movement's "cult language" (Breiner). It was a language based on the King James Bible as transmitted during British rule in the West Indies, but purged of the distortions which served to support slavery and colonial oppression. More-over, that language distinguishes the tradition as significantly as ritual performance distinguishes other black religions of the New World.

Indeed the comparative literature scholar Laurence Breiner claims that Rastafarian biblicism renders the movement unique among black religions in the Caribbean and South America. In "*vodun, shango, santeriá,* and *candomble,*" Breiner notes, spirit possession is central, women are prominent, and "traditional West African elements are preserved in a matrix of Catholicism." Rastafarianism reverses these and other African-based commonalities:

> In aggressive opposition to the norms of West Indian society, [Rastafarianism] is dominated by men. Though repatriation to Africa is its central tenet, it has no core of preserved African religion, and has not even adopted African elements from the surrounding culture to any considerable extent. Its matrix is Protestant rather than Catholic, and so at its center is the King James Bible, rather than any ritual mystery. . . . It rejects institutional hierarchy as much as it does possession by spirits or ancestors. What it establishes instead is in effect a religion of the Word, a community of prophets.[41]

Breiner argues convincingly that the prominence among Rastafarians of prophetic texts drawn from the English Bible displaces the centrality of spirit possession as found in other Caribbean and South American black religions. Instead of possession we find a distinctive emphasis on revelatory discourse and poetic biblicism. That emphasis, I suggest, renders Rastafarianism closer in phenomenological perspective to black North American Christianity in its Protestant, more biblicist, and prophetic variants.

Many black North American Protestant communities evince the same revelatory biblicism and prophetism that distinguish Rastafarianism farther south. Here I refer to this common feature as the iconographic or 'iconic' dimension of African American religions. In addition to, and coexistent with, ecstatic behaviors and practices, it constitutes another major aspect of black spirituality in North America. As previously employed, the term 'iconic' connotes a transparent employment of images and figural representations. Its use here suggests yet another aspect of the bicultural nature of African American religion and culture. African sources of this bicultural, imagistic propensity are found in the "ritual cosmos" or ancestral worldview discussed earlier, in which "one must see every thing as symbol" (Zuesse). Christian iconographic traditions, for their part, feature not only pictorial icons but also textual icons—most notably biblical narratives, symbols, and figures—which also mediate divine significations and transcendental meanings. As a kind of bicultural fusion of African and Christian sources, accordingly, we find a major instance of iconic expression in the figural tradition that imagistically employs biblical types to configure black experience: for example Moses (liberator), Exodus (emancipation), and Promised Land (destiny).

The Rastafarians' textual iconography focuses on the biblical book of the Psalms, the books of the prophets, and the book of Revelation. Preeminent in their religious imagery is the figure of Babylonian Captivity or

Exile, which holds the central place that the Exodus figure holds among black North Americans. Through the textual icon of the Babylonian Captivity figure, they envision their New World experience as exiles from their African homeland, Ethiopia. However, while the Rastafarians are not biblical literalists (their formation of doctrine is progressive and individualist, not frozen or constrained by a teaching magisterium or hierarchy of instruction), their interpretive tradition is nevertheless realist rather than metaphorical. "The metaphorical language of Marcus Garvey and of countless black clergy men who likened their people to the Jews and adapted the Psalms for their worship, becomes in Rastafarianism purely denotative language, and so revelatory—the revelation, in the most authoritative of texts, of the *identity* of the lost." Such figures as Ethiopia and Zion function iconically to configure and reconfigure Jamaican history and ongoing experience, even where those configurations conflict with conventional historiography and geopolitical realities.

> The Rastafarian does not speak of Jamaica as another Babylon; instead, he takes "Babylon" as the true name of the place of his captivity: Europeans may call it Jamaica, but the Bible calls it Babylon. Similarly, Zion is not the name of a hill in Jerusalem . . . [but] simply another name for Ethiopia. . . . The apparent Middle Eastern milieu of the Bible is part of [white] imposture. Zion *means* Ethiopia, and the notion of Jerusalem as the holy city is another attempt to preempt or conceal the truth.[42]

In Part I we saw that traditional conjure practitioners transform, revise, and reenvision reality by performing mimetic and medicinal operations via "material metaphors" (MacGaffey). Throughout this book I extend the traditional designation of the term "conjure" in order to encompass religious and secular applications of biblical figures as 'experiential metaphors.' Again, such applications operated as recently as the 1960s civil rights movement, in which Martin Luther King, Jr., represented himself as a Moses and configured the movement as an Exodus. More obviously secular examples include iconic uses of democratic texts and their ideals as found in the U.S. Constitution and Declaration of Independence. Together these secular and religious vectors account for the "biblical republicanism" (Bellah) that black North Americans share with their compatriots. Even black nationalist and Pan-African political movements (for example, Ethiopianism and black Zionism) derive from the iconic use of such biblical figures as Ethiopia and Egypt, Exodus and Promised Land, Zion and Babylonian Captivity. Finally, black Muslim and black militant figuration of Babylonian Exile in North America converges with the Rastafarians' poetic iconography of postcolonial oppression. The iconic dimension, it is evident, also conveys liberating and creative energies for future transformations of religion and culture.

Unlike Rastafarianism, however, black North American biblicism allows for increased transparency of communication. By contrast, the poet and novelist Derek Walcott has emphasized the exclusive nature of Rastafarian communication: "Pure Jamaican is comprehensible only to Jamaicans. . . . Within that language itself, the Rastafari have created still another . . . a grammar and a syntax which immure them from the seductions of Babylon, an oral poetry which requires translation into the language of the oppressor. To translate is to betray."[43] The differing political contexts of Jamaicans and black North Americans have resulted in less need for maintaining covert forms of communication among the latter, as already argued. In concluding this chapter it must suffice to examine a single instance of iconic transparency: the 'secular' representation of North America as the "ritual topography" of black people. Before presenting this case, however, it is helpful to clarify further my use of the term 'iconic.'

Here I draw upon the recent employment of the term by the literary theorist Christopher Collins, in *Reading the Written Image: Verbal Play, Interpretation, and the Roots of Iconophobia* (1991). Collins, following the nineteenth century American philosopher Charles Sanders Peirce, distinguishes icons from indices, on the one hand, and from symbols on the other, among the classes of signs. An icon (for example, a picture or diagram) resembles and can serve as a surrogate for the object it represents. But an index (for example, the pointing index finger, or the index of a book) insistently redirects attention to the object it represents. In this regard, Collins observes, indices are semiotically situated between icons and a third class of signs called symbols. "A symbolic sign—a word or number—does not look like, sound like, or in any other sense resemble what it represents: it is an arbitrarily designated cue that in a given conventional code evokes a particular concept. Since its immediate sensory presence is not *naturally* associable with its meaning, the symbolic sign must be strongly coded." Symbols, on this view (distinguished here from poetic or literary symbols), are the signs most abstracted from the material reality that they signify. Symbols like words and numbers are therefore best able, among all the types of verbal signs, to deceive, mislead, and conceal.

> Of the three classes of signs . . . symbol is the preeminent mode of human social play and offers, in the form of speech and writing, the readiest and most adaptable of all play-equipment. In primate evolution the emergence of this class of signs must have immediately separated the human from those related species that still had to rely on a repertory of iconic gestures and recognitions and indexical problem-solving methods. . . . The human could now name and classify . . . but, most important, it could *deceive*.[44]

By contrast a nondeceptive, iconic use of signs—that also displays conjurational or incantatory features—can be observed in the "art and imagination of W.E.B. Du Bois" (Rampersad). In this regard my evaluation of Du Bois, as a conjurational artist in the African American tradition, counters the self-estimation imputed to him by the literary critic Arnold Rampersad: "He knew that he could never be an active participant in [black] religious drama; he stood outside the folk, a psychic mulatto, an intellectual."[45] It may be that the label of "psychic mulatto" accurately represents the scholar's self-image. It is nonetheless possible to find in Du Bois's verse and fiction, as well as his scholarly essays and 'race propaganda', an incantatory figuralism that links the scholar more profoundly to the deep structures of black religious expression. Preliminary evidence for this possibility can be found in Du Bois's novels, in which he "reveals a semi-mystical respect for the black preacher, wizard, and conjure woman, who are always commanding figures of power wielding tremendous authority either to preserve or destroy those who fall within their power."[46] Further corroboration of the conjurational dimension of Du Bois's work comes from the literary theorist Robert Stepto. Stepto discloses the outlines of a black North American ritual topography, embedded in the figural discourse of DuBois's *The Souls of Black Folk* (1903). On the basis of Stepto's literary analysis of *The Souls*, I propose a reevaluation of the scholar's literary craft as that of a conjurational performer in the black North American tradition of the 'open secret': the performer who renders his coded representations with artful transparency.

Stepto distinguishes black North American literature, beginning with slave narratives, in terms of "ascension" narratives and "immersion" narratives. Ascension narratives originate with the slaves' tales of escape to freedom. "Up North" is the spatial expression that clearly designates this topography. It is a ritual topography as well, for certain escape routes (for example, the Underground Railroad) and destinations (for example, Canada) constituted a pattern of experiences to be mimetically repeated in the successive exploits of runaway slaves. Moreover the fixed points in such forms of experience constituted, in Stepto's terminology, "ritual grounds . . . those specifically Afro-American spatial configurations within the structural topography." Examples of such grounds include the slave quarters, that "prototypical ritual ground," and extend to the "Black Belt"—geographical regions of concentrated black population, and also to specific, black-identified locations such as Harlem in the North and Nashville or Tuskegee in the South. But the North-South structure of American experience also features the "down South" configuration. Therein the black North American experience acquires two ritual movements: "ascension" to the North for freedom, industry, and culture, but also return to the South in the form of "immersion" experiences and ritual journeys. Du Bois himself provided the literary prototype, the first such journey inscribed in black American letters, for subsequent ritual immersions by black Americans. In this way he participated, whether consciously or unconsciously, in the mystique of the South as a surrogate Africa for displaced African peoples.[47]

In *The Souls* we find the autobiographical description of Du Bois's passage from a New England childhood in the Berkshires to a career of teaching and service in the South. Stepto represents this prototype of the immersion ritual in figural terms—terms suggested by Du Bois's own transparent depictions of the South's "crimson soil," its "dull red hide-ousness," and its bricks, "red with the blood and dust of toil." By contrast, Du Bois's North appears in Stepto's terms as "brightly green with hope, private memory, and a certain mobility, and yet beshadowed, mottled, and green with an infection, the cure for which is immersion in the darker and yet more vivid hues of ritual ground." Thus Du Bois's 'cure' for Northern duplicity and capitalist avarice is immersion in the black American South: an incantatory, figural journey from the color green to the color red.

> Du Bois's journey south is, figuratively, a journey into the color red: red for the heat and blood of the War, red for the clay soil of Georgia, red for the deposed and dispossessed Indian, red for the relentless sun upon the plantation field, red for the spilt blood and enduring bloodlines of black brethren under assault. The structural topography of *The Souls* is thus strung between two poles, the green and shadowy hills of the Berkshires and the red dust "full of history" of the Black Belt.[48]

The major locus, however, of Du Bois's personal immersion experience is Atlanta and, more particularly, his own study at Atlanta University. There he sat and launched a career of scholarly activism in a context which Stepto describes as self-isolating—"a circuitous route to union with a race and culture!" It was nonetheless in that context, Stepto also acknowledges, that this alienated black scholar continued to attempt figural representations and iconographic transformations of black American experience: "symbolic spaces in which such matters would be resolved."[49] Finally, however, Du Bois's success as literary iconographer and figural conjuror was limited by his inability to discover and fully celebrate the South as his people's "tribal" and ritual ground—as a desirable telos of immersion. His personal misapprehensions about the South were never fully transcended or superseded by a more collective apprehension. "Even when the literal landscape from Massachusetts to Georgia becomes a figurative space, it is still *Du Bois's* space . . . [if tribal,] it is tribal *for him*."[50]

In *The Souls'* ritual topography of black experience, as also in his other figural and mythopoeic writings,[51] Du Bois enlisted his readers in an "enactive interpretation" (Collins) of that experience. Enactive interpretation differs from the naive, realist, or denotative use of figures and metaphors that we find in the case of Rastafarianism, on the one hand. It also differs from the "critical interpretation" of conventional scholarly approaches, which endeavor "to raise to conscious awareness the covert motivation of linguistic work generally." Rather, enactive interpretation consists in the transparency of "simulative play": the reader's awareness, shared with that of the author, that each is *coperforming* the fiction of iconic representation. Such authors avoid deceiving their readers because they knowingly enlist them in the "willing suspension of disbelief that constitutes poetic faith" (Coleridge).[52] Thus the reader too is cooperative in establishing the figural power of the text to configure reality (if only temporarily, for the moment of willingly suspended disbelief). Poetic faith, in this regard, precludes deception because the author and reader are "dual personae of the performer."[53] Each persona is instrumental in, and self-aware of, its role in the performative enactment of the text as a verbal icon according to embedded cues and codes that are readily accessible.

In conclusion, we may describe the iconic dimension of African American spirituality as a tradition of enactive interpretations or performances; the tradition features, preeminently, interpretive appropriations of *religious* figures and biblical narratives. But we may also observe the efficacy of secular texts like Du Bois's *The Souls*, or the Declaration of Independence. In enacting such texts the reader "reinstates a ritual showing place, a place of agonistic combat, of sacrifice, of offerings, and of witnesses to the powers that rule our lives . . . in other words . . . an interior 'seeing place,' a *theatron*."[54] Such performances can be visionary, contemplative, and even revelatory insofar as a reader always discovers new dimensions of a classic text for which any one-dimensional script is inadequate. This inherently revelational feature of enactive interpretation (a feature operating irrespective of the religious or secular nature of the text) is all the more intensified in traditions of revelation such as prophetic Christianity. Perhaps it is the acclimation to, and anticipation of, such revelation that renders 'religions of the Book' more contemplative in character than ecstatic religions of 'the spirit'. In the next two chapters I explore in further detail some of the cognitive structures at work in the kind of enactive and iconic performances introduced here.

Notes

1. Jimmy Stewart, "Introduction to Black Aesthetics in Music," in *the Black Aesthetic*, ed. Addison Gayle, Jr. (Garden City, N.Y.: Anchor Press/Doubleday, 1972), pp. 81–82.

2. Albert J. Raboteau, *Slave Religion: The "Invisible Institution" in the Antebellum South* (New York: Oxford University Press, 1978), pp. 58–59. The locus classicus of this issue in the field of black studies is the Frazier-Herskovits debate about African "survivals" or "retentions" among black North Americans. In the late 1930s the black American sociologist E. Franklin Frazier declared that his kinspeople had been "stripped" of their African heritage. It was against this view that the white American anthropologist Melville J. Herskovits argued vigorously in his *The Myth of the Negro Past* (1941), claiming that certain "Africanisms" had endured past the slavery period in North America. In his review of that debate Raboteau states a modified version of Herskovits's view, arguing that some elements of American evangelical Protestantism were sufficiently similar to the African background of early black Christian converts to allow their indigenous beliefs and practices to continue in modified form.

Thus a hypothesis of "continuity" mediates the extremes of both Frazier's and Herskovits's positions. It moves us beyond the impasse of their debate, because it does not require that we subscribe to the view that black North American culture evinces either the complete atrophy of African traditional elements or the overt kind of retention that we find among black cultures in the Caribbean and South America. See pp. 48–87 of Raboteau's *Slave Religion* for his extended discussion of the issues.

3. Lucius T. Outlaw, "Black Folk and the Struggle in 'Philosophy,' " *Radical Philosophers' Newsjournal* (April 1976): 29.

4. Gayraud S. Wilmore, *Black Religion and Black Radicalism: An Interpretation of the Religious History of Afro-American People* (Maryknoll, N.Y.: Orbis Books, 1983), p. 24.

5. James Baldwin, "Many Thousands Gone," in *Black Expression*, ed. Addison Gayle, Jr. (New York: Weybright and Talley, 1969), p. 325.

6. For a discussion of the irreducible "opacity" of black American experience, and the need to acknowledge that feature in black religious studies and theological formulations see Charles Long, *Significations: Signs, Symbols, and Images in the Interpretation of Religion* (Philadelphia: Fortress Press, 1986), 185–200.

7. Lerone Bennett, Jr., *The Negro Mood and Other Essays* (Chicago: Johnson Publishing Co., 1964), p. 57.

8. Houston A. Baker, Jr., *Blues, Ideology, and Afro-American Literature: A Vernacular Theory* (Chicago: University of Chicago Press, 1984), p. 74. Cf. Addison Gayle, Jr., ed., *The Black Aesthetic* (Garden City, N.Y.: Anchor Press/Doubleday, 1972).

9. Morton Marks, "Uncovering Ritual Structures in Afro-American Music," in *Religious Movements in Contemporary America*, ed. Irving I. Zaretsky and Mark P. Leone (Princeton: Princeton University Press, 1974), p. 67.

10. Ibid., p. 64.

11. Ibid., pp. 91, 114f.

12. Ibid., pp. 61–62, 115–16.

13. Morton Marks, "Exploring *El Monte*: Ethnobotany and the Afro-Cuban Science of the Concrete," in *En Torno a Lydia Cabrera*, ed. Isabel Castellanos and Josefina Inclán (Miami: Ediciones Universal, 1987), p. 227. Cf. Marks, "Uncovering Ritual Structures," p. 64: "The Africanization of musical and linguistic behavior is one of the performance rules underlying a number of communication events in a variety of New World settings." See also Walter F. Pitt's discussion on the "binary structure" of Afro-Baptist worship language, trance-induced behavior, and ritual performance—featuring, for example, the alternation between "Standard English (SE)" and "Black Vernacular English (BVE)," in "Keep the Fire Burnin': Language and Ritual in The Afro-Baptist Church," *Journal of the American Academy of Religion* LVI:1 (Spring 1988): 82, 85, 92, and the recent book-length treatment in Pitts, *Old Ship of Zion: The Afro-Baptist Ritual in the African Diaspora* (New York: Oxford University Press, 1993), pp. 132–45 and *passim*.

14. Ralph Eillison, *Shadow and Act* (New York: Vintage Books, 1972), pp. 9, 11–12. These themes were especially current during the Harlem Renaissance and are fictionally represented in James Weldon Johnson's novel of 1912, *The Autobiography of an Ex-Colored Man* (New York: Hill and Wang, 1960). For a sociology of religion theory of such "signals of transcendence" in ordinary experience see Peter Berger, *A Rumor of Angels* (New York: Doubleday, 1969), p. 52f.

15. Ibid. Cf. W.E.B. Du Bois's expression "Of Our Spiritual Strivings," in which he declares that "this, then, is the end of [the African American's] striving: to be a co-worker in the kingdom of culture, to escape both death and isolation,

to husband and use his best powers and his latent genius." W.E.B. DuBois, *The Souls of Black Folk* (New York: New American Library, 1969), p. 46.

16. Jahnheinz Jahn, *Muntu: The New African Culture* (New York: Grove Press, 1961), p. 225.

17. Alain Locke, "The Negro Spirituals," *Black Expression*, ed. Addison Gayle, Jr. (New York: Weybright and Talley, 1969), pp. 48–49.

18. Harold Courlander, *Negro Folk Music, U.S.A.* (New York: Columbia University Press, 1963), pp. 37, 43.

19. "The spirituals were born of suffering. Yet Zora Neale Hurston is right when, thinking of their rendition, she refuses the inclusive title 'Sorrow Songs.' Negro folk singers, certainly today, sing spirituals with great gusto. There is much more than melancholy in their singing; there is a robustness, vitality, a fused strength. The singing serves as a release; the fervor of the release indicates something of the confining pressure that folk Negroes know too well and have known too long." Sterling Brown, "The Spirituals," in *The Book of Negro Folklore*, ed. Langston Hughes and Arna Bontemps (New York: Dodd, Mead, 1958), p. 288. See also the illuminating discussion of the Fisk Jubilee Singers and their rehabilitation of spirituals in Lawrence Levine, *Black Culture and Black Consciousness* (New York: Oxford University Press, 1977), p. 166f. The most comprehensive treatment of spirituals remains John Lovell, Jr., *Black Song: The Forge and the Flame: The Story of How the Afro-American Spiritual Was Hammered Out* (New York: Macmillan Publishing Co., 1972).

20. James H. Cone, *The Spirituals and the Blues: An Interpretation* (New York: Seabury Press, 1972), pp. 108–42. Cone cites for corroboration the assertion by the African American cultural commentator Leroi Jones (Amiri Baraka) that "the blues issued directly out of... the spiritual" in his *Blues People* (New York: William Morrow and Co., 1963), p. 62.

21. Ibid., p. 115.

22. Ibid., p. 111.

23. Bennett, *The Negro Mood*, pp. 50–51. In addition Bennett criticizes both black and white commentators for missing this wholistic dimension of the folk tradition. "This has caused no end of misunderstanding, even among Negro mythologists who elaborate a blues mystique as opposed, say, to a spiritual mystique. ... What is lacking in most white interpretations of Negro reality is a full-bodied evocation of the entire spectrum. By seizing on one element to the exclusion of the other, by making an artificial separation of Saturdays and Sundays, white interpreters and white imitators of the Negro deform themselves and the total ensemble of the Negro tradition which stands or falls as a bloc." Ibid.

It is no coincidence that Bennett's representation of the folk traditional worldview, as a wholism of sacred/secular perspectives and performances, is consistent with the traditional "sacred cosmos" (Sobel) of West African peoples. For my part I will return to these observations after concluding the discussion of spirituals and after turning, in the analysis of wisdom, to the conjunctive or bipolar nature of cognitive approaches in black culture. Already, however, we anticipate such conjunctions with Bennett's declaration that "the blues are the spirituals." More reservedly, I have suggested that the spirituals are like the blues in that they both conceal their true intention: to transcend their ostensible effect as "sorrow songs."

24. Jahn, *Muntu*, p. 220.

25. Against the conventional view of possession as a form of individual or collective pathology, Shelia Walker charges that "authors who have judged pos-

session abnormal have done so in general because of the similarity of its behavioral manifestations to Western pathological symptoms." Walker's study of ceremonial spirit possession among nonurban peoples in Africa and Afro-America (the Caribbean and South America) seeks a more balanced view:

"Possession may be a very normal culturally determined phenomenon, although some instances of possession may be manifestations of pathology, but even then of controlled pathology, in order to properly situate it and understand its role and function in a given society. In some societies or segments thereof, possession seems to be extremely highly motivated and involved role playing, whereas in others it may be a refuge for people who are really disturbed and who learn to control their disturbance through participation in a possession cult which is actually an institutionalized way of coping with such problems in society."

Shelia S. Walker, *Ceremonial Spirit Possession in Africa and Afro-America: Forms, Meanings, and Functional Significance for Individuals and Social Groups* (Leiden: E. J. Brill, 1972), p. 2.

Walker's work establishes that trancelike states can be induced in normal people by a variety of processes. Rather than a one-dimensional view of the phenomenon, she cites a number of elements that function to create this state: neurophysiology (through which trance can be induced by drugs, sensory deprivation, or physiological stress), hypnosis and autohypnosis, and socialization or cultural determinism (see pp. 1–4). Cultural determinism, Walker argues convincingly, is the most important element in patterning the distinctive content of possession. This is because the underlying physiological or psychological condition may be described in every case as a "hypnotoform state." If the physiological determinants of possession are indistinguishable, as she argues persuasively, then cultural determinants must account for the distinctiveness of varying traditions. Cf. I. M. Lewis, *Ecstatic Religion: A Study of Shamanism and Spirit Possession*, 2nd ed. (London and New York: Routledge, 1989).

26. Locke, "The Negro Spirituals," pp. 48–49.

27. Ralph Ellison, *Shadow and Act* (New York: Random House, 1964), p. 78.

28. Quoted in Cone, *The Spirituals and the Blues*, p. 115. See his p. 151f. for his sources.

29. James Baldwin, *The Fire Next Time* (New York: Dell, 1963), p. 60. Other commentators extend Baldwin's inclusive claim regarding all the traditional genres of black music (jazz and blues, the spirituals and gospel). Sacred and secular genres together may be described also as transformative, ecstatic, cathartic, or healing: "The communal creativity of an instrumental, vocal, or dance 'soul session' in a night club still has some of the same creative possibilities for human healing that a church service might have. . . . The same kinds of psychophysical phenomena . . . help in the achievement of the state of consciousness in both cases. Religious and jazz forms of possession are based on the same African-influenced cultural expectations." Henry Mitchell, *Black Belief: Folk Beliefs of Blacks in America and West Africa* (New York: Harper & Row, 1975), pp. 145–46.

Another commentator analyzes jazz specifically in terms of "the ancient notion that music is in some sense sacred or magical" or "supernatural," a source of "expurgation or revitalization" (cf. catharsis). Here we find practitioners corroborating Mitchell's conjectures: "Veteran pianist Art Hodges once explained, 'The blues heal you. Playing the blues is like talking trouble out. You have to work the blues out of you.' Still other players . . . found it similarly revitalizing or medicinal. Pianist Mary Lou Williams occasionally interrupted her performances to implore inattentive audiences, 'Listen, this will heal you.' " Neil Leonard, *Jazz: Myth and Religion* (New York: Oxford University Press, 1987), pp. ix, 58.

30. Jahn, *Muntu*, p. 223.

31. Howard Thurman, *Deep River* (New York: Harpers, 1945), p. 36. Cf. this more recent, colloquial formulation of the same transformation: "The man systematically killed your language, killed your culture, tried to kill your soul, tried to blot you out—but somewhere along the way he gave us Christianity, and gave it to us to enslave us. But it freed us—because we understood things about it, and we *made it work in ways for us that it never worked for him.*" Reverend Calvin Marshall, *Time*, April 6, 1970, 71, as quoted in Grace Sims Holt, "Stylin' outta the Black Pulpit," in *Rappin' and Stylin' Out: Communication in Urban Black America*, ed. Thomas Kochman (Urbana: University of Illinois Press, 1972), p. 189.

32. Marks, "Uncovering Ritual Structures," p. 110. Emphasis mine. In this regard consider Robert Blauner's insightful remark about the mystique of "soul" in black culture (e.g., soul food, soul music): "The central thesis of the idea of soul is precisely that Afro-Americans have always maintained their unique spiritual life." Robert Blauner, *Racial Oppression in America* (New York: Harper & Row, 1972), p. 134.

33. Holt, "Stylin' outta the black pulpit," p. 190. Emphasis mine.

34. Ishmael Reed, ed., *19 Necromancers from Now* (Garden City, N.Y.: Doubleday, 1970), introduction. *Open Secret: Versions of Rumi*, trans. John Moyne and Coleman Barks (Putney, Vt: Threshold Books, 1984), p. 27.

35. The etymology of the word "conjure" is discussed in the Introduction.

36. Courlander, *Negro Folk Music*, p. 43. Cf. similar caveats by E. Franklin Frazier—the spirituals "were essentially religious and otherworldly and to abstract a political message solely is a suspect venture"—and by Calvin Bruce—"the spirituals bear a crucial spiritual message, not merely a political appeal in disguise;" in Calvin E. Bruce, "Black Spirituality, Language and Faith," *Religious Education* LXXI: 4 (July–August 1976): 374.

37. On the relationship between High John the Conqueror and the sacred trickster figure in Yoruba traditions, namely the *orisha* Eshu, see Diedre L. Badejo, "The Yoruba and Afro-American Trickster: A Contextual Comparison," *Présence Africaine* 147:3 (1988): 3–17.

38. Lawrence W. Levine, *Black Culture and Black Consciousness: Afro-American Folk Thought from Slavery to Freedom* (New York: Oxford University Press, 1977), pp. 385–86.

39. Martin Marty, "America's Iconic Book," in *Humanizing America's Iconic Book*, ed. Gene Tucker and Douglas Knight (Chico, Calif.: Scholars Press, 1982), p. 17.

40. M. G. Smith, Roy Augier, Rex Nettleford, *The Rastafari Movement in Kingston, Jamaica* (N.p., University College of the West Indies, 1960), pp. 5, 18.

41. Laurence A. Breiner, "The English Bible in Jamaican Rastafarianism," *The Journal of Religious Thought* 42:2 (Fall–Winter 1985–86): 30.

42. Ibid., p. 33.

43. Walcott in ibid., p. 39.

44. Christopher Collins, *Reading the Written Image: Verbal Play, Interpretation, and the Roots of Iconophobia* (University Park: Pennsylvania State University Press, 1991), pp. 4–5. Here we should note, with Charles Long, literary and

poetic uses of symbol that are non-arbitrary: "But all is not signification. There is a long tradition in the interpretation of symbol that defines symbol as an intrinsic relationship between the symbol and that which is symbolized. 'One characteristic of the symbol is that it is never wholly arbitrary; it is not empty, for there is a rudiment of a natural bond between the signifier and the signified' (Saussure, *Course in General Linguistics*" Charles Long, *Significations: Signs, Symbols, and Images in the Interpretation of Religion* (Philadelphia: Fortress Press, 1986), p. 2.

45. Arnold Rampersad, *The Art and Imagination of W.E.B. DuBois* (Cambridge: Harvard University Press, 1976), p. 14.

46. Wilson Jeremiah Moses, *Black Messiahs and Uncle Toms: Social and Literary Manipulations of a Religious Myth* (University Park: The Pennsylvania State University Press, 1982), p. 20.

47. Cf. the notion of the South as "neo-African" in Blauner, *Racial Oppression*, p. 136.

48. Robert B. Stepto, *From Behind the Veil: A Study of Afro-American Narrative* (Urbana: University of Illinois Press, 1979), pp. 71, 73, 76.

49. Stepto, *From Behind the Veil*, pp. 73–74.

50. "On one hand, DuBois cannot be faulted: tribal images and 'race-messages' must be tethered to a symbolic geography, and so he constructs one based on what he has experienced, imagined, and seen. But on the other he must be scored, since once again, as the final symbolic space of the narrative instructs, his autobiographical impulses dominate to the point where certain other narrative strategies are left undone, or perhaps done in. . . . This becomes an issue when we realize that, despite all the narrative energies that have been directed toward fashioning a narrative of cultural immersion in an Afro-American ritual ground (the Black Belt), the hero-narrator's final positioning is elsewhere, in Atlanta, in what may be a ritual ground of only a private sort. What does this final posture signal? Is it meant to suggest a final phase to the immersion journey, and hence an affirming of DuBois's personal ritual into Race? Or is it a new development, a removal, an act of distancing if not outright rejection? It is clear, I think, that Du Bois intends a positive and affirmative conclusion, but that his construction of a private ritual ground as a spatial expression of self- and race-consciousness conflicts to a degree with his intentions." Stepto, *From Behind the Veil*, p. 78.

51. For example, Wilson J. Moses observes that "Du Bois' early work struggles to fuse two complementary but substantially different mythological traditions. The first of these is 'Ethiopianism' . . . the other is the European tradition of interpretive mythology . . . the medieval practice of examining Greco-Roman mythology with the intention of either discovering within it, or assigning to it, Christian meaning." Thus in his *The Quest of the Silver Fleece* Du Bois "created a universe in which the ideology of progressive socialism and the traditionalism of Christian black nationalism work harmoniously within the framework of a Greek myth." Wilson Jeremiah Moses, "The Poetics of Ethiopianism: W.E.B. Du Bois and Literary Black Nationalism," *American Literature* XLVII:3 (November 1975): 411, 417.

52. From Samuel Taylor Coleridge's *Biographia Literaria*, xiv, as cited and discussed in Collins, *Reading the Written Image*, p. 139.

53. Ibid., pp. 169–70.

54. Ibid., p. 171.

5

Wisdom

"Old Satan am a liar and a conjurer, too—If you don't watch out, he'll conjure you."

Slave song, Rawick collection, *The American Slave:
A Composite Autobiography*

The Negro tradition, read right, recognizes no such dichotomy. The blues are the spirituals, good is bad, God is the devil and every day is Saturday.

Lerone Bennett, Jr., "Ethos," *The Negro Mood*[1]

In this chapter I intensify my efforts to explore, in theoretical terms, African American cultural referents involving conjure. We will see how conjure and its related phenomena can constitute, in the words of one observer, a "new source of intellectual fermentation . . . [involving] magic, precognition, imagination," but which "the experts of facts do not know or recognize." Because they exclude magical categories from their cognitive universe, such "experts" inevitably preclude, "at the level of discourse," serious consideration of "the nature of black cultural reality."[2] Precisely because magical categories are so vexed or problematic in contemporary academic discourse, I have chosen the expedient of framing this philosophical discussion in terms of the wisdom traditions of African and African American oral cultures. Before proceeding, however, it is illuminating to note how the African American philosopher Cornel West has addressed such problems of discourse in black studies. For his part, West is not as optimistic as many 'Afrocentric' scholars about efforts "to articulate a competing Afro-American philosophy based principally on African norms and notions." In West's judgment, "it is likely that the result would be theoretically thin," not, however, because African cultures lack the materials for discursive depth, but because African *American* thought is ineluctably more American than African.

Philosophy is cultural expression generated from and existentially grounded in the moods and sensibilities of a writer entrenched in the life-worlds of a people. The life-worlds of Africans in the United States are conceptually and existentially neither solely African, European, nor American, but more the latter than any of the former. In fact, ironically, the attempt by black intellectuals to escape from their Americanness and even go beyond Western thought is itself very *American*.[3]

Notwithstanding the irony that progressive American intellectual movements typically strive to transcend America, the more profound, underlying issue is the interpretive problem of African American self-identity—as West also acknowledges. He then provides a helpful formulation of that identity problem with the query, What is the relationship between the African, American, and European elements in this experience? This section contains my own response to that query in terms that show the convergence of Afro-American and Euro-American trajectories in black America's cognitive styles and sensibilities.

Conjunctions

While discussing Christian theological traditions, the Catholic theologian Robert Schreiter presents four cognitive approaches that can also apply to (1) variations on a text; (2) wisdom, or *sapientia*; (3) sure knowledge, or *scientia*; and (4) *praxis*, or practice in-concert-with-theory. In this chapter I am most concerned with the first two demarcations: with cognitive "variations on a sacred text" and with wisdom or "sapiential" cognitive forms, in black folk tradition. Finally, note that I am deferring to the final chapters the more directly theological issues of this study. In the present chapter, accordingly, I do not intend any direct theological applications of Schreiter's schema, since its Christian or theological framework is formally incidental to my interest in black culture's cognitive styles and sensibilities (aspects of which are, in any case, extra-Christian and pretheological). Schreiter's treatment is commendable because he highlights wisdom and oral traditions in ways that suggest their primacy for cultures such as black America, and because he avoids subordinating such traditions either to scientific or to praxial approaches.

Wisdom traditions, from ancient Greece to contemporary black America, display one of the features which Schreiter highlights in particular for wisdom traditions in Christian cultures: the cosmological imperative "to see the world, both the visible and the invisible, as a unified whole."[4] The achievement of a unified vision that joins disparate realms, such as "the visible and the invisible," is also operative in black America's apprehension of reality as a 'pharmacosm' or a ritual cosmos (Zuesse). I have introduced these related concepts to provide indigenous referents for an African American worldview.[5] In his exploration of biblical hermeneutics, Paul Ricoeur has further elaborated the desire of wisdom traditions for a unified vision of reality. Ricoeur emphasizes the *nonnarrative* aspect of the Bible's wisdom literature. Thereby he accents my interest in this chapter in orality over against cognitive traditions informed by literary structures. He underscores, for example, that wisdom literature is not interested in history and historical narration, but rather in uniting or conjoining the opposed realms of "the everyday," on the one hand, and "the immemorial" on the other.

> I am first of all struck by the way in which the proverbs, in spite of their modesty, conjoin in a striking way the everyday and the immemorial. The everyday is the time of works and days. It is punctuated by those maxims that tell how to conjoin a righteous heart and a happy life. This time of the everyday ignores the great events that make history.... And this time without events does not get narrated. It is spoken of in proverbs.... And yet, it is by way of the everyday that wisdom brings to light the immemorial, that is, what, as ageless, has "always existed."[6]

In conjoining the everyday with the timeless and immemorial, biblical literature begins to shift from its proverbial form of discourse. It does so first by means of a "hypostasis" (found particularly in Proverbs 8.22–32): the personification or reification of wisdom. "At this stage, the immemorial is not just what wisdom says but Wisdom itself when it speaks." Next, biblical wisdom addresses the universal experience of suffering and failure in the human condition—what Ricoeur calls the "radical questioning" of existence (for example: why do the righteous suffer and the wicked prosper?). The texts that convey this second, nonproverbial form of wisdom are the radical critiques of religious experience constituted by the books of Job (theodicy) and Ecclesiastes (resignation). However, the same motivation underlies both the proverbial and the nonproverbial, more critical forms of biblical wisdom: the conjunctive impulse to unite everyday experience with immemorial truths, insights, and precepts. More useful at this point, for distinguishing Afro-American wisdom traditions, is Ricoeur's alternative way of framing the conjunctive nature of wisdom traditions: as the conjoining of "ethos" and "cosmos."

> Wisdom fulfills one of religion's fundamental functions which is to bind together *ethos* and *cosmos*, the sphere of human action and the sphere of the world. It does not do this by demonstrating that this conjunction is given in things [cf. *scientia*], nor by demanding that it be produced through our action [cf. praxis]. Rather it joins *ethos* and *cosmos* at the very point of their discordance: in suffering and, more precisely, in unjust suffering. Wisdom does not teach us how to avoid suffering, or how magically to deny it ... [but] how to endure, how to suffer suffering.[7]

This formulation, in which the conjoining of ethos and cosmos is correlated with suffering in human experience, is advantageous for understanding the conjunctive wisdom of black folk culture in North America. However, the contents that black America's conjunctive tradition comprises are all-encompassing. That is, black folk wisdom conjoins many more conventionally opposed categories than Ricoeur's everyday and immemorial, on the one hand, and his ethos and cosmos on the other. The wisdom proficiencies of a folk ethos that is conjurational, and of a cosmic vision that is pharmacopeic, feature the conjoining of opposites in every context of human experience. Recall here the kind of perception required within a ritual cosmos: "One must see *every thing* as symbol" (Zuesse). A partial list of such conjoined 'things as symbols' is sufficient to suggest an entire universe composed of binary opposites like everyday/immemorial and ethos/cosmos. The reader recognizes immediately, and even anticipates, the culturally privileged terms among the elemental pairs one/many, same/other, truth/error, presence/absence, good/evil, male/female, white/ black, and so on, indefinitely throughout our Western cognitive constructs. The crucial issue in displaying such pairs, however, consists in whether or not one typically conjoins or disjoins them.

The wisdom tradition of black North American folk culture dissents from the predominant Western form of disjunctive thinking—that conventional "either/or" in which rationalism insists on unambiguous, univocal meanings for things. Instead this tradition prefers the conjunctive "both/and" of archaic and oral cultures, in which ambiguity and multivocity are taken for granted (even promoted). In this regard one observer, the black studies scholar Vernon Dixon, derives a conjunctive predisposition for black Americans from a social history of oppression and biculturality and an ancestral African heritage. Using the term "diunital" to denote conjunctive thinking, Dixon claims that

> American Blacks rather than Whites are more deeply attuned to a diunital [both/and] existence for two reasons. First, we live in a dual existence. We are American citizens, yet we are not. American institutions are ours, yet they are not. We have one identity that are two identities.... Secondly, we may embody a predisposition to diunity that arises from our African identity. Jahnheinz Jahn points out that the union of opposites characterizes traditional thinking throughout Africa, for African philosophy as such.[8]

We should note here, in addition, that Dixon is careful to distinguish his use of the term "diunital" from contemporary use of the term 'dialectical'. In conventional dialectical approaches (for example, Hegel's), "opposites are conceptualized as coexistent, but only antagonistically."[9] Rather than the thesis-antithesis disjunction of such formulations, conjunctive approaches are able to affirm both elements in a dyad. This dual affirmation of opposites is the crucial aspect of wisdom traditions that feature conjunctive forms of cognition. Its consequences are most dramatic in the area of human relationships, as Dixon also indicates.

In the course of conflicts between cultures, disjunctions typically rigidify (in Girardian terms, "mimetic rivalries" create a "crisis of distinctions"). Then ensue not only distinctions, but also devaluations of rejected elements existing in binary opposition to preferred elements: male over female, white over black, middle class over working class. In such cases a cognitive-spiritual temperament impels groups to prefer certain elements in opposition to others—impels them (during a "sacrificial crisis"—Girard) to reject certain categories and then deny validity, even existence, to the rejected terms. When those devalued terms are attached to human beings, as in cases of gender or ethnicity, caste or class, disjunction can become dehumanization, even death and mass death. Reflecting on ethnic conflicts in the United States between Afro- and Euro-Americans, Dixon describes the nature of conflict with reference to such disjunctive thinking: "Theories and policies that lead to racial polarization and racial conflict are the rational and logical development of the application by Blacks and Whites of the either/or conceptual framework to American race relations. Conflict is not caused by color itself, but by people of different colors viewing each other in either/or terms."[10] To emphasize further: conflict is not inherent in difference per se but in the conceptual—disjunctive or antagonistic—framework by which we view difference. By contrast Dixon calls for a multiethnic reconstruction of society on the basis of Afro-America's diunital predisposition and cognitive heritage. He wants to retrieve an ancestral or traditional ethos that features the conjunction and affirmation of both elements in a dyad: in this case, to be explicit, the dyad of black and white ethnic identities and communities.

However, we are now in a position to see that Dixon is attempting to foster in the social-political arena, and with analytic terms and discourse, an achievement that black culture has effected in aesthetics (for example, in the ironic duality of the spirituals and the blues) and in folk medicine and magic (for example, in the toxic-tonic duality of conjure and of herbal pharmacopeia). I have already examined certain dualistic dynamics in Afro-American musical expression. In the context of the current discussion we have the opportunity to probe more closely the cognitive structure that undergirds this dualism. Corroborating the "double system" (Marks) or the "dual communication system" (Craig)[11] observed in black aesthetic and linguistic phenomena, we also find a conjunctive approach in the folk tradition's wholistic view of experience and reality. I have referred to this conjunctive feature as a wisdom tradition in black culture which, commonly with other wisdom traditions, readily conjoins categories such as visible and invisible, everyday and immemorial, ethos and cosmos. Other observers add to this series the overarching categories of the secular and the sacred. Gayraud Wilmore, for example, locates in the African heritage of black America "a profound sense of the pervasive reality of the spirit world above and beneath the artifactual world; the blotting out of the line between the sacred and the profane; the practical use of religion in all of life."[12] But the historian Lerone Bennett has provided perhaps the most evocative statement of this conjunctive proclivity.

Bennett articulates the complementarity of the sacred and the secular in terms of two root metaphors in black culture: one drawn from the culture's seminal aesthetic forms—the spirituals and the blues—and a second taken from its ethos of everyday life—where life comprises both Sundays and Saturdays. The wisdom of complementarity, which conjoins

Saturdays and Sundays, or a "blues mystique" and a "spiritual mystique" in Bennett's terms, also conjoins two of the major terms of this study: conjure and Christianity. In this regard Bennett's criticism of the conventional wisdom, which dichotomizes or bifurcates the blues and the spirituals, also applies to the bifurcation of conjure and Christianity in most interpretations of black religion. "What is lacking in most white interpretations of Negro reality," Bennett concludes the passage, "is a full-bodied evocation of the entire spectrum . . . [instead of] seizing on one element to the exclusion of the other."[13] Indeed, here I must acknowledge that such flawed perspectives also characterize black theological and scholarly interpretations. Many traditional black religionists also insist on treating black Christianity and conjure as thoroughly dichotomous. In this regard black church communities have conventionally condemned conjurational practices as uniformly 'the devil's work'.

Conjure certainly carries negative connotations for orthodox Christianity. "The variety of illnesses, injuries, and misfortunes blamed on conjure was endless," Albert Raboteau reminds us. Other commentators like Newbell Niles Puckett have observed a moral ambivalence in the practice of conjure. "The same spirit can be persuaded to work indifferently good or evil . . . the same power being turned to different ends, just as fire may be used for warmth and protection or for burning a neighbor's barn." But this "moral ambivalence" means that conjure cannot be reduced, in such one-dimensional terms, to evil magic, demonic machinations, or "malign occultism."[14] Rather, the same phenomena viewed as morally ambivalent can also be viewed as nonmoral (neither immoral nor amoral but simply without moral categorization). In this regard Raboteau reminds us of the folk "refusal to dichotomize power into good and evil—a refusal which Herskovits and others see as African." Moreover, he concludes:

> The conflict between Christianity and conjure was more theoretical than actual. . . . Christian tradition itself has always been attuned to special gifts (charisms) of the Spirit as they are manifested in prophecy, healing, and miracles. As a result, Christianity, especially on the popular level, has a certain tendency to appropriate and baptize magical lore from other traditions. In an important sense, conjure and Christianity were not so much antithetical as complementary.[15]

Because of this conjunctive feature in folk cognitive patterns, conjure intentions can readily modulate from harming or toxic modes of operation to healing or curative modes, *without thereby ceasing to be conjurational* (i.e., mimetic and pharmacopoeic). For this "double-edged nature" of conjure intention, the folklorist Michael Bell offers a suggestive example from the field research of Harry Hyatt. Hyatt asked an informant why he recommended burying under someone's door a bottle containing a woman's hair and horsehair. The informant replied that the charm would prolong the illness of an opponent or, on the other hand, procure a woman's love for the client. Bell remarks: "Causing lingering ill health and attracting a woman's love obviously are two very different kinds of intention, yet, according to this informant, either may be accomplished by the same performance."[16] Analogously, I postulate a conjunctive wisdom tradition in which Christian practitioners of conjure readily modulate from the intentions popularly ascribed to their craft—imputedly malign intentions, to include intentions popularly identified with orthodox Christian aspirations—imputedly benign intentions. "Everything in hoodoo," Hyatt asserted, "bows before intention and desire."[17]

Intentions

Intention is, in summary, a fundamental component in the hoodoo system, if not the fundamental component. . . . Actions, objects, and other components in the hoodoo performance complex provide the means for sending a desire on its way to fulfillment, but it is the formulation of intention which guides the performance, insuring that it reaches its destination.

Michael E. Bell, *Pattern, Structure, and Logic in
Afro-American Hoodoo Performance*[18]

Conjurational intention need not be explicitly expressed in oral form—for example, in the overt form of a spoken command. At the other extreme, intention may be mentally implicit in the form of an opinion or desire. In this regard, Michael Bell's research, has discovered a "continuum" of intention. "Between these two extremes are wishing, praying, cursing, and speaking or writing charms or names. By means of these formal incantations, frequently used in combination, intention may become explicit." It is intriguing that Bell's continuum extends from voiceless intention, to spoken forms of desire, and finally to written incantation. Moreover, he claims that calling or writing the name of a conjure client or "target" predominates as a means of formulating and communicating intention in conjure practice. Finally, in view of preceding and subsequent comments on orality and literacy in black culture, I note Bell's contrast between "voiced" and written intentions. "While incantations such as wishes, charms, commands, and prayers usually are voiced rather than written, names . . . most frequently are written . . . they are a very significant part of formulating intention."[19] In this connection I redirect the reader's attention to incantatory writing in black literary-religious expression.

We have already examined incantatory writing in the tradition of Ethiopianism, along with other forms of figural discourse. Recall such naming performances as David Walker's invocation of "the God the Ethiopians":

Though our cruel oppressors and murderers, may (if possible) treat us more cruel, as Pharaoh did the Children of Israel, yet the God of the Ethiopians, has been pleased to hear our moans in consequence of oppression, and the day of our redemption from abject wretchedness draweth near.[20]

In this passage Walker named or 'targetted' the biblical God by means of a figure representive of all Africans, "the Ethiopians." Because God is herein named, targetted, or invoked as "the God of the Ethiopians," I have designated Walker's rhetorical device as a literary form of conjuring God ("God-conjuring"). But in addition Walker embeds that name within the narrative context of the Exodus story, and within the prophetic context of Psalm 68.31, thereby compounding the convergences and identifications operating between biblical texts and black experience.

A similar "strategy of inducement" (Burke), bearing similar incantatory effect, is evident in another nineteenth century Ethiopianist passage:

The following will show what God is doing for Ethiopia's sons in the United States of America.

"But ye are a chosen generation, a royal priesthood, and an holy nation, a peculiar people...[who] in time past were not a people, but are now the people of God" (1 Peter, ii.9,10).[21]

In this passage the author also 'names' with the figure of Ethiopia, but here he targets his client community, black slaves and free persons "in the United States of America." Moreover he embeds the significant name "Ethiopia's sons" within a New Testament text (1 Peter 2.9–10)—a text which in its turn quotes significantly from Exodus (19.6) and also from a prophetic Scripture (Hosea 2.23). Note too that a subtextual element in this strategy, nearly subliminal in effect, is the Exodus figure of ancient Israel as God's 'chosen people', which the author hereby invokes for identification with enslaved and oppressed black Americans. In this connection it is illuminating to return to Bell's treatment of hoodoo performance: "If I were to assign your name to a bottle of ingredients (or include your name in the bottle) and throw it into running water, then, in accordance with the [mimetic] logic of hoodoo, I would be throwing you into running water as well."[22] In the preceding Ethiopianist texts the author assigns to his target (God) or his client (black Americans) a figural name ("the God of the Ethiopians" or "Ethiopia's sons," respectively). Then—to borrow a metaphor from the description of the hoodoo performer in Bell's example—he 'throws' or introjects that designation into a biblical and a prophetic text (Exodus, and Psalms or Hosea, respectively). The final intent is both mimetic and incantatory, so that whatever happens to the name may also happen to the target or client. To be explicit, the figural recontextualization of the name is intended to effect a transformation of reality (a conjured culture) for the client.

The intention of such literary-religious expressions is to induce material transformations of reality for enslaved people in America. An examination of the conjurational provenance of these expressions, I submit, offers a comparably precise and enlightening analysis alongside the

more familiar, narrative category of applied story or myth. Recall Raboteau's cogent formulation of mythic appropriation: that the slaves
"kept hope alive by incorporating as part of *their* mythic past the Old
Testament exodus of Israel out of slavery. . . . [They] applied the Exodus
story, whose end they knew, to their own experience of slavery, which
had not ended."[23] This view of a (hermeneutic) *application* of mythic
narrative, among preliterate slave believers, is compatible with both oral
and literary processes of narrative appropriation. A conjurational account,
on the other hand, includes such narrative applications but frames them
as constitutive elements within a larger performance. That performance
(1) targets a subject (whether client or victim), (2) names that subject,
and then (3) affects that subject by way of (4) a mimetic operation in
which whatever happens to the name is also (5) intended or desired for
the subject.

In concluding this section I want to stress the transition in black
America from a primarily oral culture in the slave period to its current
situation as a formally literate but still orally oriented culture. Alongside
my hypothesis of complementarity between conjure and Christianity, I
posit conjure as a covert tradition still operating in black America's religious, aesthetic, and political formations. Though hidden, conjurational
processes are peculiarly disclosed in the incantatory naming—calling and
writing—of biblical figures. As already demonstrated, black America's conjure tradition endures in its religious and political uses of biblical typology,
featuring such figures as Exodus, Egypt, Ethiopia, Moses, Promised Land,
Captivity, and Diaspora. Moreover this ongoing conjure heritage is conveyed or empowered by oral modes of incantatory discourse, even when
expressed in literary form. Because of their oral patterns and ethos, conjurational impulses and strategies remain opaque to uninitiated observers
of black culture, most of whom are inundated by the transparency and
hegemony of literary processes. For literacy, as Ong tells us, "consumes
its own oral antecedents and, unless it is carefully monitored, even destroys
their memory."[24] Forgetting conjurational processes, accordingly, has accompanied black America's development into a fully literate society. But
as Ricoeur also reminds us, it is possible to retrieve religious categories
from the erosion of modern consciousness and technical developments,
and thus overcome such forgetfulness. A "second naivete" (Ricoeur) can
be achieved or, in Ong's terms, literacy can restore memory.[25] I continue
the task of restoration in the following section.

Significations

Signifying *can be a tactic employed in game activity—verbal dueling . . .
however, [it] also refers to a way of encoding messages or meanings which
involves, in most cases, an element of indirection. . . . Viewed as an*

alternative message form, selected for its artistic merit, [it] may occur
embedded in a variety of discourse . . . [yet] not define the entire speech
event.

Claudia Mitchell-Kernan, "Signifying"

My community was a community that knew . . . it was a community
signified by another community. This [arbitrary] signification
constituted a subordinate relationship of power expressed through custom
and legal structures. While aware of this fact, the community undercut
this legitimated signification with a signification upon this legitimated
signifying . . . on the other hand, the very fact that the relationship was
arbitrary was the source of its terror.

Charles Long, *Significations*[26]

Earlier I referred to Robert Schreiter's fourfold typology of cognitive styles
and approaches, the first two categories of which are variations on a
(sacred) text and wisdom approaches. I have already discussed African
American wisdom traditions in terms of their conjunctive features and
intention as a cognitive feature of conjurational performance. We are now
in position to take up Schreiter's first category of textual variations. This
is an appropriate emphasis for an oral culture in which a sacred text, the
Bible, functions as an iconographic source of figures and incantations to
be mimetically appropriated and replicated—as a conjure book, in short.
In Schreiter's terms such variations on a text consist of coded messages—
messages composed of signs, on the one hand, and codes that govern the
use of those signs on the other. Each "message" of this kind, he asserts,
varies "the signs originally employed, yet maintains the same codes.
Through this expansion of the sign system, the message is invested with
a wider semantic capacity than was originally the case."[27] For the purposes
of this study, however, it is more instructive to refer to the African American verbal convention called "signifying."

From the Signifying Monkey to the Signifying King

Signifying, according to the literary theorist Henry Louis Gates, Jr., "is a
uniquely black rhetorical concept, entirely textual or linguistic, by which
a second statement or figure repeats, or tropes, or reverses the first."[28]
Gates locates the folk roots of the verbal art in African storytelling and
particularly in the trickster characterization found in the "Signifying Monkey tales" of black folklore in North America. Signifying can also be
regarded as a folk hermeneutic—as an interpretive art or discipline for
eliciting meanings beyond the dictionary or lexical signification of the
words involved.[29] Furthermore, manipulating this surplus of meanings

involves games of reinterpretation and counterinterpretation, as well as devices for subtly tricking, misleading, or outwitting others. But perhaps most interesting for present purposes is the usefulness of the device for understanding the "intertextual relation" (Gates) between Afro- and Euro-American systems of meaning. Two examples of such intertextuality in African American literature can be mentioned briefly. In his 1911 novel *The Quest of the Silver Fleece*, W.E.B. Du Bois interwove black American history with Greco-Roman mythology. In a parallel manner Zora Neale Hurston, in her 1939 novel *Moses: Man of the Mountain*, rendered black American experience through the allegory of the biblical story of Exodus. Using the Afro-American vernacular term, we may say that in his novel Du Bois 'signified on' Greek mythology, whereas Hurston signified on the Bible. Indeed, together these novels constitute African American 'significations on' a received European American literary tradition, by virtue of their intertextual relation to foundational texts of Western culture.[30]

Under the various modes of signifying Gates gives a rather technical list of tropes or rhetorical devices: metaphor, metonymy, synecdoche, irony, hyperbole and litotes, metalepsis, aporia, chiasmus, and catechesis. All of these forms, he adds, "are used in the ritual of signifying." He thereby acknowledges the performative contexts in which the device occurs, and then also catalogues such performances by their vernacular nomenclature: "marking, loud-talking, testifying, calling out (of one's name), sounding, rapping, playing the dozens, and so on." Beyond such cataloguing, however, Gates pursues more closely the trickster impulse operating in ritual performances that employ signifying. Indeed, he locates the source of that impulse in an indigenous West African spirituality whose contemporary ambience is Pan-African. That spirituality features, among other elements, the sacred trickster of the Yoruba people, "Esu-" or "Eshu-Elegbara," who is also found in the Americas. In Brazil he is known as "Exu," among the Haitians and the Cubans as "Legba," in Vodun as "Papa Legba," and among practitioners of Hoodoo as "Papa LaBas." These are all manifestations of one deity, Gates explains, representing the type of spirit who is a messenger of the gods:

> he who interprets the will of god to people ... [who is] guardian of the crossroads, master of style and the stylus ... of the mystical barrier that separates the divine from the profane worlds ... the divine linguist, the keeper of *ase* (logos) with which Olodumare created the universe.[31]

These characteristics Esu shares with the Greek messenger god, Hermes, and his Roman alter ego, Mercury. With these references Gates displays the functional similarities between two gods of interpretation, one Pan-African and the other Pan-Hellenic. He thereby makes a case for positing a parallel between Esu and the deity who gave his name to the Western philosophy of interpretation, "hermeneutics." As the Pan-Hellenic Hermes is to hermeneutics, Gates declares, so Esu is to Afro-American traditions of tropological significations. But in addition to these multicultural parallels, a sacred-profane convergence operates here as well. "Esu's functional equivalent in Afro-American profane discourse is the Signifying Monkey." The monkey, like Esu, is a trickster character, traditionally celebrated in the retelling of a folk story in which he outwits a lion, "the king of the jungle." The monkey succeeds by speaking figura- ∠ tively or symbolically in a code that the lion interprets literally, to his own embarrassment and misfortune. This "trick" of discourse, which effectively overturns or "reverses" the lion's status as the jungle king,[32] has become paradigmatic in the slave tradition of "puttin' on ol' massa." In this connection, as previously discussed, the most familiar tradition of outwitting slavemasters is the coded use of slave spirituals to signify escape strategies. "As many writers have argued and as some former slaves have testified, such spirituals as the commonly heard 'Steal away, steal away, steal away to Jesus!' could be used as explicit calls to secret meetings."[33]

Another observer has reflected on comparative aspects of Esu and an Afro-American trickster called High John the Conquerer. " 'Trickster' in the *diaspora* context should include transcendence as well as wit over force," insists Diedre L. Badejo. "The term 'trickster,' " Badejo continues, "suggests not only a literary figure who dupes for self-aggrandizement but also one who challenges an established social order and probes the question of fate." This last comment applies especially to the trickster hero who survives into the twentieth century: High John the Conqueror. "High John knows the society in which he operates. He is a student of its strengths and weaknesses and his actions manipulate them."[34] In this regard we should recall the trickster aspects of signifying in black oral and literary tradition. "Transcendence as well as wit over force," Badejo claims, are the essential features of the trickster impulse in the context of the African diaspora. Can we locate these features in the Afro-American art of signifying as well? I think so. For this purpose recall Henry Louis Gates's formulation, in which signifying can be either a "textual or linguistic" performance, "by which a second statement or figure repeats, or tropes, or reverses the first."

It is remarkable how closely Gates's formulation of signifying, which he calls "a uniquely black rhetorical concept," converges with the Euro-Christian hermeneutic of biblical figuralism as presented by Erich Auerbach and quoted in the Introduction. In Auerbach's terms biblical typology or "figural prophecy," to be more precise, "implies the interpretation of one worldly event through another; the first signifies the second, the second fulfills the first."[35] In the foregoing treatment of black religious uses of this figural hermeneutic we have seen how a "worldly event," such as Lincoln's emancipation of American slaves, has been interpreted through "another" event: the biblical Exodus of Hebrew slaves from Egypt. Here ∠ I propose that such figural interpretation rightly claims two sources: a Euro-Christian source in the figural tradition of biblical typology, and an Afro-Christian source in the signifying tradition of black folk culture. In brief explication of this richly textured, multicultural convergence of sources, we may say that slave interpretation of the 1860s emancipation 'signifies on' the Bible story by rendering the emancipation as an Exodus event.

To cite a more contemporary example we may say that by his self-representation as a Moses figure, Martin Luther King, Jr., 'signified on' the Exodus story in terms of his catalytic role in the 1960s black freedom movement. In both the slavery instance and the contemporary example of King, we find the elements highlighted in Badejo's comment: the trickster elements of transcendence operating in the context of a society of domination, with related displays of "wit over force." Indeed, in this regard we are able to compare the public persona of King with that of other African American activist heroes who have operated as trickster figures, such as Adam Clayton Powell, Jr., and even Eldridge Cleaver. Such leaders have functioned as agents of transcendence, in Badejo's terms, and models of direct manipulation of the strengths and weaknesses of the society around
> them. However, they display trickster skills not simply because they employ guile or deception, but because they must confront and outwit society's use of force and deception. The situations confronting these leaders range from the most brutal to the most sophisticated forms of racism, political subjugation, and economic exploitation. In this regard it is fitting that one writer has compared King to the best known Afro-American trickster animal, in a study titled "Brer Rabbit and Brer King: The Folktale Background of the Birmingham Protest."[36]

Mother wit and conjure from High John de Conquer

Mother wit is a popular term in black speech referring to common sense.
Mother wit is the kind of good sense not necessarily learned from books or
in school. Mother wit [bears the] connotation of collective wisdom
acquired by the experience of living and from generations past [and] is
often expressed in folklore.

Alan Dundes, *Mother-Wit from the Laughing Barrel*[36]

In this chapter I am concerned to display not only Christian or religious forms of African American signifying in the folk wisdom tradition. There are extra-Christian and nonreligious developments in the tradition which also bear important conjurational features. For the purpose of displaying those features I return to the trickster figure who emerged during the slavery period, High John the Conquerer. Again the folklore research and literary craft of Zora Neale Hurston provide a rich resource here. Hurston's article, "High John de Conquer," has preserved for us a magnificent folk character who represents the best of black America's signifying wit, trickster lore, proverbial wisdom, and conjurational expertise. High John, Hurston begins, was not originally a human being but rather "our hopebringer": a kind of spiritual material emanation that emerged from an enslaved people in the extremity of their suffering and their need for a champion. "He was a whisper, a will to hope, a wish to find something worthy of laughter and song. Then the whisper put on flesh. His footsteps sounded across the world in a low but musical rhythm as if the world he walked on was a singing-drum. The black folks had an irresistable impulse to laugh." This "impulse to laugh," in the midst of gross dehumanization and abuse, constitutes the ironic and double-edged element also found in the spirituals and the blues. That laughter-in-the-midst-of-suffering appears here as the very emblem of a people's transcendence and will-to-survive.

As Hurston remarks (betraying an ethnographer's semiotic interest), "the *sign* of this man was a laugh, and his singing-*symbol* was a drum-beat."[38] She then proceeds, most eloquently and in quite simple language, to elaborate first the subjective dimension of a folk spirituality, and then the intersubjective or communal, the transcendental and the ironic, aspects of High John—this legendary spirit of laughter and song.

> It was an inside thing to live by. It was sure to be heard when and where the work was the hardest, and the lot the most cruel. It helped the slaves endure. They knew that something better was coming. So they laughed in the face of things and sang, "I'm so glad! Trouble don't last always." And the white people who heard them were struck dumb that they could laugh.[39]

Hurston's plantation locale for High John's exploits and endeavors recalls Robert Stepto's notion of the slave quarters as black America's "prototypical ritual ground." In addition, Stepto's category of genius loci, or spirit of place, provides a useful designation for High John as a "locale integrity" in which the "imagery of the tribe are given bounding outline" (Geoffrey Hartman).[40] Hurston gives us densely textured and earthy depictions of such imagery.

> [High John] was treading the sweat-flavored clods of the plantation, crushing out his drum-tunes, and giving out secret laughter. He walked on the winds and moved fast. Maybe he was in Texas when the lash fell on a slave in Alabama, but before the blood was dry on the back he was there...then somebody in the saddened quarters would feel like laughing, and say, "Now, High John de Conquer, Old Massa couldn't get the best of him. That old John was a case!" Then everybody sat up and began to smile. Yes, yes, that was right. Old John, High John could beat the unbeatable. He was top superior to the whole mess of sorrow. He could beat it all, and what made it so cool, finish it off with a laugh. So they pulled the covers up over their souls and kept them from all hurt, harm and danger and made them a laugh and a song.[41]

The pathos of this depiction is nearly overwhelming, and equally powerful is its rendering of the slaves' spiritual and therapeutic transcendence of their vicissitudes. Here we find a trace of a psychosocial folk therapy that employed cathartic laughter and singing as a strategy of recovery from both physical and emotional injuries.

Equally compelling is High John's function as a repository of trickster acumen, folk wisdom, and signifying wit. Consider the following proverbial units, all descriptive of High John and derived from various tales of his feats and exploits:

> making a way out of no way
> hitting a straight lick with a crooked stick
> winning the jack pot with no other stake but a laugh
> fighting a mighty battle without outside-showing force
> winning his war from within......................
> He who wins from within is in the "Be" class. *Be* here
>   when the ruthless man comes, and *be* here when he
>   is gone.

In addition to this catalogue Hurston gives High John the following accolades, each of which is a signifying 'toast' or boast—a kind of "playing the dozens" by means of tribal affirmations and legendary esteem. "He had the wisdom tooth of the East in his head" is a boast that signifies on 'wisdom tooth', on the one hand, and the category of Asian wisdom on the other. Indeed this boast assimilates, through the figure of John, the folk wisdom of a despised people to the legendary and widely esteemed wisdom of Eastern figures like the Magi, for example, or the Buddha or Confucius. "Morever," Hurston's vaunt continues, "John knew that it is written where it cannot be erased, that nothing shall live on human flesh and prosper. Old Maker said that before He made any more sayings." With considerable skill Hurston herself signifies on the gospel formula for prophetic declarations in the New Testament "it is written." She thereby gains for High John the reverential stature which biblical literature reserves for prophets and holy men. Since the content of the saying is also prophetic and attributed to a creator deity—"nothing shall live on human flesh and prosper . . . Old Maker said"—Hurston effectively presents High John as an extrabiblical prophet or a spirit of divination for black folk.[42]

Finally, we are in a position to appreciate the conjurational aspects of High John's legendary character and actions. For this purpose two additional indications must suffice here. The first is quite explicit and directly relates to the lore of conjure practice and paraphernalia: "High John de Conquer" is the name not only of a folk hero, but also of a root in which are located his magical powers. "High John de Conquer went back to Africa, but he left his power here, and placed his American dwelling in the root of a certain plant. Only possess that root, and he can be summoned at any time." A second indication of conjurational power is implicit in the most exalted claim, presented by Hurston as if expressed by her informant, one Shady Anne Sutton: that it was High John who achieved the liberation of the slaves by means of the Civil War. In this connection the phenomena of conjuring culture are quite evident in the testimony of "Aunt Shady Anne." That testimony begins with a rebuke for "smart colored folks" who refuse to believe in High John as their forebears did. Hurston assures

Shady Anne that she is not one of those who are "shamed of the things that brought us through." Then her informant proceeds to give an account of High John as a type of diviner, who prophesied to slaves about the emancipation a century before the war.

> Sho John de Conquer means power. . . . My mama told me, and I know that she wouldn't mislead me, how High John de Conquer helped us out. He had done teached the black folks so they knowed a hundred years ahead of time that freedom was coming. Long before the white folks knowed anything about it at all. . . . A heap sees, but a few knows. 'Course the war was a lot of help, but how come the war took place? They think they knows, but they don't. John de Conquer had done put it into the white folks to give us our freedom, that's what. Old Massa fought against it, but us could have told him that it wasn't no use. Freedom just had to come. The time set aside for it was there. That war was just a sign and a symbol of the thing.[43]

The claim here, to be explicit, is that High John conjured "the white < folks to give us our freedom." It is a magical counterclaim to that of black Christian traditionalism. A Christian providential view claims instead that the God of history acted in American experience, and acted with the same emancipatory purpose displayed in the miraculous liberation of Hebrew slaves as depicted in the biblical Exodus. Such a discrepancy, between conjure folklore and religious traditionalism, is typical of the distinctions drawn between so-called magic and so-called religion. Magic thus seen is not intrinsically concerned with theological propriety or moral scruples, but rather with mimetic operations and processes. For "magics," another commentator has asserted, "unlike deities, make no moral demands and, above all, will operate automatically and inexorably."[44] So testified Hurston's Aunt Shady Anne Sutton in her conviction that "freedom just had to come" to the slaves because "High John de Conquer" conjured it. In the next chapter, however, we will discover, in the context of black social prophetism, the coinherence of magic, morality, and allegiance to deity.

Notes

1. George P. Rawick, ed., *The American Slave: A Composite Autobiography*, vol. 5, *Texas Narratives* (Westport, Conn.: Greenwood Press, 1972), pt. 2, p. 3. Lerone Bennett, Jr., *The Negro Mood and Other Essays* (Chicago: Johnson Publishing Co., 1964), p. 50.

2. Chukwulozie K. Anyanwu, *The Nature of Black Cultural Reality* (Washington, D.C.: University Press of America, 1976), p. 58. Cf. Willis W. Harman's discussion of "The Postmodern Heresy: Consciousness as Causal" in *The Reenchantment of Science: Postmodern Proposals*, ed. David R. Griffin (Albany: State University of New York Press, 1988), pp. 115–28.

3. Cornel West, *Prophesy Deliverance! An Afro-American Revolutionary Christianity* (Philadelphia: Westminster Press, 1982), p. 24.

4. Robert J. Schreiter, *Constructing Local Theologies* (Maryknoll, N.Y.: Orbis Books, 1985), p. 86.

5. See the section "Pharmacosm" in Chapter 1 and, for my introduction of

the notion of an African American "ritual cosmos," the section "God-conjuring" in Chapter 2.

6. Paul Ricoeur, "Temps biblique," *Archivio di Filosofia* 53 (1985): 29–35. Quoted from an unpublished English translation by David Pellauer.

7. Paul Ricoeur, "Toward a Hermeneutic of the Idea of Revelation," in *Essays on Biblical Interpretation*, ed. Lewis S. Mudge (Philadelphia: Fortress Press, 1980), p. 86.

8. Vernon J. Dixon and Badi G. Foster, *Beyond Black or White: An Alternate America* (Boston: Little Brown, 1971), p. 64. Dixon's reference to Jahn reads: "Since it could not be accommodated to European systems of thought, the African way of thinking was considered non-logical. Levy-Bruhl called the attitude of the primitives 'prelogical,' a term by which he meant to characterize a kind of thought which does not refrain from inner self-contradiction, a kind of thought in consequence of which 'objects, beings, phenomena can be, in a fashion unintelligible to us, both themselves and at the same time something other than themselves'. At the end of his life Levy-Bruhl renounced his theory of 'prelogicism' and thus furnished a rare example of scholarly integrity." Jahnheinz Jahn, *Muntu: The New African Culture* (New York: Grove Press, 1961), p. 97.

9. Dixon and Foster, *Beyond Black or White*, p. 39.

10. Ibid, p. 28.

11. E. Quita Craig, *Black Drama of the Federal Theatre Era* (Amherst: University of Massachusetts Press, 1980), pp. 19–30.

12. Gayraud S. Wilmore, *Black Religion and Black Radicalism: An Interpretation of the Religious History of Afro-American People* (Maryknoll, N.Y.: Orbis Books, 1983), p. 239.

13. Lerone Bennett, *The Negro Mood and Other Essays* (Chicago: Johnson Publishing Co., 1964), p. 51.

14. Norman E. Whitten, Jr., "Contemporary Patterns of Malign Occultism among Negroes in North Carolina," *Journal of American Folklore* 75:298 (October–December 1962): 311–25. Albert J. Raboteau, *Slave Religion: The "Invisible Institution" in the Antebellum South* (New York: Oxford University Press, 1978), p. 278. Newbell Niles Puckett, *Folk Beliefs of the Southern Negro* (Chapel Hill: University of North Carolina Press, 1926), p. 175.

15. Raboteau, *Slave Religion*, pp. 287–88. The passage ends with a summary statement: "Conjure could, without contradiction, exist side by side with Christianity in the same individual and in the same community because, for the slaves, conjure answered purposes which Christianity did not and Christianity answered purposes which conjure did not."

16. Michael Edward Bell, *Pattern, Structure, and Logic in Afro-American Hoodoo Performance* (Ann Arbor, Mich.: University Microfilms International, 1980), pp. 43–44.

17. Michael Bell elaborates further: "Frequently, all that separates two hoodoo transactions is intention, since the same procedures may be used to achieve disparate goals. 'As so often in hoodoo,' Hyatt observes, 'the same rite can be used for an opposite purpose.'" Bell, Ibid., p. 43.

18. Ibid., pp. 55–56.

19. Ibid., p. 45; p. 54.

20. David Walker, *David Walker's Appeal*, ed. Charles M. Wiltse (New York: Hill & Wang, 1965), p. xiv.

21. Daniel Coker, "A Dialogue between a Virginian and an African Minister,"

in *Negro Protest Pamphlets*, ed. Dorothy Parker (1810; reprint, New York: Arno Press/The New York Times, 1969), pp. 39–40.

22. Bell, *Pattern, Structure, and Logic*, p. 55.

23. Raboteau, *Slave Religion*, p. 311.

24. Walter Ong, *Orality and Literacy* (London: Metheun, 1982), p. 15.

25. Commenting on Ricoeur's category of forgetfulness, Walter Wink has pointed out that "*forgetfulness* of the radical questions enshrined in the language and conception of another time . . . is nothing other than what psychoanalysis means by selective repression." Walter Wink, *The Bible in Human Transformation* (Philadelphia: Fortress Press, 1973), p. 47. Here I can only pose for future consideration the possible repression involved in African Americans' forgetting conjurational strategies as their ancestors' most skillful performances and variations on biblical texts.

26. Claudia Mitchell-Kernan, "Signifying," in *Mother Wit from the Laughing Barrel*, ed. Alan Dundes (New York: Garland Publishing, 1981), p. 311. Charles H. Long, *Significations: Signs, Symbols, and Images in the Interpretation of Religion* (Philadelphia: Fortress Press, 1986), p. 2.

27. Schreiter, *Constructing Local Theologies*, p. 80.

28. Henry Louis Gates, Jr., *Figures in Black: Words, Signs, and the "Racial" Self* (New York: Oxford University Press, 1987), pp. 48–49. Coincidentally, we should note Gates's efforts to fashion, in his own field, a practice of "critical signification." For Gates this term designates "a form of critical parody, or repetition and inversion . . . of formal signifying," and provides him with an operational "metaphor" for his craft as a literary historian and theorist.

29. See Mitchell-Kernan, "Signifying," p. 325.

30. On intertextuality see Gates, *Figures*, pp. 41, 49. Cf. George E. Kent, *Blackness and the Adventure of Western Culture* (Chicago: Third World Press, 1972). On Du Bois's intertextuality, Wilson J. Moses observes that "DuBois' early work struggles to fuse two complementary but substantially different mythological traditions. The first of these is 'Ethiopianism' . . . the other is the European tradition of interpretive mythology . . . the medieval practice of examining Greco-Roman mythology with the intention of either discovering within it, or assigning to it, Christian meaning." Thus in his *The Quest of the Silver Fleece* Du Bois "created a universe in which the ideology of progressive socialism and the traditionalism of Christian black nationalism work harmoniously within the framework of a Greek myth." Wilson Jeremiah Moses, "The Poetics of Ethiopianism: W.E.B. DuBois and Literary Black Nationalism," *American Literature* XLVII:3 (November 1975): 411, 417.

Similarly, as already treated in Chapter 1, Hurston interweaves the shamanistic abilities of a conjuror, who explicitly conjures the plagues of Egypt and the parting of the Red Sea, in her representation of Moses as "the finest hoodoo man in the world." *Moses: Man of the Mountain* (Urbana and Chicago: University of Illinois Press, 1984), p. 147. Moses the conjuror is an Afro-American signification deriving from the slavery period, in which "Moses was understood to be a snake-controlling magic worker of great power." Mechal Sobel, *Trabelin' On: The Slave Journey to an Afro-Baptist Faith* (Westport, Conn.: Greenwood Press, 1979), p. 73. Cf. Lawrence W. Levine, *Black Culture and Black Consciousness: Afro-American Folk Thought from Slavery to Freedom* (Oxford: Oxford University Press, 1977), p. 57.

31. Gates, *Figures in Black*, pp. 236–37. For an extensive "portrait" of Eshu-Elegba as a major *orisha*, or spirit, see Robert Farris Thompson, *Flash of the Spirit:*

African and Afro-American Art and Philosophy (New York: Vintage Books, 1984), pp. 18–33.

32. Gates, *Figures in Black*, pp. 239–40.

33. Lawrence Levine, *Black Culture and Black Consciousness* (New York: Oxford University Press, 1977), p. 52.

34. Diedre L. Badejo, "The Yoruba and Afro-American Trickster: A Contextual Comparison." *Présence Africaine* 147:3 (1988): 10,15.

35. Erich Auerbach, "Figura," in *Scenes from the Drama of European Literature*, ed. Wald Godzich and Jochen Schulte-Sasse (Minneapolis: University of Minnesota Press, 1984), p. 58.

36. Don McKinney, "Brer Rabbit and Brer King: The Folktale Background of the Birmingham Project," Paper presented at the Southeastern Commission for the Study of Religion, American Academy of Religion, March 10–12, 1989, Atlanta, Georgia. McKinney was a doctoral candidate of Vanderbilt University. For other aspects of King's self-representation, see Keith D. Miller, "Composing Martin Luther King, Jr." *PMLA: Publication of the Modern Languages Association of America* 105:1 *(January 1990): 70–82,* and *Keith Miller, Voice of Deliverance: The Language of Martin Luther King, Jr. and Its Sources* (New York: Free Press, 1991). See especially Miller's treatment of King's use of biblical "typology" in his Chapter 9: "Becoming Moses."

37. Dundes, ed., *Mother Wit from the Laughing Barrel*, p. xiv.

38. Zora Neale Hurston, "High John de Conquer," in *The Book of Negro Folklore*, ed. Langston Hughes and Arna Bontemps (New York: Dodd, Mead, 1958), p. 93. Emphasis mine.

39. Ibid., p. 94. In elaborating the covert nature of such folk traditions, Hurston subsequently adds this comment about the dumbfounding of white people: "It is no accident that High John de Conquer has evaded the ears of white people. They were not supposed to know. You can't know what folks won't tell you. If they, the white people, heard some scraps, they could not understand because they had nothing to hear things like that with. They were not looking for any hope in those days, and it was not much of a strain for them to find something to laugh over. Old John [de Conquer] would have been out of place for them" (p. 94f.).

40. Robert B. Stepto, *From Behind the Veil: A Study of Afro-American Narrative* (Urbana: University of Chicago Press, 1979), p. 70.

41. Hurston, "High John," p. 94.

42. Ibid., p. 95. Note also Hurston's comparison of High John and the legendary King Arthur of the English people, on p. 96. On boasting rituals as verbal art in African American folk culture see: Roger Abrahams, "Playing the Dozens"; John Dillard, "The Dozens: Dialectic of Insult"; William Labov, Paul Cohen, Clarence Robins, and John Lewis, "Toasts"; and Mimi Clar Melnick, "I Can Peep Through Muddy Water & Spy Dry Land: Boasts in the Blues"; in Dundes, ed., *Mother Wit,* pp. 267–304.

43. Hurston, "High John," pp. 96–97.

44. M. J. Field, *Search for Security: An Ethno-Psychiatric Study of Rural Ghana* (Evanston, Ill.: Northwestern University Press, 1960), p. 40. Also quoted in Raboteau, *Slave Religion,* p. 288.

6

Prophecy

As there is a politics of medicine, there may also be a medicine of politics.
. . . Why should political systems have the last word? They can also be
symptoms and aspects of diseases. The political power of the shaman may
be identical to his methodology of cure.

Richard Grossinger, *Planet Medicine*[1]

In this chapter I examine the black American tradition of social prophetism
in terms of its shamanic and conjurational, and its therapeutic or homeopathic, elements. One of the most celebrated instances of the convergence of these elements was the Haitian revolution of 1791, when
Toussaint L'Ouverture led a slave revolt that successfully overthrew the
French. That victory featured the prominent use of Voodoo or Vodun to
secure supernatural invincibility for the Haitian combatants. It even led,
after the revolution, to attempts to establish Vodun as Haiti's state religion.
Moreover, the "conjuror's doctrine of invincibility"[2] was a distinctive feature of slave revolts in the United States as well, notably in the betrayed
1822 conspiracy of Denmark Vesey in Charleston, South Carolina. Vesey
preached from the Bible and espoused a radical Christianity on the one
hand, but also relied solidly on the conjurational practices of his cocon-
spirator, Gullah Jack. Furthermore we should recall the failed 1831 revolt
and massacre of fifty-seven whites led by Nat Turner in Southampton
County, Virginia, the bloodiest slave revolt in United States history.

Unusual as a literate slave, Nat Turner like Denmark Vesey was a
compelling Bible preacher. He was more self-consciously Christian than
his fellow slaves and eschewed the common practices of conjure as sub-
Christian. "I always spoke of such things with contempt," his interviewer
records his saying in Turner's *Confessions*. Nonetheless historical sources
represent him as a seer and a folk prophet, given to mystical experiences,
shamanic visions, and dreams. Turner himself described his visions in terms
of explicitly Christian symbolic interpretations; for example, blood found

on corn in the fields was Christ's blood "returning to earth again in the form of dew." Such interpretations supported his convictions of divine retribution against slavery: "he had a remarkable vision in which white and black spirits were engaged in a great battle with blood flowing in streams."[3]

Together Vesey and Turner represent two phases in the development of conjure as a form of "magical shamanism" (Joyner).[4] Phase one, represented by Vesey, features conjurational practices alongside or in concert with the use of biblical and Christian theological elements. Phase two, represented by Turner, features ostensibly the repression of conjure but (precisely thereby) the return of conjurational impulses via biblical symbolism and Christian theological discourse. Common to each of these phases, however, is the phenomenon of prophetic oratory that the historian Gayraud Wilmore attributes to the fusion of African spirituality and biblical religion. "With the mystical sense of prophecy and divine intervention that both laity and clergy found in the Scriptures—particularly in the apocalyptic writings—they were able to invoke a power that required no human justification, a power that gave them license to become oracles against the whites in vindication of their own people."[5] One social context for investigating this oracular phenomenon among Afro-American clergy and laity is the revival tradition. In this connection, the religion scholar Amanda Porterfield has distinguished three leadership roles in the American revival tradition generally: preaching, prophecy, and shamanic performance. Before reviewing her treatment, however, I begin with a brief reference to the work of the scholar who conducted the most exhaustive treatment of shamanism in the field of history of religions, Mircea Eliade.

Shaman

On the one hand Eliade stressed the centrality of shamanism in the traditional cultures where observers first came to recognize it as a distinct religious phenomenon: shamanism is foremost the archaic religion of Siberia and Central Asia (the provenance of the word is Russian). On the other hand Eliade acknowledged that "similar magico-religious" phenomena elsewhere may be "thoroughly shamanic" without exhibiting the central place that shamanism holds in the forementioned cultures. In particular his short definition of shamanism, "archaic techniques of ecstasy," has been implicitly employed to indicate certain African religious types by the Africanist scholar Luc De Heusch.[6] Also pertinent here is the historian Charles Joyner's use of the term "magical shamanism" to apply to conjure tradition in African American culture.

> Amanda Porterfield has examined the shamanic character of revival preaching among white and black preachers in North America. Porterfield criticizes Eliade for neglecting the "historical particularity" and "psychosocial functions" that characterize archaic religions. She cites his alternative interest in articulating a "transcendental conception" of spiritual beings

and processes in which they manifest "the sacred order behind ordinary reality."[7] For reasons that will become clear, Porterfield's functionalist approach is conducive to my examination of the role of prophetic and shamanic leadership in African American religious and social history.

Revival preaching, since the frontier revivals of the eighteenth and nineteenth centuries in the United States, has been vividly documented as a type of ecstatic performance. While acknowledging the ecstatic nature of revivalism, however, Porterfield offers a "psychosocial definition" of shamanism that enables her to distinguish the shamanic aspects of revival < preaching from its prophetic or moral aspects. Especially notable is her insistence on the embodied nature of the preaching. As Porterfield observes, revival preaching

> often combines the prophetic activity of making moral pronouncements with the shamanic activity of representing human dilemmas in bodily gesture. Some preachers go into trances that enable them to act out the intense pain and hope that is represented by their symbols of sin and redemption.... The compelling power of these preachers is in large part the result of their ability to dramatically embody the emotional problems and social tensions besetting their patrons.[8]

I have already treated the embodied nature of Afro-American spirituality in discussing folk appropriations of biblical cosmology. This feature has also been acknowledged in recent works by the theologians James McClendon, Jr., and Archie Smith, Jr., specifically with reference to ethics and therapy in black American Christianity. Most observers tend to focus on one or the other of these aspects. But it should be understood that "body ethics" (McClendon) and psychotherapy are integral in black social prophetism. Correlating these distinct fields is the project undertaken by Smith in his effort to "think ethics and therapy together" in the context of black church traditions. Smith claims as antecedent to those traditions < an ancestral "African character" that typically combines moral action and therapeutic practice. More immediately available for illustration however is his focus on the revival tradition in black churches. "Revivals [are] social and psychological therapy for the participants ... [providing] a form of psychic release and healing, social cohesion, and a sense of communion with the Divine ... [and linking] therapeutic expressions with the moral life."[9] Like Porterfield, it is evident, Smith focuses on the psychosocial aspects of revival religion. In addition he calls attention to the revival preacher. But where Porterfield separates the preacher's shamanic or therapeutic expertise from a prophetic or priestly function, Smith emphasizes their integral relationship in black experience.

Thus Smith represents the preacher as a folk "therapist" who is concerned not only with the psychological and spiritual wholeness of his congregation. The traditional black preacher has also felt commissioned to address the congregation's collective moral integrity, alongside its external strength in the form of social-political empowerment and freedom.

> Not infrequently, these compound tasks have united in one person the role of a theological ethicist, a social activist, and a therapeutic practitioner in the community.[10] We may speculate that, conversely, other social character types in black communities have shared the preacher's shamanic role—not only ethicists and activists but also other charismatic figures, including orators and musical performers. These social characters can also claim to induce psychosocial transformations in their audiences and constituencies. Their leadership can be regarded as shamanic insofar as their performances constitute "a symbolic means of addressing psychological and social conflict . . . [that aims] to control or remedy specific problems or bring about a transformation of reality that improves upon existing conditions" (Porterfield).[11] Shamanism in these terms may be observed not only in revival preaching but also in social revitalization phenomena like the 1960s freedom movement in the United States.

"Revitalization movements" typically feature communal healing as well as social-political reform. Archie Smith explicitly links these features by describing, as a "therapeutic relationship . . . a change in oppressive relational patterns which relieves the suffering of the oppressed and may result in constructive and supportive relations in society."[12] Such developments have occurred in black social history, Smith indicates, in contexts that range in scale from family systems, small groups like churches and self-help groups, and neighborhood organizations, to large group settings and mass social efforts like bus boycotts, labor strikes, organized demonstrations, and other forms of protest and revolutionary struggle. In these diverse contexts one can observe not only distinguished individuals but also groups that function as agencies of psychosocial cure. Here we may refer particularly to the cure of disorders generated by the oppression of racism.

> Such disorders are experienced by any social group subjected to prolonged mistreatment by a more dominant contact group. Deteriorating self-esteem, internalized feelings of inferiority, and intergroup mistrust are among the demonstrable results of such mistreatment. Insofar as revitalization movements are not only political in focus, but also therapeutic in nature, they can additionally address these interior disorders. In this regard it is instructive to note that shamans have often been the subjects of their own curative abilities.[13] This feature is corroborated in the life story of the exemplary shamanic figure discussed in the concluding sections of this chapter.

Sojourner

"You know, children, I don't read such small stuff as letters, I read men and nations."

Sojourner Truth[14]

Sojourner Truth (c. 1797–1883) was born with the name "Isabella" on a Dutch estate in Ulster County, upstate New York. She was born a slave some twenty years after the American Revolution of 1776 and would have been freed at the age of thirty according to the state emancipation laws. But in the emancipation year of 1827 her current master insisted that she serve yet another year. Isabella boldly resisted by running away with her infant daughter. Later she also reclaimed a son who had been illegally sold into Alabama, appealing in court to win his return. But Isabella was not always so zealous for freedom. As a young slave woman she evinced the typical marks of a victim's internalized domination. There was a time, if we are to credit her biographer, when "she looked upon her master as a God; and believed that he knew of and could see her at all times, even as God himself."[15] That claim is compatible with Isabella's tortured decision ∠ at one point to return herself and her child to bondage voluntarily. Her biographer records Sojourner's mature self-criticism in figural terms: "She says she 'looked back into Egypt.' " Fortunately there intervened a cataclysmic religious experience that served to bar her return to bondage.

> God revealed himself to her, with all the suddenness of a flash of lightning, showing her, "in the twinkling of an eye, that he was *all over*"—that he pervaded the universe—"and that there was no place where God was not." She became instantly conscious of her great sin in forgetting her almighty Friend and "ever-present help in time of trouble." All her unfulfilled promises arose before her . . . and her soul, which seemed but one mass of lies, shrunk back aghast from the "awful look" of Him.[16]

Although she did not return to slavery, Isabella's new religious life featured its own form of human bondage. I refer to her experience with a communitarian religious group described by one biographer as "the Matthias Delusion." The recently emancipated young woman was imperiled when after reports of other abuses and indiscretions, the group's leader, Matthias, was charged with (although cleared of) poisoning his associate. Accusation fell on Isabella also; as one of the most devoted of Matthias's followers she was suspected of being an accomplice and even a witch, but her innocence was vindicated. Again we must rely on her biographer, who describes this episode as one among others in which Isabella confronted a dangerous tendency toward self-delusion. "Having more than once found herself awaking from a mortifying delusion,—as in the case of the Sing-Sing [Matthias] kingdom,—[she resolved] not to be thus deluded again."[17] Isabella's religious development reached a climax in 1843 when she realized the vocation to preach. At that point she additionally acquired a new name. It is significant that her account of the event features the figural language typical of slave religion.

"My name was Isabella," the mature preacher recalled. "But when I left the house of bondage, I left everything behind. I wasn't going to keep nothing of Egypt on me, and so I went to the Lord and asked him to give me a new name."[18] Here we should recognize that this event in the freedwoman's life bears the marks of a personal transformation. The transformation is articulated in the figural discourse of Exodus, as a change from the condition of bondage to that of freedom. But the change also corresponds to the biblical pattern of a prophetic call or vocation. Indeed we do well to observe, in the change of name from Isabella to Sojourner, a quality of existential transformation similar to that of the young Saul in 1 Samuel 10.6—yet without any report of similar ecstatic effects: "the spirit of the Lord will come mightily upon you, and you shall prophesy ... and be turned into another man." As other biblical texts also report, the divine call to prophetic vocation is accompanied by a depth transformation in which, as in Isabella's case, one is 'turned into another woman'. Indeed, by her change of name Isabella 'signified on' the Bible's prophetic call tradition.

> And the Lord gave me Sojourner, because I was to travel up and down the land, showing the people their sins, and being a sign unto them. Afterward I told the Lord I wanted another name, because everybody else had two names; and the Lord gave me Truth, because I was to declare the truth to the people.[19]

One of the ways in which Sojourner traveled 'up and down the land declaring the truth to the people' has already been observed: She was convinced that God wanted the United States government to set aside lands in Kansas and the western states for the newly emancipated persons. Accordingly she not only prayed for that social-political transformation in the present, but also lobbied in Washington and advocated this plan throughout the nation. The terms of Sojourner's conviction deserve repeating: "I have prayed so long that my people would go to Kansas, and that God would make straight the way before them. Yes, indeed, I think it is a good move for them. I believe as much in that move as I do in the moving of the children of Egypt [*sic*] going out of Canaan [*sic*]—just as much.[20] In this regard the prophet's own mobility mimetically dramatized the content of her prophetic message. She commenced an uncommon career for a nineteenth century black woman and former slave, of travel, oratory, and political advocacy. Her promotion of a massive national land grant proposal stands in clear contrast to her former sequestering in a religious sect, in a community represented by her contemporaries as mystical and delusionary. Insofar as Sojourner had cause to resolve never to be deluded in such a way again, we may speculate that her inspired social activism provided a kind of spiritual antidote to religious excesses.

In any case, the shamanic character of Sojourner's self-presentation can be observed by attending to the signifying import of her name, "sojourn(ing) truth." That name is also emblematic of her prophetic and political vocation. Indeed, James M. Glass has correlated the two in his formulation of political vision as shamanic when it employs incantatory signs in the therapeutic effort to address a "diseased situation."

> What the shaman does, how he enters into a diseased situation, depends on his capacity to construct "signs," to devise an incantation that will reach the unconscious...[an ability] as critical to the political vision."[21]

The incantatory use of her own name was the initial performance that Sojourner Truth undertook. Her name was a sign and conveyed a signifying intention. (Recall the importance of naming in the enactment of conjurational intention.) In this regard it is evident that Sojourner did not promote her Kansas project primarily as a political orator or a social activist. Rather, her self-presentation was that of an inspired visionary. "All she does and says is, as she believes, inspired by the Almighty, and she connects with his direct agency the events and circumstances which surround and control her."[22]

Most notably Sojourner was represented by her contemporaries as a prophetic seer or "sibyl." Precisely that term was popularly applied to her as a result of the efforts of a famous acquaintance, the abolitionist and novelist Harriet Beecher Stowe. In 1863 Stowe published, in the *Atlantic Monthly*, an article, "Sojourner Truth, the Libyan Sibyl."[23] The article recounted the visit of Truth with Stowe and her celebrity brother, Henry Ward Beecher. Their visitor favored them with an extensive review of her life as Isabella and of her career as "sort of a self-appointed agency through the country." Stowe's article included, alongside a description of the visit, Sojourner Truth's own powerful autobiographical synopsis followed by the author's reflections on the significance of her visitor's life. "When I went into the room," Stowe recollected at the beginning of her article, "a tall, spare form arose to meet me. She was evidently a full-blooded African, and...I do not recollect ever to have been conversant with any one who had more of that silent and subtle power which we call personal presence than this woman. In the modern spiritualistic phraseology, she would be described as having a strong sphere."[24] Certainly Sojourner gave observers adequate cause for such impressions as Stowe has recorded. A striking example is the quotation prefacing this section in which, speaking as an illiterate freedwoman, Sojourner claimed to "read men and nations" rather than letters or books. Deciphering and interpreting the personal, social, and historical signs of existence was a key concept in Sojourner's self-understanding of her shamanic ability.

However, it is also evident that Stowe's perception of Sojourner was < colored by a romantic conceit, typical among certain abolitionists of her time, in which African peoples are represented as more spiritual or mystical than white Europeans and Americans. "The African seems to seize on the tropical fervor and luxuriance of Scripture imagery as something native," she declared. "He appears to feel himself to be of the same blood with those old burning, simple souls, the patriarchs, prophets and seers, whose impassioned words seem only grafted as foreign plants on the cooler stock of the occidental mind."[25] The literary historian Wilson Jeremiah Moses has described this perspective as a pernicious and stereotypical view of black peoples. It is stereotypical by the way in which it indiscriminantly casts all black people as somehow naturally religious. It is all the more pernicious because of the deceptively positive way in which it relegates them to the vulnerable virtues: to virtues susceptible to domination within a rapacious culture, to "the supposed characteristics of the black race— gentleness, loyalty, and the enduring power of unshakable faith." Such charges apply particularly to Stowe's representation of black "Christian heroism" in her immensely popular abolitionist novel, *Uncle Tom's Cabin* (1852), to which I turn in the following chapter. By projecting such overdetermined characteristics onto black Americans, argues Moses,

> the African race was thus assumed to have a natural predisposition to the acceptance of Christian principles. Stowe believed that the millennium would be ushered in by Christianized Africans leading their race into the messianic era. When Ethiopia stretched forth her hand unto God, the entire human race would be uplifted.[26]

These views were compelling for Stowe's peers (black and white), one of whom labeled Sojourner Truth "the colored American sibyl." But note also that writer's added caveat: that Sojourner was not only "one of the most remarkable women of our time—a true sibyl, as Mrs. Stowe calls her, but a Christian sibyl, and more devoted to good words and works than to obscure predictions."[27] That distinction, although stated as a Christian identity determination, implicitly indicates the biblical formation of Sojourner Truth as a prophetic figure in continuity with Jewish and Christian Scriptures. It implies that the earlier tendencies of Isabella toward restrictive or privatized forms of mysticism have been overcome in Sojourner's activist attention to social and historical realism. Yet there is also continuity in this new identity with the younger Isabella, for Sojourner became neither an obscurantist nor simply an activist but was recognized as "a true sibyl"—thus linking her past religious aspirations with a present fulfillment.

The final irony, however, is that Sojourner's new name did not signify a fulfillment of personal aspiration in the figural terms that she herself so often invoked: the Exodus deliverance of slaves and their entry into a Promised Land. Rather, the new name evoked the biblical figures of Wilderness or exile; her persona is reconfigured as that of a wanderer or pilgrim lacking permanent residence. Thus "sojourner" signifies one who actually lacks her promised land. It thereby evokes more appropriately—indeed, more prophetically—that generation of wilderness wanderings that preceded Israel's entry into the promised land of Canaan. Such an evocation was prophetic in context despite Isabella's personal exodus from bondage, an exodus shared with the other slaves of New York State. As a recapitulation of biblical narrative, a generation of thirty-five years intervened between her name change and Lincoln's presidential order making all of her people free. In the interim, the prophet labored as a "sojourner" obliged to go from place to place to speak the "truth." With this formulation, however, we are now positioned to conclude this chapter's larger task by exploring the mimetic and homeopathic dimensions of Sojourner's "truth."

Conjure-woman

Sometimes a whole people needs healing work. Not a tribe, not a nation. Tribes and nations are just signs that the whole is diseased. The healing work that cures a whole people is the highest work, far higher than the cure of single individuals.

Ayi K. Armah, *The Healers*[28]

Mimetic elements are featured crucially in the three roles that Sojourner's prophetic career comprised. Taken together these roles span the issues of < ethnicity, gender, and class. The social context for her multiple activist roles was also threefold: the antislavery movement, the women's rights movement, and the land grant project for disowned and destitute freed persons. Moreover, a mimetic tension involving Sojourner's identity operated in each of these contexts of social change and activism. The first two areas present an irony: tension with regard to her ethnicity was displayed in the context of a women's rights meeting, whereas a contrasting tension regarding her gender arose at an antislavery event. Of course, as emancipatory issues these movements were intimately related, but the politics of ethnicity and gender in the nineteenth century led many participants to keep the movements as separate as possible. In particular popular sentiment indicated that a women's rights movement that appeared to be in sympathy with the antislavery movement would fail. Many white women were zealous, therefore, to avoid the appearance of linking their own civil rights issue with abolitionism. Sojourner's (conjunctive) folk wisdom, however, would not allow her to maintain such myopia.

In 1851 Sojourner Truth turned up at the women's rights convention in Akron, Ohio. The meeting was presided over by the established abolitionist and women's rights activist Francis Dana Gage. Gage, heedless of the possibility that the convention might be maligned in the popular press as sympathetic to or (even worse) a covert sponsor of abolitionist agendas, granted the floor to the only black woman participant. The rhetorical skill displayed in Sojourner's brief speech shows that she was fully aware of the mimetic tension of the moment. The crisis impacted her female peers, as well as her evidently contemptuous male listeners. It consisted in their shared resistance to her presence among them, because she was not (mimetically—the same as, equal to, or like) a white woman. Her speech, excerpted here, implies an awareness of that audience resistance, an awareness that her listeners were likely to deny her gender equality with the white women present, by way of despising her ethnic identity and class status. But Sojourner refused to comply with this disjunctive thinking (one is either white or not fully a woman), this deformed cog-

nition. Rather, she resolved the mimetic tension by insisting on the audience's re-cognition of her identity as (conjunctively) both black and
female. In the following excerpt we see how she simultaneously spoke to
the issue of her ethnicity (and implicity her social class), even as she
addressed the overarching issue of gender rights.

> (1) That man over there says women need to be helped into carriages,
> and lifted over ditches, and to have the best place everywhere. Nobody ever
> helps me into carriages, or over mud-puddles, or gives me any best place!
> And ain't I a woman?
> (2) Look at me! Look at my arm! I have ploughed, and planted, and
> gathered into barns, and no man could head me! And ain't I a woman? I
> could work as much and eat as much as a man—when I could get it—and
> bear the lash as well! And ain't I a woman?
> (3) I have borne thirteen children, and seen them most all sold off to
> slavery, and when I cried out with my mother's grief, none but Jesus heard
> me! And ain't I a woman?"[29]

This feat of discourse has been preserved, and is still celebrated among
feminists, because it powerfully addressed the interrelated issues of ethnicity, gender, and class for nineteenth century women activists. As a tour
de force of signifying speech it marks a moment of solidarity for black,
white, and poor women in the face of social forces that continue to divide
them. The quest for such solidarity may be underscored, rather than compromised, by recent scholarly judgment that identifies Frances Gage as the
inventor of the now classic refrain: "And ain't [ar'n't] I a woman?"[30] If
pursued, this revision would require us to consider the implications of
signifying discourse in white American traditions also, just as we considered the possibility of conjurational practices among Puritan Americans
in Chapter 3. In any case this fragile moment was achieved by signifying,
through images of her marginalized womanhood, Sojourner's equality as
a woman among white women. The very images that worked certain
negative cognitive perceptions, because of which her peers discounted her
womanhood, were employed in a rhetorical framework that reinstated or
re-cognized her womanhood. This was a homeopathic performance therefore, because it mimicked the "diseased situation" (Glass) of racist cognitive
perceptions, precisely in order to counter those perceptions and to cure
that disease. The following is a simple schema suggesting how this rhetorical performance achieved its effect.

(1)   image--------------nobody helps me into carriages, over ditches, in best places
      perception--------this is not a 'woman of quality'
      re-cognition------"And ain't I a woman?"

(2)   image--------------I have worked like a man, ploughing, planting, harvesting,
                         hearty eating
      perception--------this is not a 'woman of property'
      re-cognition------"And ain't I a woman?"

(3) image--------------I have lived a slave's life with lashings, imposed slave (child)
                              breeding, forced unions and separations
       perception--------this is not a free/white/virtuous woman
       re-cognition------"And ain't I a woman?"

By definition, homeopathic practices consist in using a substance or <
process to affect a like substance or process. In the therapeutic context of
treating illness and prescribing medicines from a pharmacopeia, the practitioner typically uses a skillfully prepared or modified dosage of a disease
in order to cure the disease. (In the signal case of immunizations, the
intention is to mimic the disease in a manner that skillfully engages the
body's natural defenses without allowing the disease a full range of operation.) In the preceding schema we see first skillfully crafted images of
Sojourner deployed in accordance with a homeopathic intent. The images
effectively corroborated her deviation from the dominant culture's conventional model of womanhood. Thus they reinforced the (mis)perception
that she was somehow unwomanly. Subsequently, however, the author
'signified on' the images ("And ain't I a woman?") in a way that exposed
them to be distorted cognitive perceptions. The affected listener, or reader
confronted with her or his deformed view, reforms that view accordingly
and so accedes to a transformed perception of Sojourner's identity and
humanity. I defer for the moment the case of the disaffected observer who
remains impervious to such homeopathic craft and signifying skill.
    A second, related mimetic tension emerged when Sojourner Truth
was directly accused of being a man and impersonating a woman. The
charge surfaced on the occasion of her conducting an antislavery tour
through Indiana in the late 1850s. It is possible that more than once such
distorted perceptions fastened onto the tall, imposing physical presence
and forceful personality of this singular black woman. The accusation of <
gender impersonation was delivered in particular by a white physician
serving as spokesperson for the proslavery faction in that community. His
ostensible strategy was to discredit Sojourner's integrity and veracity as a
public speaker by demonstrating that her presumed gender identity was
a pretense: the doctor specifically called into question the speaker's voice,
charging that it was the voice of a man. An equally likely (if not more
plausible) objective was simply to humiliate her in the public view: he did
so most offensively by challenging her to expose her breast to the women
present so that they could assure the entire audience of her womanhood.
    In apparent compliance Sojourner proceeded, homeopathically, to
submit to humiliation precisely in order to countermand humiliation. She
did not, however, turn aside to the other women: she refused to draw
them into the doctor's net of her humiliation by forcing them to become
accomplices of another woman's public exposure. In homeopathic terms,
she refused to contaminate them by means of an involuntary, mimetic
indentification with another woman's shame. Instead, she took the shame <
solely to herself and countered it by a manifestly free act. Acceding to the

man's demand but contrary to his directive, she insisted on baring her breast to the entire audience. Thereby Sojourner used her own ostensible humiliation to humiliate even more substantially her antagonist. The homeopathic performance was consummated with a paradoxical signification directed at the doctor: "It is not my shame but yours that I do this."[31]

Each of these episodes is a brilliant instance of Sojourner's performative skill and signifying virtuosity. However, in context they were focused primarily on her individual person as an example or, at best, an exemplar. Observers were left on their own to extend the implications of the performance to apply to other issues, or (unfortunately) to fail to do so. Accordingly, from the point of view of a practitioner who seeks ever more rational means and effective strategies for her ritual performances, additional feats are desirable. It is additionally desirable for practitioners of such transformations to acquire the following improvisational skills: (1) to craft diverse performances featuring similar recognitive breakthroughs with respect to multiple conflicts (including and exceeding the issues of ethnicity, gender, and class); (2) to extend the range of such breakthroughs as far as possible, in order to increase the numbers of affected participants and, by transforming their (mis)perceptions, 're-cognize' the humanity of as large a population as possible; and finally (3) to render such breakthroughs increasingly less momentary and fragile, more enduring and compelling for greater numbers of affected participants.

The need for such improvisational extensions became more acute and demanding in the context of two homeopathic crises that Sojourner faced in her career. The mimetic tensions previously examined were focused, as already acknowledged, on Sojourner's individual gender, class, and ethnic identities. But the resolution of those tensions was elementary (a mere exercise) in comparison to the performative requirements that confronted her on the large scale of national events. From a different perspective we have already looked at one national arena of Sojourner's performance— the massive land grant project to relocate displaced and destitute freed persons. I previously treated Sojourner's advocacy of this plan in terms of prophetic inspiration and deferred fulfillment. In a moment I want to treat the matter from the perspective of her shamanic and homeopathic craft; but first I take up an earlier large scale national crisis that confronted her in the abolitionist movement: the possibility of massive slave revolts and armed insurrections on the eve of the Civil War. That possibility was imminent in 1859, when Sojourner Truth took part in her now-famous exchange with the celebrated former slave and abolitionist orator Frederick Douglass.

Frederick Douglass had been instrumental in recruiting Sojourner for the activist career of an abolitionist. However, the two activists sustained a prickly relationship. Douglass once described her in an amusing, if also aggrieved, manner as someone "who seemed to feel it her duty to trip me up in my speeches and to ridicule my efforts to speak and act like a person of cultivation and refinement."[32] On the occasion of their 1859 encounter at an antislavery meeting in Salem, Ohio, Douglass (the senior lecturer and featured speaker) had been engaged in ongoing conversations with the militant visionary and slave liberator John Brown. Indeed, Douglass was subsequently implicated in John Brown's failed raid on the federal arsenal at Harper's Ferry, Virginia (October 16, 1859), for which Brown and his captured coconspirators were executed after one of the most notorious slave revolts in American history. Douglass himself had to flee north to escape prosecution, despite the fact that he had declined Brown's entreaties to participate in the desperate venture. But that night in Salem, Douglass already knew enough to caution his supporters against expecting to succeed in their stated policy of a peaceable and nonviolent resolution of the slavery issue. Indeed, the orator spoke in fateful tones of coming calamities. He achieved his noted spellbinding effect, and his dire predictions of bloodshed especially affected Sojourner Truth. In the morbid atmosphere and dense melancholy of the moment, the "quaint old black woman, gaunt and tall, her head tied in a white turban over which she wore a field hand's sunbonnet, rose in agony and cried out, 'Frederick, is God dead?' Remembering John Brown, the speaker paused, then answered positively, 'No, dear sister, God is not dead, and because God is not dead slavery can only end in blood.'"[33]

Sojourner evidently did not 'have the last word' over her occasional rival that night. However, her words have been preserved beyond that event in the epitaph on her tombstone: "IS GOD DEAD?" In fact, she was best known by Americans in the nineteenth and early twentieth centuries for that (theotropic) interjection. It would be intriguing to compare the popularity and signifying import of the epitaph with the more recent celebration by feminists of that equally compelling (anthropotropic) interrogative, "And ain't I a woman?"[34] For earlier generations of Americans, presumably, "Is God dead?" conveyed a more compelling signifying force. It appears to have surmounted Douglass's rejoinder and remained the definitive message of that dark night in 1859. On this view the interrogative effectively 'signified on' Douglass by interjecting, into his all-too-human apprehension of reality, the reality of divine power and unfettered possibilities. However, from the perspective of preceeding analysis in this chapter, "Is God dead?" is a query that marks the limits of Sojourner's shamanic vocation in the face of imminent violence and bloodshed. There are two indications of shamanic limitations here. On the one hand the words lack homeopathic force or potency because they did not enact or otherwise represent a form of the disease that they were intended to cure. Sojourner's sometime rival did indeed 'have the last word', both that night and in history, with his own signifying assertion: "because God is not dead slavery can only end in blood." I want to address these criticisms in turn.

The absence of performance is one element rendering this event unlike the previously examined incidents, where Sojourner not only 'signifies on' her antagonists but also enacts her significations: "Look at my arm," she demands while displaying her person in the first incident; "It is not my shame but yours," she declares as she exposes her breast in the second.

> As we have seen throughout this study, concrete practices and embodied performances are crucial for conjurational and other ritual practices in African American religious traditions. The performative element is indispensable (a *conditio sine qua non*) for shamanic operations generally, and for the conjurational and shamanic aspects of black social prophetism in particular. It is true that Sojourner's query "Is God dead?" signified on Douglass's fatalism precisely by claiming to counter it. However, in the absence of a performance, enactment, or demonstration sufficient to persuade observers to the contrary, Douglass's countersignification is equally if not more compelling: the formula, 'God exists'; therefore 'there must be bloodshed' (where justice means retribution), is all too axiomatic for religious traditions throughout human history.

The mimetic aspects of the "Is God dead?" episode can be stated here in a preliminary way. For that episode confronted Sojourner with the same homeopathic crisis that she subsequently encountered in the land grant project: how to use her own embodied persona to address or cure a "diseased situation" at the mass level of the nation? We have already seen that, like a biblical prophet, Sojourner signified with her name the divine intention to circumscribe the nation in judgment. To reiterate that intention in her own words: "The Lord has made me a sign unto this nation, and I go around testifying and showing them their sins against my people."[35] Thus a specifically figural or tropic representation of truth is rendered by the name, "sojourn(ing) truth." The very name signifies her individual destiny—wandering the cities and lecture halls to denounce the evil, and advocate the abolition, of slavery. A larger frame of reference, however, would include not only her personal or biographical experience but its ethnographical or ethnohistorical significations. Indeed, "sojourn(ing) truth" acquires additional signifying force in the corporate context of black American experience. That experience has been configured in diverse and overlapping ways by hermeneutic or typological appropriations of biblical narrative. In concert with such a hermeneutic I have acknowledged the biographical reference of Sojourner's name as a proleptic signifier of her midnineteenth century personal experience. In addition I would propose "sojourn(ing) truth" as a prophetic signifier of collective black experience, both before and after Emancipation.

There were also present in this woman's career, coterminous and complementary with her biblical prophetism, certain shamanic and conjurational intentions which her peers acknowledged by labeling her a sibyl. This 'sibylline' persona is implicit in her words "the Lord has made me a sign," in which Sojourner indicated her awareness of personally embodying the truth of divine intention in nineteenth century America. Her shamanic
> features correspond to those denoted in Porterfield's psychosocial definition of shamanism: shamanism is personally *embodied* symbol production

for the purpose of psychological and social *conflict resolution*. As a shaman, Sojourner Truth offered her activist life to cure the situation of black people's wandering homeless and jobless from southern farms and plantations. As an embodied "agency" (Stowe) of that cure she sought to heal the situation of freed persons expelled by economic and political disfranchisement under retrenched southern legislatures. But her curative practice was more than metaphorical. It was also homeopathic: she used a 'dosage' of a disease to cure the disease. She used her own, individual, voluntary, < and limited form of sojourning, as a cure for her clients' corporate, imposed, and exhaustive forms of sojourning. By the praxis of her own travels and in collaboration with other activists, she acted as a counterforce to the condition of sojourning that drove black Americans out of the South— to Kansas in her own lifetime, to the urban North at the turn of the century—still seeking a promised land.

Nevertheless, Sojourner's homeopathic performance in the land grant project failed. It failed because it was not a sufficiently precise form of the disease itself. That is, it did not adequately mimic the real psychosocial disorder, of which the homelessness and dispossession of the freed persons were mere symptoms. The real disease of their nineteenth century expe- < rience was not their transitional experience as sojourners, but their perennial status as scapegoats or *pharmakoi* in United States history and culture. That is the prevailing status that has plagued black experience in the New World since the advent of the Atlantic slave trade, the status that proved intractable after Emancipation and Reconstruction, and into our own century. The scapegoat status of the slaves also constituted the crucial issue on that night in 1859 when, in the encounter between Sojourner Truth and Frederick Douglass, it appeared that the only way to end slavery was to sacrifice the lives of slaves in a bloodbath of armed insurrections. As a matter of historical record other victims were found to sacrifice their blood in creating a new social and economic order for the nation through civil war. They were the laboring and poor classes of men: men who are always available, as soldiers, to negotiate with their lives the new political order that their governments fail to produce by other means. There are always more scapegoats available somewhere.

Thus the treatment of the prophetic and shamanic career of Sojourner Truth concludes within a larger context than that of African American prophetic experience alone. For addressing this larger context, however, Sojourner has bequeathed us both a solution and a problem. On the one hand she demonstrated the power of homeopathic performance to address diseased situations involving issues of ethnicity, gender, and class. On the other her practices seemed powerless at the level of massive psychosocial transformation. The problem of how to bring the 'truth' of her successful < homeopathic performances to bear at this broader level of transformation is the way in which she 'sojourns' with us into the twentieth century and the foreseeable future. In this way of sojourning, moreover, her vocation indicates the promise of black social prophetism for human transformations on a global scale. How can prophetic performances that have been crafted to address specific issues of racism, sexism, and class oppression, and that commonly feature homeopathic performances, be generalized to address the scapegoating of human beings in any way whatsoever? Successful response to that question is the most far-reaching promise of black American prophetic performances. In its resolution consists the most rewarding future for black social prophetism.

Notes

1. Richard Grossinger, *Planet Medicine: From Stone Age Shamanism to Post-Industrial Healing* (Berkeley, Calif.: North Atlantic Books, 1980), p. xv. Cf. the explicit and suggestive subheadings in David Loye's treatment of United States race relations, *The Healing of a Nation* (New York: W.W. Norton & Co., 1971): "The Years of Sickness and the Search for Therapies"; "The Healing Urgencies"; "The Prescriptions of W.E.B. Du Bois"; "Social Physicians and the Political Imperative."

2. Sterling Stuckey, *Slave Culture: Nationalist Theory and the Foundations of Black America* (New York: Oxford University Press, 1987), p. 50. On Vodun as the state religion of Haiti see Gayraud S. Wilmore, *Black Religion and Black Radicalism: An Interpretation of the Religious History of Afro-American People* (Maryknoll, N.Y.: Orbis Books, 1983), p. 23.

3. For these representations from Turner's *Confessions* and visions see: Wilson Jeremiah Moses, *Black Messiahs and Uncle Toms: Social and Literary Manipulations of a Religious Myth* (University Park: Pennsylvania State University Press, 1982), p. 64; and also, Wilmore, *Black Religion and Black Radicalism*, p. 67. For a full text of the "The Confessions" with commentary see Herbert Aptheker, *Nat Turner's Slave Rebellion* (New York: Humanities Press, 1966.) The reader should note the reluctance of many scholars to attribute the words of the *Confessions* directly to Turner himself, because of the circumstances of the interview: an unsympathetic white interviewer, Thomas R. Gray, receiving the confession of a black man awaiting execution at the hands of a frenzied white populace.

4. Charles Joyner, *Down By the Riverside: A South Carolina Slave Community* (Urbana and Chicago: University of Illinois Press, 1984), p. 000.

5. Wilmore, *Black Religion and Black Radicalism*, p. 6.

6. Mircea Eliade, *Shamanism: Archaic Techniques of Ecstasy* (Princeton: Princeton University Press, 1964). See the discussion of "shamanic possession" in Luc De Heusch, *Why Marry Her? Society and Symbolic Structures* (Cambridge and New York: Cambridge University Press, 1981), passim.

7. Amanda Porterfield, "Shamanism: A Psychosocial Definition," *Journal of the American Academy of Religion* 55 (1987): 722, 735.

8. Ibid., pp. 728–29. Perhaps the best example of American revivalism as ecstatic religion is the celebrated Cane Ridge (Kentucky) revival in 1801, which was called "the greatest outpouring of the Spirit since Pentecost." For this comment and a vivid description of bodily behaviors see Sidney Ahlstrom, *A Religious History of the American People* (New Haven: Yale University Press, 1972), pp. 432–35.

9. Archie Smith, Jr., *The Relational Self: Ethics and Therapy in a Black Church Perspective* (Nashville: Abingdon Press, 1982), p. 76. On "body ethics" and spe-

cifically "black religion as embodied ethics" see James McClendon, Jr., *Ethics*, vol. 1: *Systematic Theology* (Nashville: Abingdon Press, 1986), pp. 78–10.

10. For a contemporary renewal of these multiple, convergent interests see Henry Mitchell and Nicholas C. Lewter, *Soul Theology: The Heart of American Black Culture* (San Francisco: Harper & Row, 1986).

11. Porterfield, "Shamanism," pp. 725–26.

12. Smith, *Relational Self*, p. 75. On revitalization movements see Anthony F. C. Wallace, "Revitalization Movements," *American Anthropologist* LVIII (1956): 264–81. For a recent and groundbreaking discussion of a praxis combining personal with social transformation see Erica Sherover-Marcuse's outline "A Practice of Subjectivity" in her *Emancipation and Consciousness: Dogmatic and Dialectical Perspectives in the Early Marx* (New York: Blackwell, 1986), pp. 135–42.

13. Particularly with reference to psychic conflicts it is instructive to note that shamans have often been the subject of their own curative abilities. To repeat Potterfield's emphasis we may say that the shaman personally embodies his or her own therapy. Studies of the shaman as a neurotic, and shamanic insight as analogous to schizophrenia, have stressed the pathological aspect of the shamanic personality. But balanced observers also stress the curative dimension: "As Eliade puts it, the shaman is not merely a sick man [*sic*] but a sick man who has been cured and has become an agency of curative powers." Dennis Hughes, "Shamanism and Christian Faith," *Religious Education* 71 (1976):395–96. On the psychotic or neurotic aspects of shamanism see I. H. Boyer, "Remarks on the Personality of the Shaman," *The Psychoanalytic Study of Society*, ed. by W. Muensterberger and S. Axelrad (New York: International Universities Press, 1962); George Devereaux, "Shamans as Neurotics," *American Anthropologist* 63 (1961):1088–90; A. L. Kroeber, *Psychosis or Social Sanction: The Nature of Culture* (Chicago: University of Chicago Press, 1952), pp. 310–19; Julian Silverman, "Shamans and Acute Schizophrenia," *American Anthropologist* 69 (1967):21–321.

14. Bert James Loewenberg and Ruth Bogin, eds., *Black Women in Nineteenth Century American Life* (University Park: Pennsylvania State University Press, 1976), p. 239.

15. Olive Gilbert, *Narrative of Sojourner Truth; with a History of Her Labors and Correspondence Drawn from Her "Book of Life"* (Battle Creek, Mich., 1878; reprint ed., New York: Arno Press/The New York Times, 1968), p. 33.

16. Gilbert, *Narrative*, pp. 64–65. Compare similar experiences in the collection of Afro-American conversion stories titled *God Struck Me Dead: Religious Conversion Experiences and Autobiographies of American Slaves*, eds. Clifton H. Johnson and A. P. Watson (Philadelphia: Pilgrim Press, 1969).

17. Gilbert, *Narrative*, p. 122. For a concise description see the chapter on "Matthias" in Arthur H. Fauset, *Sojourner Truth: God's Faithful Pilgrim* (New York: Russell Co., 1971). For a contemporaneous account see the two-volume exposé of Matthias's excesses, featuring also the author's extensive interviews with and defense of Isabella, in G. Vale, *Fanaticism, Its Source and Influence, illustrated by the simple narrative of Isabella* (New York: n.p., 1835).

18. Gilbert, *Narrative*, p. 164. The dialect rendition of her speech, typically attempted by her biographers, is omitted here.

19. Ibid., p. 196.

20. Fauset, *Sojourner Truth*, p. 175.

21. James M. Glass, "The Philosopher and the Shaman: The Political Vision as Incantation," *Political Theory* 2:2 (May 1974):186.

22. Gilbert, *Narrative*, p. 238.

23. Ibid., p. 150.

24. Ibid., p. 151.

25. Ibid., p. 169.

26. Moses, *Black Messiahs*, p. 227.

27. Gilbert, *Narrative*, p. 241.

28. Ayi Kwei Armah, *The Healers* (Nairobi, Kenya: East African Publishing House, 1978), p. 100.

29. Lowenberg and Bogin, eds., *Black Women*, p. 235.

30. I am indebted to the historian Nell Irvin Painter for this counter-conventional attribution, as stated in her unpublished lecture, "Selling Truth: Slave Woman as Commodity," a paper prepared for the Rutgers Center for Historical Analysis Workshop on "Mass Consumption and the Construction of Race," New Brunswick, New Jersey, August 18, 1992. See also her full-length treatment, *Sojourner Truth, A Life, A Symbol* (New York: W.W. Norton, 1994).

31. Jacqueline Bernard, *Journey Toward Freedom: The Story of Sojourner Truth* (New York: Feminist Press of the City University of New York, 1990), p. 183.

32. William S. McFeely, *Frederick Douglass* (New York: Norton, 1991), p. 97.

33. Arna Bontemps, *Free at Last: The Life of Frederick Douglass* (New York: Dodd, Mead, 1971), p. 180.

34. Bernard, *Journey toward Freedom*, p. 253.

35. Gilbert, *Narrative*, p. 152.

III

THEOLOGICAL
PERSPECTIVES

At the root of a theological position there is an imaginative act in
which a theologian tries to catch up in a single metaphorical
judgment the full complexity of God's presence in, through, and
over-against the activities comprising the church's common life.

David Kelsey, *The Uses of Scripture in Recent Theology*[1]

In these final chapters I am concerned to retrieve the promise, and redirect
the trajectory, of contemporary black theology. I consider that this academic discipline, after little more than two decades of development, is still
a nascent enterprise and remains capable of fundamental reformulation.
When black theology emerged in the 1960s and 1970s it had common
features with other 'contextual' theologies: feminist, Latin American, and
South African. By definition contextual approaches reconstruct the traditional concerns of theology in light of the sociocultural, political, and
economic contexts within which theology is formulated. Theological reconstruction in the black American context first drew widespread attention
with the liberation theology of James Cone. Cone attracted this attention
by boldly linking terms that had been kept separate until his provocative
Black Theology and Black Power (1969): the conjunctive title connoted black
identity and social empowerment, and theological reflection on black existence and empowerment. Cone followed this manifesto with *A Black Theology of Liberation* (1970) and *God of the Oppressed* (1975), which provided
the earliest statements in the genre of systematic theology in the field.

It is a field that includes other emphases in addition to liberation as
a broad umbrella category: biblical studies (Robert Bennett); black fem-
inism or "womanist" ethics (Delores Williams; Katie Cannon); the priority
of black religious experience and divine sovereignty (Cecil Cone); womanist Christology (Jacquelyn Grant); black awareness, ethics, and theology
of hope (Major Jones); reconciliation in concert with liberation, and the

retrieval of African heritage (J. Deotis Roberts).[2] In addition I would include certain black religious philosophies as further indications of the breadth of the field, particularly the uses of process theology (Eulalio Balthazar), theodicy and humanist thought (William Jones), history of religion perspectives (Charles Long), and Marxist theory in concert with American pragmatism (Cornel West). The relation of this present study to that larger field of black theological and philosophical studies is both dependent and critical. I concur with liberation theologians, on the one hand, in the recognition of theology as a mode of reflection that is inevitably related to issues of power and empowerment. In this connection I have been greatly instructed, and spiritually disencumbered, by liberation theology's critique of Western theologies and of their systemic complicity in racism. Moreover my sense of vision and professional empowerment to pursue black religious studies, in the terms and categories of the preceding chapters, depends intellectually and institutionally on the pioneering work of black theologians, particularly that of James Cone.

On the other hand I find that, in the turn from critique to construction, black liberation theology has become enamored of, and stereotyped by, its own positivism or "ideology" of intraethnic religious empowerment.[3] In the interests of such empowerment, I judge, liberation theology as a discipline systematically abstracts and privileges selected contexts of liberation and oppression over against other forms of experience. In particular it subordinates traditions of spirituality that, as shown throughout this study, also compel attention and empower social transformation. Earlier critics too have charged that black liberation theology so privileges the experience of oppression (its import and significance) that it tends to displace or supplant people's multifaceted religious experience with its one-dimensional interest in a hermeneutic of liberation.[4] Black religious experience, however, is not reducible to the experience of suffering and oppression, nor to the quest to overcome suffering and oppression.

Admittedly Afro-American traditions of spirituality convey liberation imperatives and strategies. As wisdom traditions, however, they also contextualize and relativize issues of liberation and oppression within a wholism of social and individual change that values and includes other dimensions of human existence: aesthetic experience and creativity, practical inventiveness and improvisations, knowledge and transformations of nature, matter, and society—both spiritual and scientific knowledge and transformations. Although liberative or emancipatory conditions are invaluable for fully pursuing the other dimensions of experience, the relative lack of freedom means merely that such pursuit is more circumscribed—not nullified. In any case liberationism is too specific an interest to encompass the full range of life trajectories of a people, even a severely oppressed people. In this regard I suspect that black liberation theologians have been tone deaf to others' criticism of the ideological susceptibilities and excesses of their work, because they have proceeded with minimal benefit from the folk sources and wisdom traditions of black religion itself.

In particular[they have not incorporated, or shown sufficient regard to, the cognitive wisdom and spiritual discernment represented in their folk religious sources. That wisdom consists in acknowledging and exploring all the ways in which human creative expression generically, and black folk improvisational genius in particular, refutes the vaunted claim of oppression to dominate a people's existence and strip them of the human prerogative to transcend imposed limitations, both natural and societal.]

More recently however black theologians have begun to highlight their indigenous historical and religious sources. Later studies draw upon Afro-American cultural materials and African-identified perspectives to formulate theological insights and arguments. James Cone's useful study *The Spirituals and the Blues* (1972) is the most obvious among other efforts in this regard. For a more recent instance see the work of younger scholars influenced by Cone's work, *Cut Loose Your Stammering Tongue: Black Theology in the Slave Narratives* (1991), and *Shoes That Fit Our Feet: Sources for a Constructive Black Theology* (1993).[6] Although much more remains to be accomplished in this connection, the discipline has already surpassed earlier approaches. As critics have convincingly and sufficiently argued,[7] those approaches displaced indigenous sources with excessive reliance on formal academic theology, or on revolutionary discourse drawn from outside the religious tradition itself. The displacement of the tradition's own rightful role as an essential source of theological reflection created "the identity crisis in black theology."[8] The most problematic feature of that "crisis" was the inordinate reliance on other sources to perform the critical-emancipatory, and the recollective and authenticating, tasks to which traditional sources must also contribute. In the words of one commentator, black theology "unwittingly ignored that which gives black North American Christianity its uniqueness—vestiges of its African heritage."[9] However, from my own conjunctive perspective I would say instead that what distinguishes Afro-Christianity in the Americas is its interactive biculturality. On this view, black theologians have yet to hold Euro-American academic approaches in creative tension with ancestral and indigenous modes of thought.

Two decades ago when the new discipline was launched the times seemed propitious for liberation theologies that would proceed in direct engagement with their tradition's spirituality.[10] A spiritual agenda appeared to be coextensive with the political interests of the liberation theologies. As the Catholic theologian Robert Schreiter notes, "a spirituality for liberation was . . . high on the agenda of communities in the mid- and late-1970s after a decade's experience of action in liberation movements." But, at least with reference to black religious experience in North America, the spiritual agenda of liberation theology remains unfulfilled. In this connection Schreiter observes that liberation theology seems to work "best when acute social transformation is needed. It has been less successful as a means for sustaining identity. It still exhibits difficulty in dealing with the complexities of popular religion." He then proceeds to present alternative theological approaches that are more conducive to a people's spirituality, designating them variously as "ethnographic approaches," "theology as variations on a sacred text," and "theology as wisdom."[11]

In the preceding pages we have passed through all of these alternative approaches. Ethnographic approaches focus on cultural and religious identity—as earlier where we displayed indigenous forms of oral and literary discourse, and forms of ritual and conjurational performance. Liberation approaches, on the other hand, or theologies of "praxis" seek social transformation. The best possibility, of course, as Schreiter himself acknowledges, is to pursue both—as I endeavor to do throughout this study. Next, we have discussed traditional or preacademic black theology as a typological or figural tradition of biblical hermeneutics. There are Schreiter's variations on a sacred text, in which (1) theology varies "the signs originally employed, yet maintains the same codes"; (2) theology finds "oral over literate forms of culture" most conducive to its discourse; and (3) its "criteria of adequacy" are perhaps best articulated in terms "which would ordinarily be considered aesthetics."[12]

But it is the conjunction between theology as wisdom and a fourth type—"theology as sure knowledge" or *scientia*—that provides the concluding emphases of this essay. Wisdom theology proceeds from the desire "to see the world, both the visible and the invisible, as a unified whole." On the other hand, theology as certain knowledge or *scientia*

> tries to give a critical ... account of faith, using the tools of a discipline that can offer the most exact form of knowledge known to the culture. This has often meant the use of human reason, but also now includes disciplines in the social sciences and to some extent in the natural sciences. The knowledge gained in this kind of theological reflection is, first and foremost, *sure*.[13]

Here we have returned ironically to the dominant theological tradition against which black theology and other liberation theologies have launched their criticisms. In comparison to those criticisms Schreiter's description appears innocuous, almost vacuous. Nonetheless he does proceed to acknowledge the tendency of such theologies to dominate the discourse of other cultures, and also to displace the legitimacy of alternative approaches to theology. He finally espouses a best possible state of affairs, in which theology as certain knowledge is "but one form of theology, alongside wisdom theology, theology as praxis, and theology as more occasional variations on sacred texts." But what Schreiter's account leaves out is the double valence in all these forms of discourse: the human will-to-power operating in all speech, action, and writing with effects that are good/evil, healing/harming, nurturing/dominating, and saving/destroying. Along with Schreiter, but to more effect, we may also recognize that theology as scientia typically excludes theology as sapientia because science traditions in the West are chronically disjunctive. By contrast, wisdom theologies as conjunctive traditions are more capable, in theory if not also in practice, of attending to the interests of scientific discourse. Finally, we may observe that science orientations in theology have pursued certainty or "sure knowl-edge" in the form of rational-dogmatic or rational-apologetic interests. By contrast, what operates in the present study is a rational-ritualistic and rational-pharmacopeic interest. The final chapters of this study lay the groundwork for a reconstruction of black theology in terms of that con-junctive approach.

Notes

1. David H. Kelsey, *The Uses of Scripture in Recent Theology* (Philadelphia: Fortress Press, 1975), p. 163. I quote Kelsey here because my use of the term "conjuring culture" conforms precisely to his description of a theological project that is imaginatively informed by a "single metaphorical judgment." However, Kelsey's comment also orients the trajectory of such a judgment toward the church and its activities. In this regard the pertinence of the current study to ecclesiology, practical theology, and church life may be insufficiently articulated. It is evident that my initial efforts to elaborate the metaphorical utility of "conjuring culture" more readily reflect a cultural rather than an ecclesial theology. Indeed, the major influence on my efforts in this regard is Paul Tillich's program of a "theology of culture," which he presented as a newly developing contemporary source of sys-tematic theology. However, this initial influence does not (of course) preclude ecclesially oriented presentations of this study in the future.

Tillich described theology of culture as a discipline which 'reads styles' at the level of a culture's "ultimate concern." In these terms the present study reads the ultimate concern of African American culture in terms of its incantatory styles, conjurational traditions, and transformative practices. "The key to the theological understanding of a cultural creation is its style.... It is an art as much as a science to 'read styles,' and it requires religious intuition, on the basis of an ultimate concern, to look into the depth of a style, to penetrate to the level where an ultimate concern exercises its driving power." Paul Tillich, *Systematic Theology*, vol. 1 (Chicago: University of Chicago Press, 1951), p. 36. In recalling this program-matic formulation I acknowledge the Tillichian sources of this study, as a study in theology of *black* culture. Cf. Paul Tillich, *Theology of Culture* (London: Oxford University Press, 1959). For convergences and divergences between Kelsey and Tillich see also David H. Kelsey, *The Fabric of Paul Tillich's Theology* (New Haven: Yale University Press, 1967).

2. For a concise and informed survey of some of these authors see Josiah U. Young, *Black and African Theologies: Siblings or Distant Cousins?* (Maryknoll, N.Y.: Orbis Books, 1986), pp. 31–61.

3. On Cone's theology as ideological see Stephen Reid, "Review of *God of the Oppressed*, by James Cone," *The Drew Gateway* 48:1 (1977): 53–56. For a biblical hermeneutical critique see H. Wayne House, "An Investigation of Black Liberation Theology," *Bibliotheca Sacra* 139 (1982): 159–76.

4. See Calvin E. Bruce, "Black Spirituality and Theological Method," *Journal of the Interdenominational Theological Center* 32 (Spring 1976): 65–76; Carlton Lee, "Toward a Sociology of Black Religious Experience," *The Journal of Religious Thought* 29:2 (1972): 5–18; and Talbert Shaw, "Religion and Afro Americans: A Propaedeutic," *The Journal of Religious Thought* XXXII:1 (Spring–Summer 1975): 65–73.

5. It was the mature wisdom of the tradition in the voice of the black novelist and critic Ralph Ellison that chided the young black poet and critic Leroi Jones for his exaggerated assertion that "the slave could not be a man." Ellison queried:

> But what, one might ask, of those moments when he feels his metabolism aroused by the rising of the sap in spring? What of his identity among other slaves? With his wife? And isn't it closer to the truth that far from considering themselves only in terms of that abstraction, "a slave," the enslaved really thought of themselves as *men* who had been unjustly enslaved? . . . Slavery was a most vicious system and those who endured and survived it a tough people, but it was *not* (and this is important for Negroes to remember for the sake of their own sense of who and what their grandparents were) a state of absolute repression. Ralph Ellison, *Shadow and Act* (New York: Random House, 1964), p. 254.

6. See Dwight N. Hopkins and George C. L. Cummings, *Cut Loose Your Stammering Tongue: Black Theology in the Slave Narratives* (Maryknoll, N.Y.: Orbis Books, 1991), and Dwight N. Hopkins, *Shoes That Fit Our Feet: Sources for a Constructive Black Theology* (Maryknoll, N.Y.: Orbis Books, 1993). Coincidentally, Hopkins anticipates the trajectory of these concluding chapters in Part 3 when he asks: "Will black theology acknowledge conjurers as legitimate sources for the actual doing of theology?" (*Shoes That Fit Our Feet*, p. 73).

7. Young, *Black and African Theologies*, pp. 32–36.

8. See Cecil W. Cone, *The Identity Crisis in Black Theology* (Nashville: African Methodist Episcopal Church, 1975). Cf. Carl F. Ellis, Jr., *Beyond Liberation: The Gospel in the Black American Experience* (Downer's Grove, Ill.: Intervarsity Press, 1983), and R. L. Jordan, *Black Theology Exposed* (New York: Vantage Press, 1982).

9. Young, *Black and African Theologies*, p. 33.

10. See, for example, Gutierrez, 1973; Christ and Plaskow, 1979; and Leonard E. Barrett, *Soul Force: African Heritage in Afro-American Religion* (Garden City, N.Y.: Anchor Press/Doubleday, 1974).

11. Robert J. Schreiter, *Constructing Local Theologies* (Maryknoll, N.Y.: Orbis Books, 1985), pp. 13, 62, 80–93.

12. Ibid., pp. 80–84.

13. Ibid., p. 88.

7

Gospel

When Christ came into the world, he said, "Sacrifices and offering you have not desired, but a body you have prepared for me."

Hebrews 10.5[1]

In the following sections I examine the black Christian solution to a millennial problem in Christian praxis: the threefold problem of (1) how to initiate victims in an imitation of Jesus that navigates between the two poles, equally contrary to his persona in the gospels, of docility and enmity; and in conjunction, (2) how to induce victimizers to desire their own imitation of the God who saves victims, over against their continued practices of victimization; and finally (3) how to provide, for both prospective victims and incipient victimizers, a nonconflictual model for the resolution of rivalry and acquisitive, "mimetic desire" (Girard). In the quest for a nonviolent praxis, such a model is called upon to terminate and deter victimization, and thus regenerate culture on the basis of new and nonviolent forms of imitation and the will-to-power. The traditional Christianity of black North Americans offers an instructive case study for addressing such problems. This chapter treats black Christianity as the primary source of the civil rights activism of Martin Luther King, Jr. I investigate King's religious heritage for its power simultaneously to overturn ethnic victimization and to transform the victimizer—for its power to realize what he called "the beloved community."[2] That power resides, by hypothesis, in a performative tradition of mimesis in black religion.

The mimetic basis of such performances is twofold, consisting in (1) an imitation of Christ—*imitatio Christi*—in which Jesus as "Suffering Servant" (Isaiah 53) provides a preeminent model for the nonviolent transformation of "victimage" (Girard), and (2) 'homeopathic' applications of the imitatio Christi designed to 'cure' violence in the form of racism at the level of social change. Finally, I also examine these issues from the perspective of René Girard's theory of religion and violence. But the nature of such transformations in the King movement poses a question that remains untreated by Girard's theory: what praxis can ensure a culturewide resolution of mimetic conflict, in which scapegoating and sacrificial compulsions are defused or transcended? For black Christians before and after King this is an ongoing problem of praxis and survival, not theory alone. It is the problem of *curing* racist Christianity as a deformed cult: that is, healing Christianity itself of its deformation as a religion of sacralized violence that is contrary to its own gospel origins. This chapter examines the Afro-Christian cure.

Suffering Servants

"When I was a slave my master would sometimes whip me awful, specially when he knew I was praying. He was determined to whip the Spirit out of me, but he could never do it, for de more he whip the more the Spirit make me content to be whipt...." That contentment, it may be said, stifled outward political resistance, but it may also be argued that it represented a symbolic inward resistance... a victory of the spirit over the force of brutality.*

Albert J. Raboteau, *Slave Religion: "The Invisible Institution" in the Antebellum South.*[3]

The transformation of encounters with ethnic violence into ritual occasions for identification with Christ and his suffering or passion extend from the period of slave religion in the United States to the King movement of the 1960s. A dramatic instance in the early phase of the tradition is found in a fictional work published in 1810 by Daniel Coker, *A Dialogue between a Virginian and an African Minister*. Coker's work, probably the earliest publication by an Afro-American of his own writing, is also distinguished as one of the first antislavery texts by a black author. As a young man, Coker (1780–1835) escaped slavery and was later manumitted through the efforts of friends and benefactors. Ordained a deacon in the Methodist church around 1800, he is perhaps best known as one of the founders, along with Richard Allen and others, of the African Methodist Episcopal (A.M.E.) Church in 1816. Allen and his colleague Absalom Jones, who was subsequently ordained by the Episcopal church as its first black priest, had separated from the Methodist Episcopal church in Philadelphia after the establishment of segregated seating during Sunday worship.

The *Dialogue* is cast as a defense of emancipation by "an African minister," and as a condemnation of those slavemasters who forbid the religious instruction of their slaves. Its story form is narrated by an African minister, Coker's alter ego, who begins by relating the occasion of a visit from a Virginian slaveholder. The slaveholder politely, even graciously, requests to interview the minister. After opening civilities on both sides this "Virginian" confronts the minister with a charge which has been popularly circulated: he is accused of having "imbibed a strange opinion ...repugnant to reason and justice," and "wrong in the highest degree," that the slaves should be emancipated. Yet by the end of the dialogue the minister has not only succeeded in defending that "strange opinion." He has also, albeit fantastically, persuaded the Virginian to manumit all of his "fifty-five negroes."[4]

The slave as "suffering servant"

In the course of his biblical argumentation the African minister purports to tell the slavemaster a true story of one of his enslaved kinsmen, "in his own words." Within this framework we hear about a slave who is unjustly stretched and lashed with a whip because, as we learn in due course, his master considers him too devout and fervent in his religious life. In the first person voice of the slave we read the following account:

> "First I am chained, and kept back from my public [church] meetings; secondly, I am chained in and out of the house for thirty and some times forty hours together, without the least nourishment, under the sun; thirdly, I am tied and stretched on the ground, as my blessed master was, and suffer the owner of my body to cut my flesh, until pounds of blood, which came from my body, would congeal and cling to the soals [*sic*] of my shoes, and pave my way for several yards. When he would have satisfied his thirst in spilling my blood, he would turn from me to refresh himself with his bottle."[5]

At this point in the narration the Virginian, evidently offended by the stark brutality in the depiction, interrupts and entreats the minister to stop the story. But the narrator refuses to capitulate to decorum, insisting instead on fidelity to "the experience of one of those sufferers." However, he does not continue to dwell on the atrocities committed by the drunken slaveholder. Rather, he turns to celebrating the irony that the lashing of the slave actually redounds to an increase of the victim's religious zeal. The narrator represents that zeal in the slave's declaration, that "when I looked and saw my blood running so free, my heart could not help praising my Savior, and thanking God that he had given me the privilege, and endowed me with fortitude sufficient to bear it without murmuring."

Here is one of the first instances in Afro-American literature of the slave as a type of the preeminent figure who endured the lash "without murmuring." The locus classicus for this figure is the Suffering Servant passage from Isaiah 53—a passage still celebrated by Christians as a messianic prophecy of the Hebrew Scriptures that was fulfilled in the "passion" of Christ (53.7):

> He was oppressed, and he was afflicted,
>   yet he opened not his mouth;
> like a lamb that is led to the slaughter,

and like a sheep that before its shearers
is dumb,
so he opened not his mouth.

Already the narrator has represented the slave in accordance with Christian mimetic tradition as a suffering servant. Yet a deeper dimension of representation subsequently appears: "Finding this [abuse] a great means to make me more fervent in prayer, [the master] bethought himself of another diabolical strategem to put me to shame." The slave is carried to a nearby blacksmith, fitted with an "iron collar," and then dragged away to labor in a field "although scarce able." But this final torment ushers in the climactic moment in the slave's ordeal: the moment of mimesis in which, by becoming a figure or type of Christ, he experiences his suffering as redemptive.

> When I recollected that my dear Lord and Master had commanded me to bear my cross, and take his yoke upon me, my soul, my heart was elevated. I thought I could have flown, and I went to work with more submission, and with more apparent love than I had done heretofore.[7]

The slave's return to work "with more submission, and with more apparent love" than before is the action that he undertakes in imitation of Christ. For most contemporary readers that action and its accompanying sentiment may seem an extreme form of docility, an acquiescence in one's own victimization that amounts to self-abuse and masochism. However, before we dismiss Coker's representation prematurely, let us read it in the context of Coker's authorial strategy as the writer of an antislavery tract.

If we are attentive readers we will not fail to penetrate the authorial presence of Coker in the guise of his alter ego, the "African minister" who is narrating the slave's story. But it is neither the slave nor the slave's advocate and spokesperson, the minister, who is the central performer. Rather it is Coker himself who is engaged, by hypothesis, in performing a literary-magical feat. The essential and conjurational purpose of his work is to induce the formation of antislavery convictions in his white readers. His strategy for achieving that feat is to simulate (or model or 'mimic') such a transformation in the person of the Virginia slavemaster in the narrative. Thus the climactic moment of the *Dialogue* is the Virginian's response after hearing this story. Recall that it is the story of a slave who returned to work for an abusive master, while rejoicing in the commandment of his other Master "to bear my cross, and take his yoke upon me." Marvelously, simply by hearing this rendition of a slave's imitation of Christ, the Virginian slavemaster is himself induced to obey the gospel commandment. In the words of Jesus, he determines to "deny himself and take up his cross and follow me" (Matthew 16.24). He then promises to do so, fantastically and in defiance of all the strictures of realistic expectation, by emancipating his fifty-five slaves. Therein he 'denies himself': by dispossessing himself of other human beings as 'property' he denies himself as master. Moreover, he experiences such dispossession as an act of allegiance or obedience to the commandments of Christ, as his own performance of the imitation of Christ.

In Coker's *Dialogue* the author's magical or conjurational intent is to induce similar transformations among his white American readership. Through his writing Coker summoned white readers to perform their own imitation of Christ by similar self-denial: that is, by the dispossession of their property and by the general abolition of slavery. Of course, with respect to historical reality this literary-conjurational strategy was inadequate to the task. The *Dialogue* was ineffectual for inducing such large scale transformations as the abolition of slavery. But this ineffectiveness resulted, I suggest, not so much from a defect of intent or power (the inability to publish and compel affective reading, for example) but more substantively from a deficiency of mimetic application. It was the same deficiency that curtailed the prophetic efficacy and mimetic performances of Sojourner Truth. Indeed, it is a mimetic deficiency that has persisted in black social prophetism and transformational practices until the recent performances of Martin Luther King, Jr. That deficiency consisted in the inability to affect and enlist antagonists directly in the enactment of their own mimetic performances. In the absence of such engagement practitioners of mimetic transformation have had recourse only to the persons and bodies of the victims themselves.

In the absence of direct engagement with the embodied persons of antagonists (as a correlate of the embodied practitioner), any transformation such as that which Coker portrayed in the character of his Virginian slavemaster lacks reality. The claim of transformation, therefore, necessarily appears fantastic: lacking adequate cause, it is mysterious and even occult. Did Coker project that transformation out of sheer hope or desperate fancy? What act, process, or power could sufficiently account for such a change? To borrow a problem of causality from an older era in the physical sciences, what theory can explain such 'action at a distance'—in this case, the psychosocial distance between victims and their oppressors? As a matter of literary criticism Coker's text provides no adequate cause. He could not account for the Virginian's transformation but he represented it anyway, beyond any evident basis in reason, as the expected result of confronting a slaveholder with a slave who embodies the Christian gospel and its truth. Thus Coker projected what he could not explain, but what *should* work according to his theological convictions and his literary-shamanic intentions. That is to say, an authentic encounter with Christ in the person of his ministers or (suffering) servants should be converting.

Indeed, in the theological nature of Coker's expectation lies a phenomenological indication of his role of a shamanic performer. Negatively stated, the text is magical-conjurational precisely because Coker projected such a conversion without adequate rationale, and in defiance of commonsense expectation and ordinary experience. (Ordinary experience is represented, for example, in Stowe's *Uncle Tom's Cabin*, where Tom's purity of soul inspires a former master's piety but does not affect his final master or, in the end, secure Tom's freedom.) Indeed, the evidence against expecting such a transformation is immediately available in Coker's story itself. The master of the abused slave is depicted as a direct witness to his slave's Christoform performance (unlike the Virginian slavemaster, who only hears the story told), yet there is no reported change in his intoxicated and abusive state. Positively speaking, however, Coker's text is shamanic because it effects or intends a psychosocial transformation based on embodied symbol production (Porterfield). On this view the slave is an embodied type of Christ (a suffering servant), whose *imitatio Christi* indirectly transforms the Virginian—in his role as Coker's (conjure) 'target' or (shamanic) 'client.' Now let us examine the two foci of transformation: that of the victim and that of the victimizer.

The transformed victim. Coker represents the slave as the medium of his own transformation: he encounters violence and victimization at the hands of his earthly master, and then counters that violence by treating it as a type of his *heavenly* Master's victimization. The spiritual, or ritual, formula for such a transformation is given in Matthew's gospel in the words of Jesus spoken to his followers and is quoted by the slave himself in his explicitly stated intention "to bear my cross, and take his yoke upon me" (Matthew 11.29). Author Coker plays upon the congruence between 'the crucified God' and the victimized slave and makes the joy of that identification the basis of the slave's elation. In this manner the brute fact of undeserved suffering is transformed in a Christological mode, based on the slave's 'Christoform' experience and response. That is, the slave experiences his own suffering as conformity to, or as mimetic participation in, the sufferings of Christ (as "the fellowship of his sufferings"; Philippians 3.10). He effectively reconfigures his suffering as Christian suffering ("if you suffer, it should not be as a murderer or thief . . . however, if you suffer as a Christian, do not be ashamed, but praise God that you bear that name"; 1 Peter 4.15–6). On that basis he virtually redeems—gets back restored or even exalted—his formerly mangled humanity.

The gospel ethic of reconciliation to enemies finds expression in the slave's return to his master's service, as he says, with "more submission" and with "more apparent love" than before. Although incredible in conventional terms, such reconciliation is normative for Christians in terms of the model or rule of Jesus as stated in the beatitudes. We may refer specifically to the commands "Do not resist one who is evil" (Matthew 5.39) and "Love your enemies, do good to those who hate you, bless those who curse you, pray for those who abuse you" (Luke 6.27–8). The substance of reconciliation here consists in regarding an enemy (not as a friend but) as an intimate participant in the effort to obey those commands. Is reconciliation on that basis merely fantastic, a self-delusion or pretense?
> On the contrary, I observe in such reconciliation a phenomenon of 'mimetic intimacy', in which an enemy functions intimately as an accomplice in the process of one's own transformation. Insofar as one's enemy facil-

itates such transformation that enemy is an intimate in the process itself (albeit unwitting or alien in relation to its intent). I return to the concept of mimetic intimacy in treating Girard's theory of mimetic desire later. First we must take up the risk involved in such intimacy: that the practitioner may involuntarily, despite transformative effects and precisely because of the invasive nature of violence, internalize the negation of one's humanity that is inherent in such an ordeal. In consideration of this risk it will be necessary to examine another version of suffering-servanthood in black religion: the slave as Uncle Tom. First, however, I turn to Coker's Virginian.

The transformed victimizer. The Virginian's transformation is indirect because he does not himself experience the slave's victimization. Nonetheless what he does experience, and to great effect (albeit for occult or occluded reasons) within the structure of Coker's narrative, is an oscillating correspondence between himself and two forms of alter ego. The Virginian is confronted with a moment of decision, indeed, whether to identify himself with either the drunken, abusive slavemaster, on the one hand, or the slave's "Dear Lord and Master" on the other. As potential versions of himself these alternative masters constitute tensions, tendencies, or potencies within the Virginian's nexus of desire. The decision that confronts him consists in determining which master to desire as a model of his own actions: the conventional model of domination—the slave's "owner of my body"—or the narrator's proffered alternative—the slave's "blessed master." We know from the Virginian's earlier protestations while listening to the African minister's story that he finds the former model repugnant; that he is predisposed to identify his own character with that alternative Master. The literary craft of Coker consisted in his ingenious representation of the folk religious (conjurational) strategy: to foster, exploit, and augment such a predisposition in white Americans, so that they would be induced to prefer the latter model to the former.

The components of that transformative practice include compelling the intended 'target' of the performance somehow (here, via storytelling and despite resistance) to 'experience' a victimizer as his alter ego and therein to prefer a nonvictimizing alternative. The effect is catalytic, but not apart from the accompanying element of the slave's imitation of Christ. Indeed, the cumulative effect of all the dynamic elements prompts the following phenomenological schema. The logic of the Virginian's transformation reflects a synergy in which (1) a portrayal, representation, or dramatization of (2) a victimizer's scapegoating behavior, alongside (3) a victim's imitation of Christ (4) induces, catalyzes, or augments the observer's ability to identify with the latter and to dissociate himself from the former. Every element of this schema, I venture, is crucial for its transformative effect. Accordingly, for the dramatization to be compelling or efficacious, a participant or observer must be confronted with both alternatives: with a clear choice between a victimizing and a nonvictimizing role. Further, the dramatization must induce or catalyze the observer's ability to enact in current circumstances the latter role over against the former. I take up an analytic treatment of this transformative prescription after reviewing Girard's theory of religion and violence.

The slave as "Uncle Tom"

The reality of slave existence was brutal; a small assertion of one's humanity might result in death. The phrase, "No, massa, me no want to be free" (despite the risk of believing it oneself and internalizing white values) . . . is both a depravity and . . . can be a good, a means of survival.

James H. Cone, *The Spirituals and the Blues: An Interpretation*[8]

Debates about self-disesteem in the Afro-American psyche usually include the stereotype of "Uncle Tom." The Uncle Tom figure emerged in popular culture from the literary-religious craft of Harriet Beecher Stowe, the nineteenth century author and slavery abolitionist. From a balanced reading of Stowe's celebrated antislavery novel, *Uncle Tom's Cabin* (1852), we can retrieve her original, intended image of Uncle Tom as "the saintly, subservient slave who loved as he suffered; of the dutiful, abused servant who forgave unto seventy times seven, not because of truckling instincts but because he was a true Christian." The Christian integrity of Stowe's character, however, deteriorated under subsequent popularization in the theatrical entertainments of the late nineteenth century. In minstrel shows across the country a caricature of Christian longsuffering emerged, in which "Uncle Toms groaned biblically under the lash . . . Tom himself became a bobbing, bowing flunkey, a toady, a grotesque saint capable only of drugging his race into genteel apathy."[9] A discrepancy exists, therefore, between the servility of this caricature and the nobility of character of many Christian slaves. In this connection we do well to recall the character of Stowe's reputed model for Uncle Tom, Josiah Henson, an escaped slave, antislavery activist, and author.

Henson was, according to the commentator Walter Fisher, "a man many times larger than his fictional counterpart and many times nobler than the mean connotation which his name bears in the American language today."[10] So great a discrepancy between fiction and reality may indicate one form of the victimization of black people during the post-Reconstruction era in the United States. In that period Afro-American communities in the South were regularly subjected to lynchings, ritual immolations, and random, unpunished murders. Character distortion in the popular imagination was symptomatic of more lethal forms of socially sanctioned violence. Yet this distortion can also be linked to a deficiency in Stowe's psychological foresight. Her strategy for overcoming ethnic prejudice in the minds of her reading audience was to stake everything on a idealized representation of Tom as the good Christian slave. Slaves were regarded as "good" if they behaved submissively to their masters in accordance with the apostle Paul's injunction in the New Testament: "Slaves, be obedient to those who are your earthly masters, with fear and trembling, in singleness of heart, as to Christ . . . rendering service with a good will as to the Lord and not to men" (Ephesians 6.5). Inevitably the religious virtue of obedience was exploited by masters, missionaries, and the slavery system to reinforce the slaves' docile compliance to their exploitation. Heedless of reinforcing this pernicious use of Christian Scripture, Stowe's literary strategy depended on her readers' acclamation of such subservience as exemplary behavior in a slave.

It must also be admitted that slaves themselves were constantly at risk for internalizing this manipulative standard of behavior. A stunning example is provided in Henson's slave narrative, *Father Henson's Story of His Own Life*. In that narrative Henson recalls an incident in his role as overseer of the other slaves on his master's plantation. One night the slavemaster went to Henson's cabin, woke him from sleep, and, in a great show of distress, entreated him to oversee the transporting of all the slaves from Maryland to Kentucky. "There were eighteen negroes, besides my wife, two children and myself, to transport nearly a thousand miles, through a country about which I knew nothing, and in mid-winter."[11] Henson knew that his master had mismanaged the estate, that his brother-in-law had initiated proceedings to take over, and that the master's character had deteriorated under the conditions of drunkenness, desperation, and the loss of all other resources. One sign of his extremity appeared in the master's condescending to beg his own slave to save him, as Henson recounts. "For two or three hours he continued to urge the undertaking, appealing to my pride, my sympathies, and my fears, and at last, appalling as it seemed, I told him I would do my best."

Here Henson acknowledges quite explicitly that it was not his Christian virtue or Christ-like love for his master that prompted his assent. In contrast with those of Coker's slave as portrayed in the preceding section, Henson's natural passions of self-pride, fellow-feeling and pity, and finally fear for himself and his family's future were the motive forces of his action. During the prolonged entreaty the master was clever enough to arouse those passions by various devices. He alternately praised Henson as the best of overseers, apologized for former abuses, and threatened that all the slaves might be separated or, worse, sold south to Georgia or Louisiana if Henson did not comply. And throughout this emotional barrage there recurred that unnerving event of solicitation "with urgency and tears, by the man whom I had so zealously served for over thirty years, and who now seemed absolutely dependent upon his slave."[12]

Given this cauldron of influences we are not surprised to learn that Henson was confronted with a dilemma when, during the journey north, the opportunity arose to free himself, his family, and the other slaves. Inadvertently crossing the Ohio River near Cincinnati, they were informed by the black residents that this was free territory and thus they need not return to bondage. Henson, suppressing his own desire for freedom and that of his party, commanded them all to continue: he successfully completed their transferral to Kentucky. Later Henson did indeed carry out his own escape and that of his family and over a period of years participated in the underground railroad and assisted in the escape of numerous other slaves. Additionally he improved his learning and mastered literary expression. It is to that mastery that we owe the eloquence and self-penetration of this confession regarding that infamous day in Ohio:

> Often since that day has my soul been pierced with bitter anguish at the thought of having been thus instrumental in consigning to the infernal bondage of slavery so many of my fellow-beings. I have wrestled in prayer with God for forgiveness. Having experienced myself the sweetness of liberty, and knowing too well the after misery of numbers of many of them, my infatuation has seemed to me the unpardonable sin.[13]

This disavowal of his former deed is precisely the basis for regarding Henson, in the terms of the commentator cited, as "many times larger than his fictional counterpart and many times nobler." For a person who repudiates former actions in such terms has acquired a capacity that is not represented in Stowe's characterization: the capacity to discern the difference between obedience as a theological virtue and obedience due to psychosocial conditioning.

On the one hand Henson remembered thinking that releasing the slaves would be like stealing, and that ministers and religious speakers had often insisted upon "the duties of the slave to his master as appointed over him in the Lord." He also recalled that, confronted with the enticements of freedom under such circumstances, "still my notions of right were against it."[14] Later he learned to console himself "with the thought that I acted according to my best light, though the light that was in me was darkness." In the final analysis, however, it appears that Henson came to acknowledge the very real conditioning of his mind under racist domination. That acknowledgment is most striking where, recounting the start of his journey with unusual autobiographical transparency, he revealed that "to all who asked questions I showed my master's pass, authorizing me to conduct his negroes to Kentucky, and often was the encomium of 'smart nigger' bestowed on me, to my immense gratification." Finally, Henson acknowledged the degree to which the anticipated approbation of his master confirmed his decision to be obedient.

> I had undertaken a great thing; my vanity had been flattered all along the road by hearing myself praised; I thought it would be a feather in my cap to carry it through thoroughly; and had often painted the scene in my imagination of the final surrender of my charge to master Amos, and the immense admiration and respect with which he would regard me.[15]

Here Henson has provided us with a self-analysis that confirms contemporary studies of individuals and groups who undergo the rigors of social domination. "They are at one and the same time themselves and the oppressor whose consciousness they have internalized," claims the South American philosopher Paolo Freire, in his *Pedagogy of the Oppressed* (1970). Freire's "consciousness they have internalized" has elsewhere been called "internalized oppression." Internalized oppression is an increasingly familiar psychosocial phenomenon, examined in differing terms by diverse studies that treat the colonized mentality, the abused child or spouse, the concentration camp victim, the victims of cult conditioning or brainwashing, women and ethnic minorities undergoing social oppression, and so on.[16] Again, such internalization appears to be an involuntary or reflex phenomenon of the human psyche, to which a victim inevitably succumbs by assenting, to some degree, to a subordinate identity or role as imposed by an oppressor.

A promising course of inquiry for a future study would correlate the data of internalized behavior with René Girard's theory of mimesis and violence. Particularly fruitful would be Girard's concept of mimetic doubles, in which the desire to imitate each other establishes rivalry between two persons who alternate as model for each other's desires. (Consider here Hegel's dialectic of the master-slave relationship in his essay on "Lordship and Bondage.") Such reciprocity of desire more authentically reflects the Afro- and Euro-American relationship, which involves complex alternations of love and hate, imitation and rejection, and cooperation and rivalry (as in Henson's account of his conflicted feelings for his desperate master). A theory of imitation under the conditions of social domination might usefully account for Freire's observation that oppressed classes of people "have a diffuse, magical belief in the invulnerability and power of the oppressor."[17] Probably more familiar is the mimetic behavior of prisoners and captives as noted, for example, in the case of Nazi concentration camp victims: "Survivors are often accused of imitating SS behavior. Bruno Bettelheim has argued that 'old prisoners' developed 'a personality structure willing and able to accept SS values and behavior as its own.' "[18] Once again, however, I would stress the involuntary nature of such imitation.

> Having internalized the norms and values of the dominant group, members of an oppressed group often mistreat each other in an *unconscious* imitation of their own suffering. A dialectical perspective understands that *no oppressed group can remain immune* to the institutionalized and socially empowered untruths which purport to "justify" its oppression.[19]

All of these issues are raised, but then avoided, by Harriet Beecher Stowe in her figural creation of Uncle Tom. However well-intentioned Stowe's portrayal, the relative lack of complexity in her creation is exposed by the actual life story and anguished self-consciousness of Josiah Henson. Remarkably, Henson dared to expose his self-awareness before an often unsympathetic public, but Stowe lacked a corresponding courage in her fiction. To his credit Henson interrogated himself: "Did I willingly serve and obey out of love of God, or due to an insidious domination to which

I succumbed?" It was precisely that complexity of self-interrogation that Stowe omitted in her central character, despite the testimony of Henson's narrative and her reported encounters with the man himself.[20] Perhaps Stowe's characterization of Uncle Tom, by flattening out or fleeing from the complexity of such testimonies as that of Henson, bears significant responsibility for the false image that we retain today of black Christian virtue. It is the image of the docile slave and, now irrevocably, attaches to all black persons "whose behavior toward whites is regarded as fawning or abjectly servile."[21]

Finally, Stowe's literary strategy was flawed in its assumption that a straightforward display of the undeserved suffering of the good and innocent is itself sufficient to counter their violation. For such an assumption does not take into account the pernicious and contagious nature of scapegoating as a principal form of victimization. From a more informed point of view, evidence of the innocence or virtues of a victim are far from constituting a bar to victimization; they can ironically activate and even attract scapegoating attacks. It happens not infrequently that the innocent and virtuous offer more malleable targets for the accusations to be fabricated against them. On this view Stowe's representation of black Christian heroism is inherently ambiguous: both valorizing and inciting to new abuse. For she did not, and probably could not, anticipate how the appeal to pity can be worse than ineffectual: how the effort to establish a field of human sympathy, based solely on a portrayal of the suffering of the good or innocent, can backfire.

It may be anachronistic to expect that Stowe could foresee the degree to which her representations of pious submissiveness would attract and even augment victimization. But for our part we need not remain so naive. At least we should not be surprised that eventually the dominant response to Stowe's pious Tom became contempt, ridicule, and further abuse. Indeed, within a few years of the international acclaim of her work the public entertainment tradition of minstrelsy, launched just a decade earlier,[22] seized upon the Tom character and exploited the novel's popularity for profits and comic effect. Ironically, it was precisely because of its popularity and success in aiding the antislavery movement that the novel "perpetuated stereotypical images of the black as surely and effectively as did any minstrel performance." To be sure, this result more directly derived from the songs that accompanied the stage versions of the novel. "Uncle Tom on stage had little in common with his characterization in the novel. The story was twisted to suit the purposes of the adaptors, who in many cases recognized and exploited comic features as commercial windfalls."[23]

But in addition to their profitability, we may surmise, the minstrel shows constituted ritual occasions for unanimity and catharsis through the public ridicule and laughter generated by the "Tom" caricature. That unanimous laughter still resounds at the end of the present century, more than a hundred years later. Moreover we are obliged to suspect a correlation between that laughter and its social context of racist violence. "Perhaps the best measure of Stowe's novel is the violent reaction it produced in her Southern contemporaries. Not only was the book banned throughout the South and anti-Tom literature issued, but on one occasion Mrs. Stowe received a package containing an ear severed from some unfortunate slave."[24] Such vengeful and capricious acts of violence were succeeded by, and consistent with, the nation's paroxysm of lynchings and live-body burnings decades later. The complexity of such violence is indicated by the routine nature of the attacks against black people, attended by their pervasive stereotyping in the comic tradition of minstrelsy—a stereotyping that continues in minstrelsy's successor forms of entertainment in twentieth century vaudeville, film, and television portrayals. More insistently: the convergence of the nation's socially sanctioned murders, with its socially cathartic laughter, requires a kind of analysis that understands and antic-ipates such complexity. Such an analysis is readily available in René Girard's theory of ritual violence and scapegoating.

Saviors

I have used the phrase "violence and the sacred"; I might as well have said "violence or the sacred." For the operations of violence and the sacred are ultimately the same process . . . [involving] order as well as disorder, peace as well as war, creation as well as destruction. . . . The theory of generative violence permits us . . . [to see] why the sacred is able to include within itself so many opposites.

René Girard, *Violence and the Sacred*[25]

For René Girard, collective violence and scapegoating behavior are grounded in a primal aspect of anthropology and psychology: mimetic desire leading to rivalry and conflict. On this view—the view from Girard's "fundamental anthropology"[26]—characteristically human forms of conflict and violence have emerged in our species integrally. That is, ritual violence in the forms of sacrifice and scapegoating has become embedded in human interaction alongside the process of "hominization" that occurred among our prehistoric ancestors. In this connection Girard recalls Aristotle's aphorism "Humanity differs from the other animals in its greater aptitude for imitation."[27] Desire is imitative when, ostensibly attracted to the ob-ject(s) that belongs to another, it is more profoundly activated by the desire of the other person or group. Simply stated, a first person's desire imitates that of a second person when it is motivated primarily by the fact that the second person is desiring something (anything). Such behavior, of course, renders the second person a model for the first person, the subject. But adequate imitation requires that the subject also have access to, or be able to appropriate and dispose of, the model's objects or pos-sessions. (Hence Girard's hallmark symbol: the 'triangle of desire' that comprises the subject, the other as model/rival, and the other's objects or assets.) The subject is thus related to another person or group in terms of a fundamental affective disposition: the desire to imitate by means of acquisition. With this notion of acquisitive or appropriative mimesis Girard deliberately, and with full awareness of the theoretical implications, adds a new category to Western philosophical reflection.

Although Aristotle recognized the higher aptititude for imitation in human beings, he typified the intellectual perspective of his age and subsequent periods in considering only one kind of mimesis: imitation as representation. Since the treatment of mimesis by Plato and Aristotle in the classical period of Greek philosophy, imitation has been regarded in terms of representation exclusively (1) as imitation in the orders of nature and being (imitation of ideal, metaphysical forms by material and intellectual existents); (2) as imaging in human consciousness (imaging in thought, imagination, dreams); or (3) as representation in the arts, crafts, techniques, and sciences (representation in poetry, painting, or sculpture; pottery or cooking; architecture or astronomy). Accordingly one commentator has described the classical view of mimesis as "ontoepistemological" (Spariosu), because it locates representation in the ontology or metaphysics of being, on the one hand, and in human cognitive processes of understanding and knowing on the other. Girard's innovation, by contrast, is "bioanthropological": he claims to discover what Western philosophers (unlike some writers, poets, and playwrights, notably Shakespeare) have overlooked, that imitation also functions acquisitively in such areas as human instinct and behavior, in individual and group psychology, and in socialization and cultural formations.[28] One of the most familiar phenomena of human behavior, conformity in fashion, offers a clear example of acquisitive mimesis among myriad other instances of mass behavior based on imitation by means of appropriation. Imitation, thus construed, characterizes all behavior and practices, traditions and institutions, social movements and crises, and determines the structure of entire historical periods. Its most significant sphere of 'bioanthropological' operation, however, is defined by the intersection of religion and violence.

Curing violence by violence

According to Girard the dynamics of mimetic desire are never simply suspended; rivalry and 'sacrificiality' are never merely terminated. "Here, as in every case, we have to return to what one might call the primary mimeticism. This is a mimeticism which cannot fail to arouse conflicts. It is therefore disruptive and dangerous, but it is also indispensable to the cultural process." It is so indispensable, indeed, that none of us can properly situate ourselves in our own culture without acquiring attitudes based on desire and its "highly developed mimetic capacity."[29] But how do cultures deal with the toxic occurrence of rivalry and conflict that such desires and attitudes inevitably entail? In addressing that question Girard treats in detail the structures of prohibition and taboo that societies fashion in order to deter imitation and the escalation of rivalry and conflict that it fosters (for example, the outlawing of incest, or infanticide in the case of twin births). He also describes as prophylactic, and remedial or corrective, the rituals of execution, expulsion, dismemberment, or other punishments that a community imposes to anticipate or redress inevitable incidents of rivalry and conflict.[30] But nowhere does he treat a community's efforts to transform a situation of mimetic violence into nonviolence without recourse to the mediation of the scapegoat mechanism or a sacrificial victim. The task remains for us to examine the possibilities and actualities of nonconflictual and nonviolent means for overcoming violence.

Just as mimesis need not be representational (as in the thought of Plato and Aristotle), so also it need not be acquisitive. There is a *tertiam quid*, or third thing: the form of desire that Girard calls the "mimesis of apprenticeship." Consider relationships involving apprenticeship in the form of education or discipleship, and those involving friendship including diverse forms of love, affection, and intimacy. In such relationships the model willingly accedes to the subject's imitative desire by readily sharing ('objects' or 'possessions') information or knowledge, wisdom or skill, experiences or sensibilities, perhaps even social position or status. "The master is delighted to see more and more disciples around him," Girard comments on such relationships, "and delighted to see that he is being taken as a model." Next, however, Girard depicts the transition from this benign, nonacquisitive mimesis of apprenticeship, to a more conflicted and malignant form.

> Yet if the imitation is too perfect, and the imitator threatens to surpass the model, the master will completely change his attitude and begin to display jealousy, mistrust and hostility. He will be tempted to do everything he can to discredit and discourage the disciple.[31]

In this scenario the unaware disciple will experience a "double bind." He will experience, on the one hand, the master-as-model saying, "Imitate me." But on the other hand the master-as-obstacle or rival is saying, "Do not imitate me." A disciple who is confused by such a mixed message "can only be blamed for being the best of all disciples. He admires and respects the model; if he had not done so, he would hardly have chosen him as model in the first place. So inevitably he lacks the necessary 'distance' to put what is happening to him 'in perspective'. He does not recognize the signs of rivalry in the behaviour of the model."[32] The question that remains unaddressed in this description, however, is the possibility of resolving the double bind and reestablishing a nonconflictual mimesis—perhaps on some new basis that transforms the disciple-master relationship.

A similar question arises with reference to Girard's category of "mimetic doubles." Doubles are rivals whose mutual desire to displace the other renders them increasingly like one another as their conflict escalates. At the furthest extreme of mutually desired homicide the essential identity between antagonists is complete: each desires sole survival on the condition

of the other's annihilation. In such situations violence has become the
"great leveler of men and everybody becomes the double, or 'twin,' of his
antagonist."[33] The operation of rivalry here, between mimetic doubles
who are structural peers (whatever the relative difference in social status),
can of course be distinguished from our former case of the master-disciple
relationship, in which rivalry arises unilaterally from the side of the master.
However, the same question is pertinent in each case: on what basis could
there intervene a process for countering rivalry and replacing it with nonconflictual mimesis? The question arises particularly from the perspective
of a homeopathic approach, in which a practitioner employs a mimetic
form of a disease in order to cure that disease. Girard's theory already
includes such a homeopathic formulation, but only in terms of its negative
operation in scapegoating and sacrificial behaviors.

At the social level Girard's theory identifies the 'cure' for mimetic
rivalry, a cure which, it is claimed, archaic societies discovered and which
all subsequent societies have replicated: sacrificial violence or scapegoating.
By agreeing to terminate all other sources of mimetic conflict and violence
in the community and unanimously turn on a single victim or class of
victims as scapegoat (pharmakos), a society can effectively resolve an out-
break of the contagious, reciprocal violence in which each person or group
is turned against another.

> The cure must depend on the identification and expulsion of the individual
> whose presence pollutes the community. . . . Everybody must agree on the
> selection of the guilty individual. The surrogate victim plays the same role
> on the collective level as the objects the shamans claim to extract from their
> patients play on the individual level—objects that are then identified as the
> cause of the illness.[34]

In this negative homeopathy of violence, victims may or may not be
guilty of the crimes lodged against them. It is sufficient that the community
is convinced they are so guilty—sufficiently convinced that unanimity can
occur and thus enable the redirecting or introjection of all other conflicts
onto the resolution of this single, all-encompassing conflict. Ritual sacrifice
of human victims in archaic societies had precisely that "cathartic" effect,
the theory claims, and contemporary societies continue to require such
catharses even though the effects are more attenuated and ambiguous
(inasmuch as they are normally mediated through civil society and its
distancing institutions of law and order).[35] Such a homeopathy—the homeopathy of violence-curing-violence—raises the question of an alternative
homeopathy as practiced by a nonviolent type of the "shaman" referred
to in the preceding excerpts. To that possibility I turn next.

Homeopathy of a nonviolent cure

*The sacrifice of God is a troubled spirit; A broken and contrite heart, O
God, you will not despise.*

                                                      Psalm 51.17

The Christian gospels provide for personal and social transformations on the basis of Jesus' identity as a divine Savior. Consistent with our new "non-sacrificial reading of the gospel text,"[36] Jesus' identity as Savior is grounded in his will to save rather than create victims. Over against a cure of violence which generates culture by-means-of-violence, the gospels aim to regenerate culture on the basis of a salvific will—the will to save or 'make well' (Latin *salvus*: safe, unhurt, well). In such terms the will to save victims is alternative to the acquisitive desire that operates in the case of the 'primitive sacred' (Girard): in the case of the God(s) who sanctions human sacrifice and thus requires victims. The Jesus of the gospels saves by virtue of constituting the will-to-save as the preeminent divine attribute, in contradistinction to other representations of deity by other religious cultures—including the (mis)representation of deity in biblical religion itself wherever it claims that God requires sacrifice for the sake of atonement or forgiveness of sins. In this nontraditional reading, Jesus' divinity consists in his virtual identity with nonviolent deity. This is a deity whose operation in the history of consciousness and culture may well be ubiquitous, but who becomes most luminous and compelling in the life and death of this one figure.

How does Jesus as model constitute a new basis for regenerating culture and therein countering, limiting, or redirecting acquisitive desire? In this formulation Jesus-as-model may also be represented, ironically, as a 'model-obstacle'. As Western culture's preeminent exemplar/rival he both exemplifies and counters the most exalted human expressions of the will-to-power. He is exemplary in his multiple persona as (1) cosmic lord and divine *logos*—the word of God by whom all things were created and subsist; (2) an earthly (Davidic) king, heir of an ancient dynasty and of a messianic destiny of imperial power and rule; and (3) a religious leader of miraculous and prophetic powers who represents and mediates divine favor and judgment, beneficence and justice, to the acclaim and prosperity of his people and all nations. (Familiar instances of mental illness or megalomania, in which the subject claims or parodies the identity of Jesus Christ, may be derived from the double bind of intensely aspiring and simultaneously failing to displace Jesus as our premier culture hero and hence a model/obstacle. Consider in this regard Girard's discussion of mental illness and double-bind phenomena in his treatment of "interdividual psychology."[37])

On the other hand it is evident that Jesus' persona exists in radical contrast to the preceding portrayal. His exercise of the will-to-power is internally balanced: he intrinsically anticipates and negates the ego excess or self-aggrandizement which human beings inevitably experience in the process of desiring (acquisitively) our own good at another's expense. In contrast with excesses which are normative in human desiring, "he did not consider equality with God a thing to be grasped, but emptied himself, taking the form of a slave" (Philippians 2.6–7). In apprenticeship to such a model, one discovers that we all are welcome to share his attributes but not acquisitively—just as he did not acquisitively participate in divine attributes. To be a disciple on the model of Jesus therefore is to rule by serving, and to be lord by becoming a slave, as his disciples learned often to their consternation. These inversions are fundamental to Jesus' full persona as presented in the gospels. Accordingly, to emulate him with a disciple's desire is to be contradicted by him wherever one exceeds the limits of another's well-being: it is to be in conflict with this most exacting model who eschews self-empowerment at the expense of another person or group.

If the will-to-save is to be realized in contexts of violence and conflict, then the imitation of such a model is requisite for both (prospective) victims and victimizers. This requirement consists in the fact that in antagonized situations of mimetic rivalry, and in agonized conflicts of mimetic doubles, there is no clear criterion available to the parties themselves for discerning whether or not one's actions are yet another expression of acquisitive desire. It is typical that conflicted parties are irretrievably enmeshed in conflict and require the imposition of some climactic or dispersing event to free themselves. The familiar phenomenon of enmeshment in violence is rooted (by hypothesis) in the primal emergence of human consciousness via the generative mechanism of acquisitive desire. So abysmally are human beings fixated on such generative processes that even our attempts to counter rivalry, conflict, and violence are themselves infected with the germs of future violence. "When [violence] breaks out it can happen that those who oppose its progress do more to assure its triumph than those who endorse it. There is no universal rule for quelling violence, no principle of guaranteed effectiveness. At times all the remedies, harsh as well as gentle, seem efficacious; at other times, every measure seems to heighten the fever it is striving to abate." The compulsion to revert to violence plagues even nonviolent strategies and philosophies, if only in the form of voluntary victimization—that is, imputedly 'virtuous' self-sacrifice or martyrdom. But just as often a retribution against oppressors is called for, so that a new class of victims arises to replace the old. Inevitably in either case, Girard observes, "the moment comes when violence can only be countered by more violence. Whether we fail or succeed in our effort to subdue it, the real victor is always violence itself. . . . The very weapons used to combat violence are turned against their users. Violence is like a raging fire that feeds on the very objects intended to smother its flames."[38]

Whether we use the metaphor of fire or the medicinal metaphor of disease,[39] an intrinsic, specieswide contamination in violence is the import of Girard's analysis. We must be careful to observe, however, that violence as the generative mechanism of culture is not so-called violence for its own sake, but rather that lesser violence to which humans resort in the effort to cure a greater violence. For sacrificial violence in Girardian terms is double valenced: it is healing/harming, good/evil, beneficial/destructive. Scapegoating is not simply a malign phenomenon; it is not a purely negative, primitive, or barbaric evil with no positive rationale in the structures of human existence. While such violence is destructive for victims, Girard contends that it has been generative and even beneficial for cultures.

> Men do not worship violence as such. Primitive religion is no "cult of violence" in the contemporary sense of the phrase. Violence is venerated insofar as it offers men what little peace they can ever expect. Nonviolence appears as the gratuitous gift of violence; and there is some truth in this equation, for men are only capable of reconciling their differences at the expense of a third party. The best men can hope for in their quest for nonviolence is the unanimity-minus-one of the surrogate victim.[40]

On this view, normative cultural interactions are indelibly marked by that generative process. There is no way to reconstitute the actual history of consciousness on some other empirical basis. Even granting this view (which is highly debated by Girard's opponents, but conceded here because of its thoroughgoing mimetic approach), I nonetheless continue the quest for an alternative formulation for the regeneration of culture. What my formulation concedes to Girard's view is the necessity of a "third party" for dissolving human violence—a *tertium quid* operating outside the nexus of the conflicted parties. What I deny is the necessity of an act of violence with respect to this third party.

The quest for a nonviolent homeopathy for curing violence leads back to the concept introduced earlier: mimetic intimacy. Mimetic intimacy denotes the corelationship of victim and victimizer in their mutual *imitatio dei*—their imitation of the nonviolent God as the third party to every conflict. Both parties require such an imitation, because the operation of acquisitive desire in both victim and victimizer enmeshes and implicates each one in rivalry and conflict. Moreover, both (prospective) victim and victimizer are dependent on one another in the effort to perform their mimesis of the nonviolent God. According as each does or does not recognize and treat the embodied other, so each fulfills or repudiates the divine model, to "love the neighbor as oneself." This mutual dependence means that the victimizer needs the victim: one must reclaim, restore, and save the embodied victim in order to perform one's own imitatio dei. "For he that does not love his brother whom he has seen, how can he love God whom he has not seen?" (1 John 4.20). This dependence renders both parties as intimates in the process of imitation. "Insofar as you have done it to the least of these my brethren, you have done it to me" (Matthew 25.40). Even the victim must renounce any reactive victimization of the oppressor—to create a new victim is also to repudiate the nonviolent God. With a convergent logic Girard cautions against 'taking sides' in every conflictual situation. To take sides, especially in political contexts in which groups are polarized into classes of oppressors and their victims, is to risk reinvoking the 'primitive sacred': the primal compulsion to find a scapegoat and thereby establish a 'righteous' community on the basis of the scapegoat's sanctioned demise (sanctioned on the basis of theological, ethical, or political 'correctness'). By "taking sides . . . we inevitably ignore the true centre of gravity of the process—the scapegoat mechanism, still religiously transfigured, at least up to a point."[41]

The problem of nonviolent social praxis is precisely the problem of how to terminate victimization without taking sides. The phenomenon of mimetic intimacy, together with a homeopathy of nonviolence, enables us to address this issue. In mimetic intimacy the (prospective) victim and victimizer are intimates in the mimesis of a nonviolent model who saves victims and obstructs domination. But this mimesis also functions homeopathically: the form of the disease that cures the disease is our endemic, specieswide rivalry with the nonviolent model. That model rivals our human expressions of the will-to-power and obstructs our acquisitive desire to survive and advance by means of the domination and destruction of other beings. To exercise mastery in the world on the basis of such a model means to accede to this form of rule or else effectively repudiate the model. Acceding to the model constitutes apprenticeship to a non-

> violent God. Accordingly the mimesis of rivalry between antagonized parties is displaced, first, by a mimesis of rivalry with a preeminent nonviolent model and then, where effective, by a mimesis of apprenticeship to that model. This is precisely the kind of nonviolent homeopathic transformation we have been looking for. To restate: a mimetic form of a disease—here rivalry with a nonviolent model—cures or deters an outbreak of the disease: rivalry and violence between conflicted parties. Finally, in the interests of a universalist formulation (and to anticipate and defuse interreligious and theistic versus atheistic or nontheistic rivalries), I would say that discipleship to the Jesus of the gospels as a nonviolent model is coincident with apprenticeship to any other model of comparable clarity, potency, and intention to save victims.

Here I conclude this theological construal of Girardian theory and, in the next section, return to the black religious quest to cure racism. That quest is 'pharmaconic' and homeopathic in the terms discussed. The quest remained the unfinished agenda of slave religion in the United States, which failed to find a mimetic formula sufficiently potent to effect a cathartic or shamanic transformation of racist violence. There were two major symptoms of that failure: (1) the persistent phenomenon of white Americans unaffected or untransformed by black Christian performances of the imitation of Christ and (2) the accumulation of negative reactions from victimizers and covictims (ridicule and contempt, disesteem and abuse) inadvertently internalized by black practitioners attempting to apply traditional forms of the imitatio Christi to their situation. That twofold

> problem persisted from the slavery period until the King movement. King's success as a practitioner of nonviolent civil disobedience consisted precisely in discovering a type of social performance that addressed the two problems within a single format: a format that simultaneously transformed prospective victimizers into allies, and conventional victims into victors. How did the King movement achieve such mimetic intimacy that, among

black and white Americans, formerly antagonized parties became heroic practitioners of nonviolent mimesis? To that question we turn next.

Notes

1. This verse quotes Psalm 40.6. In this chapter I pursue a nonsacrificial reading of this verse taken from perhaps the most sacrificial text in the New Testament; thus I deliberately lift it out of context and reverse its import from Christ as the preeminent sacrificial victim within a cult of transformative violence, to Christ as an embodied model of nonviolent transformation.

2. For an exposition of King's concept of the beloved community see Kenneth L. Smith and Ira G. Zepp, Jr., *Search for the Beloved Community: The Thinking of Martin Luther King, Jr.* (Valley Forge, Penn.: Judson Press, 1974). The term itself, "the beloved community," was coined by the American philosopher, Josiah Royce. See Royce, *The Problem of Christianity*. Reprint of 1913 ed. (Chicago: University of Chicago Press, 1968), p. 125f.

3. Albert J. Raboteau, *Slave Religion: The "Invisible Institution" in the Antebellum South* (New York: Oxford University Press, 1978), pp. 307–8.

4. Daniel Coker, "A Dialogue between a Virginian and an African Minister," in *Negro Protest Pamphlets*, ed. Dorothy Porter (1810; reprint, New York: Arno Press/New York Times, 1969), pp. 4–5, 31.

5. Coker, "Dialogue," pp. 34–35. Note the contrast embedded in the slave's speech between "my Blessed Master" and his human master, whose power and authority are relativized or diminished by the expression "the owner of my body." I return to this contrast in discussing the Virginian slavemaster's cathartic response in the decision to free his slaves.

6. Ibid., pp. 34–36.

7. Ibid.

8. James H. Cone, *The Spirituals and the Blues: An Interpretation* (New York: Seabury Press, 1972), p. 28.

9. Walter Fisher, Introduction to *Father Henson's Story of His Own Life*, by Josiah Henson (New York: Corinth Books, 1962), pp. vi, 5.

10. Ibid.

11. Henson, *Father Henson's Story*, pp. 48, 47.

12. Ibid., p. 46.

13. Ibid., p. 53.

14. Ibid., pp. 51–54

15. Ibid., pp. 49, 52.

16. Paolo Freire, *Pedagogy of the Oppressed* (New York: Herder & Herder, 1970), p. 32f. On specifically Afro-American variants see Suzanne Lipsky, "Internalized Oppression," *Black Re-Emergence* 2 (Seattle: Rational Island Publishers, 1977): 5–10, and the discussion of the "universality of the experience of submission" in addition to the mimetic aspects of "the myth of Uncle Tom" in Wilson Jeremiah Moses, *Black Messiahs and Uncle Toms: Social and Literary Manipulations of a Religious Myth* (University Park: The Pennsylvania State University Press, 1982), pp. 58–60. Cf. the alternative figure of "Sambo" discussed in Stanley Elkins, *Slavery*, 2nd ed. (Chicago: University of Chicago Press, 1968)—a discussion which generated controversy over the stereotypical and fictive status of this personality

type; see also Elkins's comparison of slavery to concentration camp experience. For related analyses of the phenomenon of internalized oppression among colonized subjects see Albert Memmi, *The Colonizer and the Colonized* (Boston: Beacon Press, 1967), p. 187f., and also Memmi's *Dominated Man* (New York: Orion Press, 1968). On internalized oppression among concentration camp victims see Terence Des Pres, *The Survivor: An Anatomy of Life in the Death Camps* (New York: Oxford University Press, 1976), p. 116f. Finally, for a definitive statement of the phenomenon and an extensive bibliographic citation see Erica Sherover-Marcuse, *Emancipation and Consciousness* (New York: Basil Blackwell, 1986), pp. 134f., 144f. See also Sherover-Marcuse's references to Herbert Marcuse's representation of internalized oppression as a "psychic Thermidor," and her reference to Anton Pannekoek's formulation of internalized oppression as a "secret power" or *"geistige* [spiritual] power" of the bourgeoisie within the German proletariat, who, after the revolution of 1918, "rebuilt bourgeois domination with their own hands after it had collapsed" (p. 134).

17. Freire, *Pedagogy of the Oppressed*, p. 52f. Freire's comment suggests an exploration which I am unable to pursue here: applying to the issue of internalized oppression my correlation of mimesis and magic in Chapter 1.

18. Des Pres, *The Survivor*, p. 116f.

19. Sherover-Marcuse, *Emancipation and Consciousness*, p. 4. Emphasis mine.

20. According to Fisher, Stowe's novel was based on a fragment of Henson's autobiographical narrative, first published in 1849 and thus prior to Stowe's 1852 publication of *Uncle Tom's Cabin*. Fisher also reports that Stowe engaged in two conversations with Henson before writing her novel, in 1849 and again in 1850, and finally supplied a preface to the 1858 edition of *Father Henson's Story of His Own Life*. See Fisher's introduction to the 1962 edition of Henson's *Story*, previously cited. On the other hand, Wilson J. Moses argues that "it is very doubtful that the historical Josiah Henson really was the model for Uncle Tom," and cites the following sources for his view: Stowe's own publication, *The Key to Uncle Tom's Cabin: Presenting the Original Fact and Documents upon Which the Story Is Founded, Together with Corroborative Statements Verifying the Truth of the Work* (1854; reprint, New York: Arno Press, 1968), pp. 42–45; and Robin Winks, ed., *An Autobiography of the Reverend Josiah Henson* (Reading, Mass.: Addison-Wesley, 1969), pp. xviii–xxii.

21. Fisher, introduction to *Father Henson's Story*, p. v, quoting *Webster's New World Dictionary of the American Language* (1951).

22. Sam Dennison, *Scandalize My Name: Black Imagery in American Popular Music* (New York: Garland Publishing, 1982), p. 88.

23. Ibid., p. 172.

24. Ibid.

25. René Girard, *Violence and the Sacred*, trans. Patrick Gregory (Baltimore: The Johns Hopkins University Press, 1977), p. 258.

26. "Fundamental Anthropology" comprises Book I of René Girard, *Things Hidden Since the Foundation of the World*, with J.-M. Ourgoulian and G. Lefort (Stanford: Stanford University Press, 1987).

27. Aristotle, *Poetics* 4, as quoted in Girard, *Things Hidden*, pp. 1. Cf. the discussion of "The Process of Hominization" on p. 84ff.

28. Mihai Spariosu, ed., *Mimesis in Contemporary Theory* (Philadelphia: John Benjamins, 1984), pp. x–xvii, 79ff. Girard's most recent work is *A Theater of Envy: William Shakespeare* (New York: Oxford University Press, 1991).

29. Girard, *Things Hidden*, p. 290.

30. See Girard, *Violence and the Sacred*, pp. 18–19, 112–14, 211–16, 219–22.

31. Girard, *Things Hidden*, p. 290.

32. Ibid., pp. 290–91.

33. Girard, *Violence and the Sacred*, p. 79.

34. Ibid., p. 83.

35. On "catharsis" and violence see Girard, *Violence and the Sacred*, pp. 30, 81, 271–73. Note, however, this view of "the crisis of modern society" according to Girardian theory as represented by an otherwise unsympathetic critic: "Lacking any belief in the sacred, it lacks any basis for sacrificial rites. Lacking any basis for sacrificial rites, it lacks any principle by which to deflect violence outside the group. And lacking this last principle, it also lacks any ordering principle. . . . 'Purification [Girard says] is no longer possible and impure, contagious, reciprocal violence spreads throughout the community . . . the religious framework of a society starts to totter . . . institutions lose their vitality; the protective facade of the society gives way; social values are rapidly eroded, and the whole cultural structure seems on the verge of collapse.' The crisis of modern society, then, is a crisis of 'sacrificiality.' " Hayden White, "Ethnological 'Lie' and Mythical 'Truth,' " *Diacritics* 8:1 (Spring 1978): 4.

36. Girard, *Things Hidden*, pp. 180–223.

37. Ibid., p. 291f.

38. Girard, *Violence and the Sacred*, pp. 30–31.

39. Girard's analyses are replete with medicinal metaphors. One instance is explicitly homeopathic: "And what about the modern practice of immunization and inoculation? . . . The physician inoculates the patient with a minute amount of the disease, just as, in the course of the rites, the community is injected with a minute amount of violence, enabling it to ward off an attack of full-fledged violence." Girard, *Violence and the Sacred*, p. 289. Yet Girard also cautions that "we must be careful not to push our metaphor too far" in the expectation of "identifying the microbe responsible for the dread disease of violence." Ibid, p. 33.

40. Ibid., p. 258–59.

41. René Girard, *Job: The Victim of His People* (Stanford: Stanford University Press, 1987), p. 59.

8

Praxis

*One who believes in me will also do the works that I do; and greater
works than these will you do.*

John 14.12[1]

In this chapter I derive an Afro-Christian praxis from a tradition beginning
in the slavery period that includes in its purview the King movement. As
we will see, however, King's praxis refines the tradition in a qualitatively
new way. I begin by reflecting on the transgenerational history of forgiveness and reconciliation in black religious experience. On the basis of
his research, the historian Timothy Smith has argued that forgiveness and
reconciliation toward enemies were the central doctrines of slave theology.
The elements of their theology, Smith claims, were "first, forgiveness, awe
and ecstasy . . . the experience of forgiveness and the doctrine of reconciliation were primary."[2] Smith then observes how theological doctrine converged with the psychological and existential needs of the slaves. In his
view, their theology of forgiveness addressed the existential agony of impotence and outrage they experienced in the face of their severe oppression.
Slave converts to Christianity in particular realized and acknowledged their
need to be 'saved' from an internal cauldron of violent impulses. "Black
converts knew they had a lot to forgive. A long stream of testimony and
reminiscence records their outrage at the injustice and hypocrisy of Christians who held them in bondage." Indeed, in the following testimonies
Smith shows instances of an extraordinary disposition of forgiveness directed toward white Americans. Here we should recall that fictional representations of the same disposition can also be found in the two portrayals
of slaves that I have reviewed: Stowe's *Uncle Tom's Cabin* and Coker's *A
Dialogue between A Virginian and An African Minister.*

[A]nother fugitive preacher witnessed open discrimination against Black people at successive church services conducted by Baptists, Presbyterians and Methodists in the little town of Coxsackie, New York. He declared he had suffered more in spirit that day than at any time since ridding himself of slavery's chains. Nevertheless, "in pity and tears of sorrow," he wrote, "I commend them to the blood in which they must be cleansed if they ever reign in glory, and like Jesus I say from my soul, Father forgive them for they know not what they do."

We see here that, in acceding to the doctrine of reconciliation with enemies, black believers endured internal struggles in order to practice forgiveness toward their white oppressors. In addition, wherever they discovered the efficacy of such forgiveness they reflected upon it, celebrated it, and proclaimed it to others:

> Sometime after 1812 the members of the African Baptist Church in Savannah, Georgia, recorded on the tombstone of their first pastor, Andrew Bryan, that when at the outset of his ministry thirty years before he had been "imprisoned for the gospel without any ceremony and . . . severely whipped," Bryan told his [white] tormentors that "he rejoiced not only to be whipped, but . . . was willing to suffer death for the cause of Christ."

Smith argues that such forgiveness of enemies became central in the practice and theology of early black Christians, but not because of abstract idealism or sheer religious zeal. Rather, this doctrinal formation was impelled by "the psychic necessity of finding means to resist inwardly injustices they could neither condone nor for the moment curb." Therefore, he quotes in conclusion, they "thought it 'a great mercy to have all anger and bitterness' removed from their minds."[3]

Smith's correlation of theological doctrine with existential situation and religious psychotherapy is plausible, even convincing. But I consider it a partial account; alongside it, or by way of extension, I propose a further correlation. I observe here—as elsewhere in African American folk religion—a correlation of doctrine with ritual and incantatory (or conjurational) practices, and with pharmacopeic or healing strategies. Observe for example the expressions of Christian conviction that Smith quotes and highlights (and which in turn evoke the Christian Scriptures): "willing to suffer death for the cause of Christ"; "Father forgive them for they know not what they do"; "a great mercy to [remove] all anger and bitterness." Such phrases, ritually intoned and repeated in hymns, sermons, and prayers, function in ritual contexts as incantations. If internalized by the victimized believer who is struggling to achieve an interior disposition of forgiveness toward enemies, their effects are indistinguishable from the efficacy of spells. Furthermore, like fragmentary moments of initiation and reinitiation in the ritual experience of conversion or baptism (where the believer swears or vows allegiance to Jesus' commands and teachings), such phrases re-present and reinforce the mimetic requirements for the imitation of Christ. In the following section I examine the operation of this nonviolent mimesis in the King movement of the 1960s.

Pharmakeus

An analytics of nonviolent transformation in history and culture is not yet available in Girardian theory. This is most disappointing given that the theory calls for such transformations and implicitly requires such analyses.[4] It provides a brilliant analytics of victimization, and a transformative (nonsacrificial) hermeneutic of the Christian gospels, yet without a complementary treatment of ameliorative social and historical transformations based on those analyses or the gospels. Moreover, as an analytics of social violence Girardian theory does not seem disposed to address pragmatic or 'praxiological' questions. That indisposition, which some regard as an intellectual strength for a theory, on this view indicates the theory's most telling defect. Rather than belabor the matter we must treat deficiency as our own incentive to supply what is lacking: in this case to advance, in terms of a praxis of transformation, the project Girard has begun in terms of an analysis of sacrifice and a repristination of the gospels. With this turn toward praxis I attempt, in effect, to 'stand Girard on his head' in the way that Marx inverted Hegel's philosophy from speculative to praxial priorities.[5]

The pharmakeus *as* pharmakos

Our English cognate forms "pharmacy" and "pharmaceutic," "pharmacology" and "pharmacopeia," all derive from the Greek word *pharmakon*. Most recently the double-valenced use of this word, to mean either medicine or poison or both at once, has been brilliantly expounded by Jacques Derrida in his long essay, "Plato's Pharmacy."[6] Indeed, Derrida's essay skillfully elaborates the three variant meanings of the Greek form: not only *pharmakos* (victim or scapegoat) and *pharmakon* (medicine/poison), but also *pharmakeus*, sorcerer or magician. In this section we see how all three variants are highly significant for understanding African American experience. We have already traced the applicability of the pharmakos, or scapegoat, variant to black experience. Now it is illuminating to recall the second meaning, pharmakon, or medicine/poison, and to develop a third variant, pharmakeus, sorcerer or magician. To begin with, in the black cultures of North America the pharmakeus is most commonly known by the vernacular terms "conjuror," "conjure doctor," or "hoodoo doctor." Indeed, a consideration of the threefold variation of the Greek root suggests similar variations of the Latinate word "conjure." A correlation of the Greek words *pharmakos, pharmakon*, and *pharmakeus*, with parallel aspects of Afro-American conjuration, would be highly advantageous. (Because the English word "conjure" has been appropriated and indigenized in Afro-American culture, it is crucial to retain it here. However, the

Greek root and its variant meanings provide a useful threefold schema of conjure phenomena.)

The first variant proposed is (1) the conjure client, that is, a client who may be seeking redress from, or anticipating injury from, someone else (someone else's conjure), and hence a prospective or current victim—a pharmakos or where relevant a scapegoat. This formulation retrieves the Latin etymology of the term: the now-obsolete meaning of conjuring as conspiring against someone in the sense of the word *conjurare*: to swear (conspire) together as does a 'jury' or as a group of witches or sorcerers casting spells against a victim. The second variant in the meaning of "conjure" is double valenced, precisely like the curative/toxic meanings of the word *pharmakon*. Just as a pharmakon is both medicinal and poisonous, so too are (2) conjure prescriptions, featuring a pharmacopeic repertory that includes both healing and harming substances (herbs and grave dirt, for example) and practices (both benign and malign tricks and spells).

Finally, the third variant follows from the first: in the etymological sense the *conjuror* is a 'juror,' one who swears by uttering an oath or spell for the purpose of sorcery, or who conspires with a client to aid or harm another by sorcerous utterances. Contrary to English conjure or sorcery traditions, however, and consistent with the pharmacopeic reference, (3) the Afro-American conjuror is as much a healer or "doctor" as a sorcerer. Even malign conjure involves materia medica—natural and artificial materials such as herbs or whiskey—that are treated as medicinal prescriptions but for purposes of harming or poisoning. Rather than solely a system of magic spells, witchcraft, or sorcery, African American conjure traditions are also healing traditions. Conjure is therefore conjunctive or wholistic in the sense of joining together elements of a whole (a gestalt) that are conventionally bifurcated and opposed. To summarize: the three pharmacoconjurational categories are (1) the pharmakos, or conjure victim or client; (2) the pharmakon, or conjure prescription; and (3) the pharmakeus, or conjure practitioner. It is notable, in this regard, that the pharmakos (victim) and pharmakeus (practitioner) may coincide in the same person in those cases where the practitioner is also a target of malign phenomena requiring conjurational transformation. Notable as well is the coincidence of the pharmakos (victim) as an embodied pharmakon (tonic/toxin), discussed later.

Martin Luther King, Jr., provides just such a case where the practitioner of transformation is also a victim. It must first be acknowledged, however, that according to Girardian theory the scapegoat is already an agent of transformation. This power is reflected in the apotheosis or divinization of figures who were formerly victims and scapegoats. Such an apotheosis is a recurrent pattern in both mythological and historical sources and is based on the scapegoat's miraculous power to unite individuals and factions formerly at each other's throats. The transition of the same victim from criminal to hero, demigod or god, resides in the transformative power of the scapegoat. Precisely as the most accursed and toxic figure in the community, the scapegoat is able by dying to focus and purge the most virulent fears and animosities of the community.

> The effect of the scapegoat is to reverse the relationships between persecutors and their victims, thereby producing the sacred, the founding ancestors and the divinities. The victim, in reality passive, becomes the only effective and omnipotent cause in the face of a group that believes itself to be entirely passive. . . . If the relationships at the heart of these groups can deteriorate and then be reestablished by means of victims who are unanimously despised, obviously these groups will commemorate these social ills in conformance with the illusory belief that the scapegoat is onmipotent and facilitates the cure. The universal execration of the person who causes the sickness is replaced by universal veneration for the person who cures that same sickness.[7]

Examples of the sacralization of the scapegoat abound (see Girard's *The Scapegoat*). In this context I cite only the most relevant case. The celebration of Martin Luther King, Jr., as a black American culture hero conforms to the pattern of the apotheosis of a victim. Admittedly, the aspects of criminality and divinization in the case of King cannot be as extreme or as developed in contemporary societies as they were in ancient or traditional cultures. (Precisely because we are more self-aware of scape-goating as well as sacralizing processes, such processes are less compelling for us.) Nevertheless, our increasing commemoration of King as a civil rights martyr sufficiently confirms the terms of this discussion. Jailed os-tensibly for breaking laws as a practice of (nonviolent) civil disobedience, he also endured continuous harassment and investigation by agents of the Federal Bureau of Investigation (FBI) for supposed communist alliances and treasonous activities. The FBI also charged King with sexual transgres-sions or misconduct (adulterous and crossethnic) not befitting a married man and a religious leader. In this connection he was similarly accused, years after his death, in the memoir of his former colleague and sometime rival, the Reverend Ralph David Abernathy.[8] Indeed, he was during his lifetime the object of intense intraethnic rivalry and public criticism among black church religious and political leaders, particularly in his hometown of Atlanta.

King's experiences therefore conform to those of the typical scapegoat who is accused of crimes most dangerous to the state and to civil society (communist allegiance during the cold war), and who is the object of sexual suspicions and rumors if not convicted of transgressing the strongest taboos (here not incest or bestiality but southern proscriptions barring black male and white female erotic relations). Of course we easily forget King's criminal persona as our (black and white) communities' scapegoat, because we are part of the culture that now apotheosizes him as the sacral agent of our social harmony. His eventual assassination culminated an extended sacrificial career (arrests, jailing, harassed marching, bombings, stoning in Chicago, a knifing incident, and so on), a career that enabled his society to consolidate civil rights legislation and egalitarian practices in a democratic culture that valued and espoused them. Even King's Nobel

Peace Prize served a salvific function for his country: by honoring the 'prophet' himself, the prize 'saved the appearances' of the prophet's people, who could now be represented before the world as the sponsors of a national (Afro-*American*) will to fulfill vaunted democratic aspirations. In such terms we have yet to come to the end of our celebration and exaltation of this martyr to the nation's most sacred self-image and ideological commitments.

The pharmakos *as nonviolent* pharmakon

However, King was not only a pharmakos (victim) but also functioned as a pharmakeus (practitioner) and pharmakon (tonic/toxin). By the term "pharmakeus" I refer to his shamanistic ability to achieve the nonviolent amelioration of conflicted situations for his client communities, specifically black communities undergoing legalized discrimination and civil rights violations. The two features of shamanism most prominent in King's vocation are the shaman's self-cure as a sine qua non of curative ability in relation to clients, and the shaman's use of his or her own body as a site of "symbol production" in the curative process.[9] As a practitioner of nonviolent religion in the Afro-American tradition, Martin Luther King, Jr., reenacted the gospel story in a salvific rather than a sacrificial mode. More precisely: the King movement induced participants and observers alike to perform their own imitation of a nonviolent model who saves rather than sanctions new victims. The movement achieved this distinction by crafting, on the streets and before the public media of the nation, ritualized reenactments of the gospel story in which a victim secures salvation for covictims and victimizers alike, without thereby requiring or initiating the creation of new victims. Indeed, civil rights history attests that such performances were catalytic and widely transformative for current and future communities of victims and victimizers.

As we have seen, a crucial aspect of King's black Christian heritage was its mimetic or imitative features—its derivation from a long tradition of Christians practicing the imitation of Christ. In contradistinction to the conflictual mimesis presented in Girard's theory, King's imitation of Christ constituted a "good mimesis" that is nonconflictual and nonsacrificial.[10] On the one hand black participants in the movement faced the difficult problem, How do I overturn my victimization without creating a new class of victims (that is, how do I resist nonviolently)? But the freedom movement also addressed an equally agonizing problem in black religious history: How do I transform victimizers without incurring the internalized effects of abuse? The solution that emerges from this study is a conjure prescription that consists of the embodied person as a tonic/toxic *pharmakon*. The cure prescribed is homeopathic on the basis of a nonviolent model, in which the victim's own body and person serve as materia medica—toxic to the victimizer if violation ensues, but tonic if victims are saved. To this end public displays of victimization in which any violation of the victims' bodies would clearly reveal their scapegoat status were devised. In such displays precisely what is toxic for victimizers if abuse occurs—that is, public recognition of their identity as persecutors and scapegoaters—can become tonic if scapegoating is terminated and prospective victims are saved. Embedded in this model of transformation is the phenomenon of negation, to which I turn next.

Negations

The major objection to the orthodox Marxist analysis of culture and religion is not that it is wrong, but that it is too narrow, rigid, and dogmatic. It views popular culture and religion only as instruments of domination, vehicles of pacification. It sees only their negative and repressive elements... [and] refuses to acknowledge the positive, liberating aspects of popular culture and religion, and their potential for fostering structural social change.

Cornel West, *Prophesy Deliverance!*[11]

In his incisive treatment of Afro-American prophetic thought, Cornel West has collated the criticisms that Christianity and Marxism lodge against each other. Some of those criticisms are immediately pertinent to the foregoing discussion. A prior point of interest however is the commonality, rather than the controversy, that obtains between two otherwise disparate traditions. "Their prophetic and progressive wings share one fundamental similarity: *the negation of what is and the transformation of prevailing realities.*"[12] A third source of cultural negations and transformations, which West neglects, is constituted by the wisdom traditions of diverse folk cultures. Like the prophetic or progressive traditions in Christianity and Marxism, the African American folk wisdom tradition also features practices for negating and transforming imposed realities. A principal mode of negation and transformation operating in this wisdom tradition is homeopathic, however, and not dialectical as in West's example of Marxist methodology. Dialectical method is typically disjunctive. Its processes of negation and transformation juxtapose opposites in an antagonistic manner.[13]

In this regard West too is critical of conventional Marxist dialectical method, because its rigid Hegelian philosophy of negation and transformation rejects the possibility of conjunctive formations between opposed forces. "The Hegelian notion of determinate negation subscribes to the idea that only one genuinely new alternative emerges from the clash of contradictory elements within a specific dialectical configuration ... [precluding] combinations and amalgamations of the two contending systems, as has occurred...."[14] In this connection we should not underestimate the importance of historical developments in providing an effective praxis of negation. Black activism in the freedom movement followed World

War II, with its exposure of scapegoating processes on a staggering scale in Nazi Germany. An important factor in the sensitivity of political leaders and their constituencies in the United States to charges of genocidal and oppressive policies against black people was the spectacle of a European nation that had attempted systematically to exterminate its minority population. What the black freedom movement achieved cannot be fully evaluated in the absence of that historically conditioning reality.

Homeopathy of a nonviolent cure

Within the context of such midtwentieth century realities as the genocidal scapegoating of European Jews, black leaders were motivated to *seize the time*.[15] The nonviolent activist leadership did so by learning how to craft public performances that would first expose their own scapegoat status, and then provide a remedy. A crucial element in that remedy was imitation of a nonviolent model by both victims and victimizers brought together in 'mimetic intimacy'. Moreover we may infer that, in their discovery of that remedy, King and his generation were informed by slave practitioners and their descendents acting over the course of succeeding periods in American social history. But what King and his colleagues discovered, as perhaps the most proficient practitioners of the tradition to date,[16] were ritual performances that skillfully amplified the victim-as-scapegoat *appearance* of black Americans, yet without incurring the contaminating results of full victimization.

The new feature of the King phenomenon was the crafting of homeopathic performances in which a sufficiently small instance of a social disorder is rendered efficacious for exposing and (thereby) countering that disorder. By pragmatically discovering that homeopathic formula in the context of American scapegoating traditions, King and his associates provided a solution to racist violence that had eluded generations of his predecessors—from the earliest seventeenth and eighteenth century efforts surveyed in this study (those of Judge Samuel Sewall, for example, and of numerous black petitioners appealing for redress to colonial legislatures), to the nineteenth century efforts of activists like Sojourner Truth and Frederick Douglass, and, up to the midtwentieth century, the efforts of W.E.B. Du Bois among others. The new factor was historically conditioned by the midtwentieth century context of a sufficiently receptive public. Such reception was most evident in the 1954 Supreme Court decision that declared segregated schools unconstitutional. But within that context it was incumbent on activist leaders to devise models for taking advantage of that (always ambivalent) receptivity. Models were devised, we now know, to stage a sufficiently efficacious exposé of the scapegoat status of victims but without the excesses of either martyrdom or provocation leading to a full resumption or reenactment of that status. Where successful, the models engaged the public consciousness precisely where they were optimally aware (or willing to become aware) of scapegoating as a social evil. Coincidentally, the new models reduced the contaminating side effects of such exposure for the victims themselves. It is possible to make the sacrificial misconstrual that it is the victim's suffering or even destruction that is desired by God or that is efficacious for transformation. On the contrary, King's practices avoided the conspicuous abuse or self-sacrifice of victims.

Implicit in the movement was the principle that Girard's theory states explicitly: that what is actually transformative is an exposé of the scapegoating process itself. In this regard proficient ritualists like King have crafted public performances that steer participants between the deficiency of passive displays, on the one hand, and the excesses of full victimization and internalized abuse on the other.[17] This discussion represents King as a consummate conjurational performer in the African American tradition. Although he showed that he was mindful of excesses to be prevented (and although he was constantly reminded of such excesses by his critics), he nonetheless proceeded to craft social rituals of ecstatic suffering. Such displays were ecstatic because they enabled victims to be transformed from mere sufferers (from their status quo) into heroes. Thereby he also countered their internalized self-disesteem. At the same time those public demonstrations enabled observers and even protagonists to become converts and allies rather than persecutors. The freedom movement contrived such demonstrations in the form of agonized (compare the Greek *agon*) festivals, during which masses of participants and, vicariously, an observant nation were initiated into an 'ecstatics' of self-sacrifice. Television viewing of police brutality against participants "awoke the conscience of a nation" (King), leading to federal legislation and a new public ethos that has irrevocably ameliorated the victimization of black Americans.

Of course, as with any repertory of practices, King's public performances were not always efficacious. (Such performances are as much a matter of improvisation, trial-and-error, and experiential proficiency as we find in the practice of any discipline, skill, or art. Hence the need for new and increasingly proficient practitioners is ongoing: more requisite, not less.) But where such performances succeeded, they provided alternative rites for a society that has become highly self-conscious of, and scandalized by, its own propensity for scapegoating. Indeed, this heightened state of public consciousness has become increasingly sensitive and pliable in the modern period. (This is a fact that Girard acknowledges as both hopeful and dangerous; dangerous in that, wherever we fail to provide more effective, nonsacrificial means for resolving the conflicts that scapegoating conventionally addressed, populations tend to seek more effective and catastrophically sacrificial means of resolution, such as totalitarian and reactionary societies employ.) In most Western societies it is now hypothetically possible, however intractable events may prove in practice, to influence social opinion toward terminating continued assaults on clearly designated victims. Yet this precise state of affairs also bears its dangerous aspects.

Our peculiar contemporary danger is that a reductive use of performative mimesis—the superficial imitation of King's tactics, for example—eviscerates and deforms the tradition of nonviolent religious activism that we have examined. A limited focus on techniques and strategies can become manipulative and self-serving, and hence contrary to the spirituality of Afro-Christian tradition as a source of American social activism generally. Since the black freedom movement of the fifties and sixties we have witnessed an increasingly self-interested and nontransformative employment of ritual activism. I do not mean simply that a loss of religious content, rhetoric, or observances has occurred. Indeed, protest marches and other public demonstrations can easily display an explicit religious and moral commitment. The practitioners may nonetheless lack the ability to create mimetic intimacy between the conflicted parties on the basis of their mutual imitation of a nonviolent model. Because they neglect to display for public view a model of transformation like the imitatio Christi or a mimesis of comparable efficacy, they inevitably perpetuate rivalry and conflict regardless of other measures of success.

In the years following the black freedom movement and the anti–Vietnam War activism of King and other religious leaders, the United States has witnessed similar forms of public protest that draw upon predecessor traditions. Activists involved in these successor movements have learned that ritual forms of activism involve performative and symbolic aspects that are highly effective for influencing public opinion. They may nonetheless neglect to give attention to the primal and sometimes volatile mechanisms of ritual as presented in Girardian theory, on the one hand, and to the genuinely transformative potential of ritual as displayed in Afro-Christian practice on the other. Such practitioners may, for example, realize and exploit the incantatory power of biblical or religious symbols and representations. They may still lack a more profound grasp of the efficacy of a nonviolent model like the imitatio Christi, and of its transformative power for both victims and victimizers in American history.

The prospect for more proficient performances of such indigenous traditions of 'mimetic alchemy'[18] depends on the continuing emergence of practitioners who understand the kind of pharmacopeic and conjurational dynamics examined. Cultures of nonviolence still require leaders who not only are initiates themselves in such practices, but can initiate and induce others to participate in the transformative processes. But what are the prospects for the large scale amelioration of social patterns of domination that victimize and scapegoat target groups? Here we differ from Girard, who maintains a pessimistic view of social change on such a scale. Against that pessimism (otherwise called realism) we may invoke the black religious experience of curing racism which, though not completely successful, involves a centuries-long quest for individual and mass transformations.

No one can deny this tradition its efficacy, neither radical activists nor radical theorists. The tradition still conveys the spirituality that has sus-

tained a people from its experience of chattel slavery in the past to its present survival under more sophisticated and perhaps more intractable forms of domination and victimization. Rather, on this question of prospective effectiveness it is appropriate to adopt a heuristic disposition, supported by historical and ongoing evidence from all cultures of non-
> violence. We can continue the quest for more and more efficacious means of transformation, and use this praxis to review and revise Girardian and other theories of social violence. On this view theory is supplementary to a praxis that poses the consummate human project of our time and all history. [It is credible that the cascading impact of multiple communities engaged in that praxis could transform the human prospect in the foreseeable future.]

Toward socioritual proficiency and a pharmacopeic praxis

Present-day trends are best seen as revitalizations of a traditional African world view... a return to the ancestral ontology... [to] the incantation of Black magicians such as Frederick Douglass, Marcus Garvey, Malcolm X and a host of other visionaries.... Black Power is Black incantation (magic) pronouncing a curse on the witchcraft of white racism... [and] contemporary Black theology, then, is rightfully understood as primarily a reconstruction of the collective unconsciousness of African peoples.

Leonard Barrett, *Soul-Force: African Heritage in Afro-American Religion*[19]

A conventionally modern, technocratic view of the contemporary period discounts the future vitality of ritual practices and magical traditions in both Western and non-Western cultures. A clear instance is the social science perspective of Bryan Wilson, who claims in his *Magic and the Millennium* to have observed a "rational mutation of religious responses" among third world cultures. Wilson ascribes the cause of these mutations to secularization and modernizing influences. More specifically, he sees in third world societies the rationalization of folk religion in response to the "logic" of political institutions and bureaucracies, military and educational systems, and the dominance of scientific and technological procedures. As these systematizing processes begin to supersede the "communal and affective bases of traditional societies," Wilson argues, religion gradually loses its orientation in magic and thaumaturgy (wonder-working). Replacing this traditional orientation is the "logic of economic, political, bureaucratic, and educational institutions copied [cf. mimesis] from the western world."[20]

This perspective, however, depends on a one-dimensional view of both religion and rationality. It does not, from the side of reason, allow for a rational pursuit of ritual processes adapted from indigenous traditions. Nor, from the side of religion, does it allow for initiatives that seek to extend the sphere of ritual proficiency into new areas by means of analytic and technical prowess. Because of its scientistic positivism, Wilson's perspective does not value what his subjects of study may highly value: the extension of spiritual and transformative practices alongside increasing virtuosity in the use of technical-rational approaches (or again, the use of conjunctive rather than disjunctive approaches.) Lacking a more pluralist, inclusive, or differentiated view of rationality,[21] and one less ethnocentric, Wilson fails to allow for the possibility of modulations or transmutations of indigenous traditions over against their atrophy. Even in the West a rapprochement between traditional spiritualities and the ongoing rational development of political and scientific cultures is possible, as a contemporary development in postmodernism, for example. Some postmodern indications, indeed, suggest a "return to cosmology" and the "reenchantment of science," whether those developments are positively or negatively evaluated by observers and commentators.[22]

The spirituality of African peoples in North America has survived within one of the most complex technocracies in the modern world. Yet there persist among their descendants ritually proficient traditions, still displaying incantatory and pharmacopeic practices.[23] Those practices give ambiguous evidence of withering away under modern influences. In fact, what we sometimes observe is a more skillfully organized and circumspect employment of ritual performance, in areas especially of psychosocial and social-ethical transformation (for example, church related rites of passage, group counseling events, and social and political activism). Initiatory and communal rites derived from folk traditions, and incorporating ecstatic and other affective elements, continue to be employed in complex and sometimes intractable crises (e.g., healing services and exorcisms in drug plagued neighborhoods). The kind of "rationalization" these developments reveal does not suggest the atrophy of thaumaturgy or theurgy. Rather, at best we find ongoing efforts to craft more cogent and coherent practices, with intentions that are less reactive and obscurantist, more informed and responsive, than the practice of predecessor communities.

Contrary to the prevailing cognitive and spiritual disjunctions, Afro-American critical thought is well situated between Western and traditional cultures so as to preserve both its rational and its ritual sources conjunctively. More prescriptively: in order to realize its potential and its cultural mandates within this in-between situation, black thinkers must authenticate their pharmacopeic wisdom tradition as a form of Western discourse. That is, they must become methodologically indigenous in combination with articulating their thought in terms of Western logocentric disciplines. A telling feature of genuine achievement in this regard will be the ability to include the element of negation, which appears in black folk sources as malign conjure. Rather than excluding the negative—of practices that are poisoning, toxifying, harmful, or destructive— black theology must continue homologous practices of negation in its forms of critical signifying (Gates). Let me not be misunderstood here.

To be more precise: black thinkers must retain the capacity to countermand other toxic pharmacopeia. A countertoxic proficiency is crucial within indigenous traditions. Here such proficiencies must be practiced not in the form of arcane methods but in discursive modes of critical signifying. In this connection Cornel West has emphasized that, typically, "reflection by black theologians begins by negating white interpretations of the gospel, continues by preserving their own perceived truths of the biblical texts, and ends by transforming past understandings of the gospel into new ones."[24]

West refers to this sequence of negation and transformation in his own terms as a "dialectical methodology." Such methods, he elaborates, render black theologians "sensitive to the hidden agendas of the theological formulations they negate, agendas often guided by social interests." In continuity with this tradition of negation, black theologians and other thinkers can proceed best by remembering how their intellectual tradition has previously negated what distorts, defames, and destroys the humanity and freedom of its constituent communities. Beyond that, the pharmacopeic task of black theologians includes a continuous rediscovery of how to transform the convictions of its multiple communities of reference—both Afro- and Euro-American. That task requires that the most informed, rational-analytical perspectives be employed with a homeopathic interest.

The major pharmacopeic project of black America remains the need to countermand the most toxic forces impacting its communities, both internally and externally. Those forces include, from the external side, the entire range of dominating, scapegoating, and victimizing processes at work in history and society. On the other hand, those same processes are also internalized and acted-out within black communities. So toxic are these doubled effects that communities require practices and discourses of recovery from the toxicity of their own traditions, institutions, and social practices. A double-sided view allows us to see a convergence here between the curative, healing interests of black culture and its negative tasks. Its negations can become antidotal in the same way that homeopathic medicines are curative. The use of a toxin to cure its own poisonous effects offers an intriguing trope for the kind of negation that distinguishes the tradition. It must suffice here to remind the reader of two examples previously discussed in this study: critical appropriation of the Bible in ways that transform it from a toxic to an antidotal text, and praxial uses of victimization and scapegoating as in civil rights demonstrations and activism. That activism featured mimetic enactments or rites of social change that exposed the toxicity of ethnic oppression at a public level, and thereby induced social reactions that countermanded the worst effects of oppression. In the concluding chapters of this study I will examine the cognitive and spiritual resources in black America for sustaining and expanding such forms of conjuring culture into the foreseeable future.

Praxis

Notes

1. My paraphrase.
2. Timothy L. Smith, "Slavery and Theology: The Emergence of Black Christian Consciousness in 19th Century America," *Church History* 41 (1972): 498.
3. Ibid., pp. 498, 499–500, 500.
4. "There can no longer be any question of giving polite lip-service to a vague 'ideal of non-violence'. There can be no question of producing more pious vows and hypocritical formulae. Rather, we will more and more often find ourselves faced with an implacable necessity. The definitive renunciation of violence, without any second thoughts, will become for us the condition *sine qua non* for the survival of humanity itself and for each one of us." René Girard, *Things Hidden Since the Foundation of the World*, with J.-M. Ourgoulian and G. Lefort (Stanford: Stanford University Press, 1987), pp. 136–37. How to reconstitute cultures, for "the definitive renunciation of violence, without any second thoughts," is the issue of praxis that Girardian theory invites but then neglects.
5. In the turn to praxis I trust that I advance Girard's underrepresented interests. Girard (*Things Hidden*, p. 135) alludes to those interests in various places, yet without elaborating their implicit praxial dimension, as in the following 'prophecy'.

> There will be others, in any case, who will repeat what we are in the process of saying and who will advance matters beyond what we have been able to do. Yet books themselves will have no more than minor importance; the events within which such books emerge will be infinitely more eloquent than whatever we write and will establish truths we have difficulty describing and describe poorly, even in simple and banal instances. They are already very simple, indeed too simple to interest our current Byzantium, but these truths will become simpler still; they will soon be accessible to anyone.

I must say that I detect here a laissez-faire disposition that at least is morally questionable, and at worst jeopardizes the survival of our species by leaving the attainment of nonviolent cultures to chance occurrences. In fact, we need not wait for the future to bring about the immediate accessibility of nonviolent "truths." The present book is one of those written in the context of events of the recent past that make such truths accessible. The King movement of nonviolent civil disobedience and its Gandhian predecessor movement, among others, constitute precisely such events. Are we not obliged to create a literature, as one course among all the others at our disposal, that promotes the understanding, deepens the impact, encourages more advanced forms of such events?

6. Jacques Derrida, *Dissemination*, trans. Barbara Johnson (Chicago: University of Chicago Press, 1981).
7. René Girard, *The Scapegoat*, trans. Yvonne Frecerro (Baltimore: The Johns Hopkins University Press, 1986), p. 44.
8. Ralph David Abernathy, *And The Walls Came Tumbling Down* (New York: HarperCollins Publishing, 1990).
9. See Amanda Porterfield, "Shamanism: A Psychosocial Definition," *Journal of the American Academy of Religion* LV:4 (Winter 1987): 725–26. As discussed in Chapter 6, in Porterfield's view shamanism is personally *embodied* symbol production for the purpose of psychological and social *conflict resolution*.

10. Mihai Spariosu, ed., *Mimesis in Contemporary Theory* (Philadelphia: John Benjamins, 1984), pp. ix–x, xv–xvi, 97, 100.

11. Cornel West, *Prophecy Deliverance! An Afro-American Revolutionary Christianity* (Philadelphia: Westminster Press, 1982), p. 117.

12. Ibid., pp. 101; cf. pp. 95, 99.

13. In this regard see Vernon Dixon's argument for a "diunital" rather than a dialectical approach in the section "Conjunctions," Chapter 5.

14. West, *Prophesy Deliverance!*, pp. 100–101.

15. The troubled literature of the 1960s and 1970s shows some degree of anxiety about the similarity of oppression suffered by the two minority communities in their respective host nations of Germany and the United States, from the scholarly debate by Stanley Elkins and others comparing slavery to Nazi concentration camp experiences, to the more alarmist work by Sam Yette that speculated on the readiness of the U.S. government to create extermination centers in riot-torn urban areas. See Stanley Elkins, *Slavery: A Problem in American Institutional and Intellectual Life*, 2nd ed. (Chicago: University of Chicago Press, 1968), and Samuel F. Yette, *The Choice: The Issue of Black Survival in America* (New York: Putnam, 1971).

16. "Scholars such as August Meier are certainly correct to point out that the Congress on Racial Equality (CORE) as early as 1942 preceded Dr. King and SCLC in embracing nonviolent direct action as a tool to oppose segregation. Moreover, Dr. King himself indicated that he learned much from prior nonviolent resisters such as the Reverend Theodore J. Jemison of Baton Rouge, Louisiana, who had led a successful boycott against that city's public bus system as early as 1953, and from the earlier attempts of the Reverend Vernon Johns, who preceded King as pastor of the Dexter Avenue Baptist Church in Montgomery [Alabama]." James M. Washington, ed., *A Testament of Hope: The Essential Writings and Speeches on Martin Luther King, Jr.* (New York: HarperCollins Publishers, 1991), p. 417.

17. For illuminating discussions of this and other principles of nonviolent activism see especially Joan V. Bondurant, *Conquest of Violence: The Gandhian Philosophy of Conflict*, rev. ed. (Berkeley: University of California Press, 1965); James P. Hanigan, *Martin Luther King, Jr., and the Foundations of Nonviolence* (Lanham, Md.: University Press of America, 1984); Gene Sharp, *The Politics of Nonviolent Action*, 3 vols. (Boston: Porter Sargent, 1973).

18. For corroboration of this terminology see the chapter "Alchemizing Iron into Gold," Keith D. Miller, *Voice of Deliverance: The Language of Martin Luther King, Jr. and Its Sources* (New York: The Free Press, 1992), pp. 186–197.

19. *Soul Force: African Heritage in Afro-American Religion* (Garden City: N.Y.: Anchor Press/Doubleday, 1974), p. 215.

20. Bryan R. Wilson, *Magic and the Millennium: A Sociological Study of Religious Movements of Protest Among Tribal and Third-World Peoples* (New York: Harper & Row, 1973), pp. 452, 504.

21. Cf. a related argument with different subject matter in Alasdair C. MacIntyre, *Whose Justice? Which Rationality?* (Notre Dame, Ind.: Notre Dame University Press, 1988).

22. See David Ray Griffin, ed., *The Reenchantment of Science: Postmodern Proposals* (Albany: State University of New York Press, 1988), and Stephen Toulmin, *The Return to Cosmology: Postmodern Science and the Theology of Nature* (Berkeley: University of California Press, 1982).

23. See, for example, Albert J. Raboteau, "The Afro-American Traditions," in *Caring and Curing: Health and Medicine in the Western Religious Tradition*, ed. Ronald L. Numbers and Darrel W. Amundsen. (New York: Macmillan Publishing Co., 1986), p. 560.

24. West, *Prophesy Deliverance!*, p. 109.

9

Apocalypse

It is the Apocalypse which is missing from most evaluations of black Christianity.

Donald G. Mathews, *Religion in the Old South*[1]

This final chapter completes the sequence of biblical figures which the first chapter inaugurated with a review of James Weldon Johnson's *God's Trombones*. Recall that Johnson's Afro-American classic begins with "The Creation" and ends with "The Judgment Day." It thereby recapitulates the monumental span of the Christian Scriptures, from the Book of Genesis to the Book of Revelation or Apocalypse (Greek *apo-kalupsis*: un-veiled or un-hidden; hence, revealed). Following in that same tradition—the Afro-American tradition of biblical figuralism—the present study also spans biblical books and themes, beginning with the Genesis and Exodus figures in the early chapters and concluding with the Apocalypse in the present chapter. Yet, it should be noted, concluding with Apocalypse is not merely a formal or aesthetic requirement of this project. Rather, attention to apocalyptic figuration is integral to a display of the full range and interests of African American conjurational spirituality. Despite this centrality the topic was, until recently, neglected in evaluations of black religion—as the historian Donald Mathews indicates in the excerpt quoted.[2]

Mathews's criticism concurs with the findings of the historian Timothy Smith discussed earlier: our obsessive and exaggerated interest in issues of docility versus resistance, particularly in slave religion (but also in the contemporary and intervening periods), obscures other salient aspects of black experience. In this chapter I leave behind such debates—for example, whether otherworldliness rather than realism predominates in black religious consciousness—in an effort to treat apocalyptic perspectives and sensibilities on their own terms. A troubling aspect of Christian apocalyptic traditions in general can also be found in black American apocalyptic. In each case one finds a theological irony: the irony of a religion that espouses forgiveness and reconciliation, on the one hand, and yet harbors a vigorous hope for divine wrath and retribution on the other.

> The issues of accommodation, compensation, and survival, like the figure of Uncle Tom, ignore the fact that the chosen community of love and forebearance is also that of hope, but not a flaccid expectation that everything will turn out right in the end. The hope is Apocalyptic; that is, it is based on the vision of a future, violent struggle in which Evil is destroyed.[3]

For two millennia in the Jewish and Christian traditions apocalyptic texts have fostered and sustained the anticipation of a violent climax to the cosmic and historical antagonism of good versus evil, righteousness versus wickedness, justice versus injustice. Accordingly, the central feature of the Book of the Apocalypse is divine judgment, and even vengeance, against wickedness and injustice. (Hence Johnson's emphatic title for the final chapter in *God's Trombones*, "The Judgment.") Moreover, the intensive desire for retribution in biblical religion itself appears to reinforce, in the religious imagination of postbiblical audiences, an endemic human desire to place a warrior deity alongside the God of love and forgiveness. This double construction, the God of love as a God of vengeance, is especially evident among the Bible's inheritance communities of the poor and oppressed. "Slaves hoped also for justice which could not be given by the crucified Christ who gripped Mrs. Stowe's imagination—it could come only from the Lord of the Apocalypse."[4] Other commentators as well have noted in slave religion the juxtaposition of the Lord of the Apocalypse alongside the Jesus of the gospels. As the historian Lawrence Levine reminds us:

> It was not invariably the Jesus of the New Testament of whom the slaves sang, but frequently a Jesus transformed into an Old Testament warrior . . . "Mass Jesus" who engaged in personal combat with the Devil; "King Jesus" seated on a milk-white horse with sword and shield in hand. "Ride on, King Jesus," "Ride on, conquering King," "The God I serve is a man of war," the slaves sang. This transformation of Jesus is symptomatic of the slaves' selectivity in choosing those parts of the Bible which were to serve as the basis of their religious consciousness.[5]

Levine's emphasis is illuminating, but he neglects to point out that the transformation of Jesus into a warrior god occurs already in the New Testament itself, in the Book of the Apocalypse. The sword-wielding Jesus riding a white horse is a terrifying figure portrayed in the most explicit terms in Revelation 19.11–13: "Then I saw heaven opened, and behold, a white horse! He who one sat upon it is called Faithful and True, and in righteousness he judges and makes war. His eyes are like a flame of fire . . . He is clad in a robe dipped in blood, and the name by which he is called is The Word of God."[6] Perhaps even more arresting is the juxta-position of such ferocity with another traditional image, the docile Jesus (standing "like a sheep dumb before her shearers"; Isa. 53.7). We find this combined image in an earlier representation in the text: the Lord of the Apocalypse as a warrior Lamb!

> Then the kings of the earth and the great men and the generals and the rich and the strong, and every one, slave and free, hid in the caves and among the rocks of the mountains, calling to the mountains and rocks, "Fall on us and hide us from the face of him who is seated on the throne, and from the wrath of the Lamb; for the great day of their wrath has come, and who can stand before it?" (Rev. 6.15–7)

This "wrath of the Lamb" is a stunningly ironic and composite image. Moreover its prominence here, in the generative text of Christian apoc-alyptic, suggests that the combination of conciliatory and vengeful im-pulses lies at the very core of the tradition.

A similar combination of impulses is evident when one turns to "black eschatology" (Greek *eschaton*: the last things). With this term the church historian Gayraud Wilmore refers, in his brief study *Last Things First* (1982), to black religious beliefs concerning the end of the world, the Second Coming, and the millennial reign of Christ. On the one hand, Wilmore points out, utopian dreamers like Martin Luther King, Jr., have been able to rely on black religious sensibilities to support efforts toward realizing a future victory of interethnic reconciliation. For example, Wil-more finds conciliatory eschatological symbols in King's most famous address, the "I Have a Dream" speech delivered at the March on Wash-ington in 1963, and suggests that the freedom movement could not have succeeded as well as it did without the "wide acceptance" of such symbols. King's dream "was a dream of the Kingdom of God in which marching people, black and white, Jew and Gentile, lined up to be registered to vote, challenged segregated schools and lunch counters, overturned the tables of economic exploitation and job discrimination." On the other hand, Wilmore observes, black Christianity also features "the belief that the *visible* reign of Christ does not come by social action programs, but rather by catastrophe."[7] This catastrophic view of the *eschaton* has a long history, spanning biblical apocalypticism and slave religion and extending into the twentieth century in various forms and permutations—some re-ligious or mystical; some political, revolutionary, or ideological, and some romantic, fashionable or popularist. A major conduit of African American apocalyptic expression in the twentieth century has been the religious-political tradition of Ethiopianism.

For Ethiopianism has bequeathed to our contemporary period a suc-cessor movement that also conveys apocalyptic motivations: Pan-Africanism as "the latter-day ideology of the African diaspora."[8] Accord-ingly, the next sections take us from the apocalyptic aspects of Ethiopi-anism to Diaspora themes in black religious experience.

Nemesis

Thar's a day a-comin'! Thar's a day a-comin'. . . . I hear de rumblin' ob de
chariots! I see de flashin' ob de guns! White folks blood is a-runnin' on de
ground like a riber, an' de dead's heaped up dat high! . . . Oh, Lor'!
hasten de day when de blows, an' de bruises, an' de aches, an' de pains,
shall come to de white folks, an' de buzzards shall eat 'em as dey's dead in
de streets. Oh, Lor'! roll on de chariots, an' gib de black people rest an'
peace. Oh, Lor'! gib me pleasure ob livin' till dat day, when I shall see
white folks shot down like de wolves when dey come hongry out o' de woods!'

The Greek goddess of retribution, Nemesis (from *nemo*: to give what
is due), finds a fitting medium here in this slavewoman's wrath. But my
deeper interest lies in the Christian transformation of this classical, pre-
Christian figure. Black religious tradition, as an inheritor of biblical religion
and European conventions of biblical typology, participates in that transformation. Indeed, contrary to a conventional view, there is already operating in the Book of the Apocalypse a precursory, incipient effort to
purge a community's fixation on retributive vengeance. Although the effort
fails in Western history, its intended trajectory is recoverable and (by
hypothesis in this chapter) a matter for more proficient theory and practice
in the contemporary period. At this point of initial presentation it is
sufficient to suggest that the appropriation of Western and Christian representations of divine retribution constitutes yet another instance of Afro-
American participation in Euro-American traditions. In this connection,
we have already surveyed a particular black American expression of retributive justice, in the case of the "black jeremiad."

In the analysis of the eighteenth and nineteenth century genre of the
"black jeremiad" (Wilson Jeremiah Moses), I have observed its mimetic
relationship to the Puritan jeremiad. The Puritan jeremiad emerged during
the colonization of New England as a type of sermon that reprimanded
the early colonists for failure to fulfill their theological-political covenant
as a righteous community conforming to both biblical and civil law. Subsequently the jeremiad became a standard rhetorical device for lamenting
national transgressions and warning of the consequences: divine displea-
sure and forthcoming retribution. African American writers eventually
adapted the genre to their own rhetorical strategies, principally as a means
for provoking their countrymen to fulfill the egalitarian ideals of the young
republic. But in both traditions of discourse—in both the black jeremiad
and its Puritan prototype—the deep structural intent was not to repudiate
or diverge from the covenantal heritage or the democratic ideals of the
republic. Rather, by means of criticism and doom-saying the intended
result was revitalization or repair of "the broken covenant" (Robert Bellah). In such terms the jeremiad was a consensus-building device: a transformative instrument employed by diverse practitioners of the nation's
oldest social-political rite; the "ritual of consensus."[10]

Warnings of specifically ethnic violence and catastrophe were added to the jeremiadic tradition by black writers, preachers, and orators. The context for this development was the increased repressiveness of slavery in the nineteenth century as southerners, reacting to growing criticism of their "peculiar institution," became all the more entrenched in its defense. In response black leaders added to the more spiritual and abstract warnings of divine retribution the all-too-human, more tangible prospect of slave revolts and a militant black insurgency. David Walker's *Appeal to the Colored Citizens of the World* (1829) and Henry Highland Garnet's "Address to the Slaves of the United States" (1843) were perhaps the most outspoken expressions of this response. However, to repeat Wilson Jeremiah Moses's evaluation of those works, "the rhetorical threat of violence was not based on any real desire for a racial Armageddon [Apocalypse]. Black jeremiads were warnings of evils to be avoided, not prescriptions for revolution. When Ethiopia stretched forth her hands, it would not be to take up a sword but to embrace her erstwhile enemies."[11] But how cogent is such a claim, that a text is intended to forestall or remedy evils when its rhetoric and referents ostensibly prescribe them? We have seen similar claims corroborated in the phenomena of black religion and culture: (1) in the spirituals and the blues, musical genres which employ a transformative aesthetic, and (2) in conjure practices which involve homeopathic processes.

As Chapter 4 pointed out, the spirituals and the blues feature 'sad music' expressed for the purpose of counteracting sad moods. That transformative effect is captured in James Cone's designation of the blues as a "secular spiritual."[12] However, the blues mode of transformation, in which one counteracts a melancholy mood by means of a melancholy tune, is homologous, or similar in function, to conjurational practices which are homeopathic in nature. That is to say, the blues and the spirituals perform, by means of music, the kind of transformation of experience and reality effected by conjurational strategies: a mimetic form of a disease is prescribed to cure that disease. We are now in a position to perceive how this homeopathic strategy can also operate in apocalyptic figuration, within an African American tradition that is simultaneously conjurational and Christian. To vary the emphasis, as an Afro-Christian strategy *conjuring Apocalypse* can employ toxic elements with a tonic or homeopathic intent. This apocalyptic strategy, however, precedes black Christian development as an essential feature of primitive Christian apocalyptic.

In her own terms the New Testament scholar Adela Collins has detected a homeopathic strategy operating in the Book of the Apocalypse itself—but only incipiently and imperfectly. Collins's discussion of this feature begins with the Aristotelian theory of catharsis, as signaled in her book titled *Crisis and Catharsis: The Power of the Apocalypse*. It is intriguing in this connection to note Collins's reference to the medicinal (or pharmacopeic) background of Aristotle's theory.

There is a certain analogy between Aristotle's explanation of the function of Greek tragedy and the function of Revelation. In each case certain emotions are aroused and then a catharsis of those emotions is achieved. Tragedy manipulates the emotions of fear and pity; Revelation, primarily fear and resentment. Aristotle's term "catharsis" is a medical metaphor. In its medical sense it refers to the removal from the body of alien matter that is painful and the restoration of the system to its normal state.[13]

The expurgation of fear, powerlessness, and resentment is the prescriptive intent or "medical" function of the Book of the Apocalypse, according to Collins. It its context, she claims, the book performed a cathartic function for Christian communities who were currently undergoing the brutalities of Roman persecution. The cathartic strategy operating in the text involved first intensifying fear, powerlessness, and resentment; a heightened effect was intended to purge, not to reinforce, those virulent feelings. Intensification was effected by the text's portrayal of the most extreme representations of Roman wealth and power. Roman excesses are symbolized, for example, by the perversities ascribed to the 'great whore of Babylon'. Notice the toxic impact of this depiction—a toxicity that consists not only in the extreme nature of the image but also in its misogynist overtones (Rev. 17.4–5):

> I saw a woman sitting upon a scarlet beast, full of names of blasphemy.... And the woman was clothed in purple and scarlet, and had ornaments of gold and precious stones and pearls, having a golden cup in her hand full of abominations and the unclean things of her fornication; and upon her forehead a name written, Mystery, great Babylon, the mother of the harlots, and of the abominations of the earth.

A cathartic potential may inhere in the virulence of this depiction, for example, when the image is subsequently applied to the experience of its contemporary audience: "And I saw the woman drunk with the blood of the saints, and with the blood of the witnesses of Jesus." It is reasonable to suppose that this image of cannibalism successfully "evoked and intensified" the audience's experience of persecution. Thus Collins remarks: "These vivid images are certainly designed more to evoke terror than to allay it." Indeed, such passages may have served a quasi-ritualistic function of preparing or priming the book's audience for subsequent cathartic release. However, Collins's hypothesis of cathartic release is unconvincing in the terms she employs. "The projection," she offers in explanation, "of the conflict onto a cosmic screen [of good versus evil], as it were, is cathartic in the sense that it clarifies and objectifies the conflict. Fearful feelings are vented by the very act of expressing them, especially in this larger-than-life and exaggerated way."[14] But it is obvious with a moment's reflection that intensification can just as easily lead to the reinforcement—and to the reactionary and aggressive escalation—of such feelings rather than to their purgation. Indeed, in a more critical mode Collins admits that "the long-term effectiveness of the Apocalypse's means of reducing tension . . . is questionable because of the use of violent imagery, its ambiguous effect, and the ambiguity of violence itself."[15]

Lacking a more developed theory of catharsis—or an analytic treatment of violence and religion, such as Girardian theory provides—Collins is unable to derive a principle for differentiating between the violent and nonviolent potentials in apocalyptic imagery. Nonetheless her cathartic hypothesis, if tenable, would improve upon Girard's treatment of the Book of Revelation. Collins's hypothesis would 'save the appearances' of the work as a creditable Christian text: as a text composed in strategic continuity with the nonviolent trajectory of the gospels.

> As far as we know, the book of Revelation was written to avoid violence rather than to encourage it. The faithful are called upon to endure, not to take up arms. The violent imagery was apparently intended to release aggressive feelings in a harmless way. Nevertheless, what is cathartic for one person may be inflammatory for another. . . . Norman Cohn's book *The Pursuit of the Millennium* has shown that apocalypatic imagery has been linked historically to violence under certain conditions.[16]

Here Collins (perhaps naively but nonetheless in 'good faith') assumes a nonviolent reading of the Book of Revelation that runs somewhat counter to Girardian interpretation of the text. Girard too propounds a nonviolent or "nonsacrificial" reading of Christian apocalyptic, but principally in its gospel forms rather than in the form of the Johannine Book of Revelation. "The usual version of the apocalyptic theme . . . is taken for the most part from John's Revelation, a text which is clearly less representative of the gospel inspiration than the apocalyptic chapters in the Gospels themselves (Matthew 24; Mark 13; Luke 17,22–37; John 21,5–37)."[17] In contrast to this evaluation, however, Collins's cathartic reading of John's Revelation renders the text more consistent with Girard's reading of the other apocalypses. In all four of the gospels, Girard maintains, apocalyptic violence is ascribed principally to human beings, not to a vengeful God. Contrary to conventional apocalyptic sensibilities, a nonsacrificial gospel reading locates the nemesis of our species internally rather than projecting it onto 'the sacred':

> We must realize that the apocalytic violence predicted by the Gospels is not divine in origin. In the Gospels, this violence is always brought home to men, and not to God. What makes the reader think that this is still the Old Testament wrath of God is the fact that most features of the Apocalypse, the great images in the picture, are drawn from Old Testament texts.
>
> These images remain relevant because they describe the mimetic and sacrificial crisis . . . but this time there is no longer a god to cut short the violence, or indeed to inflict it in the first place. So we have a lengthy decomposition of the city of man. . . . In the last days, we are told, "most men's love will grow cold." As a result, the combat between [mimetic] doubles will be in evidence everywhere. Meaningless conflict will be worldwide."[18]

Here Girard's analysis provides a principle for differentiating between violent (sacrificial) *and* nonviolent (nonsacrificial) appropriations of the cardinal text of Christian apocalyptic. It is, moreover, a principle that can supplement Collins's cathartic hypothesis. Nonviolent readings attribute apocalyptic phenomena to human agency and failure, not to the active will or menace of a vengeful God. The principle of violence, on the contrary, operates in a sacrificial matrix within which "violence and the sacred" are identical phenomena. In that matrix (Girard's "founding mechanism") God or the gods provide the ultimate sanction for humanity's intraspecies production of victims. 'The sacred' is constituted as the sanction for our collective, primordial recourse to victimization. When instead human beings acknowledge responsibility for violence as our own social production, we require no such divine sanction. We thus dissolve the identification of violence with the sacred and, therefore, the religious mystification of our own violence. The principle of differentiation proposed here, however, requires a theory of catharsis that proceeds in a direction contrary to Collins's formulations.

In the Book of Revelation, Collins claims, fear, powerlessness, and aggressive feelings are not pacified but first "heightened," next "projected" against the backdrop or screen of cosmic good and evil, and finally "released" by the resulting catharsis. "Intensification leads to catharsis. . . . By projecting the tension and the feelings experienced by the hearers into cosmic categories, the Apocalypse made it possible for the hearers to gain some distance from their experience. It provided a feeling of detachment and thus greater control."[19] On the contrary, a cogent theory of catharsis may require not further distance between apocalyptic terror and the audience, but *less* distance. Readers and hearers may need to experience the evils of Apocalypse as phenomena deriving from their own agency. "As long as the violence seems to be divine in origin, it really holds no terrors for anybody, since it is either an aid to salvation [violence as the sacred, as 'saving'], or it doesn't exist at all [via modernist demythologizing of the text]." In order for apocalyptic phenomena "to be truly frightening," they must convey "no remedy . . . no recourse of any sort"; they must cease to be "divine."[20] It is precisely in this vein that Girard insists on reading Christian apocalyptic in prophetic correlation with the contemporaneous threat of nuclear holocaust.

The importance of subjective distance for the effective release of emotions is a central feature in a "new theory of catharsis"—a theory that can further supplement and substantiate Collins's hypothesis. In his 1979 study, *Catharsis in Healing, Ritual, and Drama*, T. J. Scheff goes beyond Aristotle to more recent research in the psychotherapeutic uses of catharsis. Scheff first retrieves the early work of Sigmund Freud and his colleague Josef Breuer in the 1880s. Freud's initial but abandoned experimentation with cathartic release has been further developed in more recent research through the contemporary peer counseling movement.[21] On the basis of Scheff's research we may be able to extend to the Book of Apocalpyse an updated theory of catharsis that encompasses ritual healing, as well as dramaturgical performances and other performance media such as film. The point of intersection between these diverse areas of consideration is a concept borrowed from dramatic criticism: "aesthetic distance."

Aesthetic distance consists in the ideal achievement of a balance between the audience role of spectator and the role of participant in a dramatic performance. "At aesthetic distance, the members of the audience become emotionally involved in the drama, but not to the point where they forget that they are also observers." On this view aesthetic distance is achieved by an audience member's "simultaneous and equal experience" of being both participant and observer. Such equality of roles, to revert to the context of ritual healing or therapy, obtains when a client experiences "balance of attention." On the one hand, such a client attends to a (past or presenting) problem of psychic distress; on the other, he or she is aware of the present context of relative safety and detachment from distress. Here is where Scheff's theory locates the phenomenon of cathartic release. The "discharge" or venting of emotions occurs spontaneously in the moment of balance between the "reliving" of past distress and the reality of present detachment.

> Under these conditions, the repressed emotion ceases to be too overwhelming to countenance.... [But] deviation from the point of balance usually keeps discharge from occurring or stops discharge if it has started. If most of one's attention is absorbed by the distress, the distressful event is simply relived, as if it were happening again.... On the other hand, if most of one's attention is in the safe present, the repressed emotion is not sufficiently felt, and again, no discharge occurs. At the balance point, one is both reliving the event, and therefore feeling the emotions associated with it, and at the same time, observing the distressful event from the safety of the present.[22]

Not only in therapeutic 'performance,' but also in drama or entertainment, role balance can be jeopardized by either "overdistancing" or "underdistancing." Overdistancing occurs when audience members' emotions are relatively unaffected by a performance. "Total overdistancing involves responding only to the nonemotional aspects of the present environment—there is no emotional resonance at all." Underdistancing occurs conversely as an audience virtually experiences or reexperiences the emotions associated with a dramatized event. "Total underdistancing means complete immersion in some past scene and complete oblivion to the present."[23] Insofar as the dramaturgic goal is audience catharsis, the requirements of aesthetic distance will entail sufficient immersion in the emotions applicable to a participant in a given experience, but also some measure of detachment appropriate to an observer of that experience.

Turning next to a discussion of the "ritual management of repressed emotion" in media entertainment, Scheff evaluates the evocation of fear and anger in contemporary film and television. In particular he is critical of dramas that typically underdistance their audiences. "In the case of the various types of dramas based on fear ... almost all of the entertainment available to the public is underdistanced." In general, fear dramas are commercially successful in achieving their goals of shocking or terrifying audiences through depicting scenes of horror, violence, or disaster. But viewers of such entertainment seldom evince the signs of emotional catharsis of fear and terror: relatively sustained shivering and sweating followed by a reduction of psychic tension. "The audiences I have observed seem to experience the still, bone-dry fear which is indicative of distress rather than catharsis ... [they are] more, rather than less, tense at the end of the drama ... as if it were happening to them.... The fear that is restimulated is so overwhelming that no catharsis takes place." In this connection Scheff concludes this subject with a recommendation regarding "fear-anger" entertainments generally. He observes that it might be profitable, not only therapeutically but also commercially, for the entertainment industry to experiment with increasing audience distance in fear-anger dramas. Specifically there could be more use of such devices as comic relief: films that mix fear with laughter "may be sufficiently distanced to allow the catharsis of fear, as well as anger and/or embarrassment." However, it is also possible that in other performance genres, such as live theatre, the problem is not underdistancing but overdistancing. In this regard Scheff calls for a multidirectional approach to redress the impoverished state of ritual performance in modern societies. "Optimal distancing" for catharsis may require explorations in the direction of increased distance in film and television media, but decreased distance in live theatre and other performance genres.[24]

At this juncture it is useful to return to the attempt to render a cathartic reading of the Book of the Apocalypse. However, since we are now considering processes involving literacy—that is, reading a bibical text—we require not only a theory of catharsis but also an attendant theory of reading as a kind of performance. Such a view has become available in recent reader-response theory. In fact, we have already reviewed a theory of the performative reading of texts: Christopher Collins's theory of iconic reading in which texts function instrumentally as icons rather than objects. Reading written texts can be considered a ludic performance or play activity. Such performances treat a text as a verbal or literary icon by transposing its dramatic images or representations onto the psychic screen or cultural stage of one's imagination. In these terms Collins develops a reader-response theory in which the reader performs a "*poiesis*" in accordance with rule-governed sets of ludic cues. This theory claims that reading (or hearing) a text can be performative in some way analogous to acting a scene or playing a musical score. In particular, we have examined reading-as-performance with respect to Rastafarian appropriations of prophetic biblical texts, and Du Bois's crafting of "enactive interpretations.".

We are now in a position to proceed with a new, performative reading of the Book of Apocalypse. Girard's nonsacrificial reading of apocalyptic literature is based on the rubrics of a nonviolent gospel—a gospel that does not require violence-as-the-sacred. More particularly: this reading

effectively decreases the distance between the audience and the terrors
depicted in the Apocalypse by (not demythologizing, but rather) correlating it with our contemporary situation. 'Contemporizing' the text, however, is also effected without (the distancing device of) positing a divine
agent of apocalyptic violence. Girard avoids both means of distancing the
text by one of his most provocative claims: The ongoing threat of nuclear
holocaust reveals that we human beings have made our global situation
'objectively apocalyptic'.

> As for the terrors of the Apocalypse, no one could do better in that respect
> nowadays than the daily newspaper. I am not saying that the end of the world
> is at hand . . . all the elements that I draw out in my analysis have something
> positive about them. . . . To say that we are objectively in an apocalyptic sit-
> uation is in no sense to "preach the end of the world". It is to say that mankind
> has become, for the first time, capable of destroying itself, something that
> was unimaginable only two or three centuries ago. The whole planet now
> finds itself, with regard to violence, in a situation comparable to that of the
> most primitive groups of human beings, except that this time we are fully
> aware of it. We can no longer count on sacrificial resources based on false
> religions to keep this violence at bay.[25]

It is evident that this reading effectively decreases the distance (that
is, increases the proximity) between readers and the terrors of the Apocalypse. At the same time, however, Girard implicitly seeks a "balance of
attention" (Scheff) by disavowing the immediate threat of nuclear holocaust. Accordingly, he stresses our essential continuity with our primitive
predecessors in respect to their endemic recourse to violent resolutions of
conflict. "In effect, people's basic make-up has not changed in the slightest,
and that is precisely what makes our situation so dangerous." On the other
hand, labeling our situation objectively apocalyptic does not require
preaching an imminent end of the world. To the contrary, "all the elements
that I draw out in my analysis have something positive about them." The
positive side of Girard's analysis consists not in claiming any progress,
improvement, or evolution of human nature—"violence has always been
inherent in man"—but rather in the (too infrequently) stated claim that
endemic violence can be and, indeed, has been overcome: "The use of
violence does not rest on an irresistible instinct. The proof of this lies in
the fact that the ultimate violence has been at our disposal for a while,
and, up to the present time at any rate, we have not yielded to the temp-
tation to use it. . . . We are reaching a degree of self-awareness and responsibility that was never attained by those who lived before us."[26]

Thus a crucial element in the phenomenon of catharsis—balance of
attention (in therapy) or aesthetic distance (in drama)—is implicitly operative in Girardian interpretation of apocalyptic literature. Although that
interpretation brings the terrors of the Apocalypse home to humanity
where violence originates, it allows, in *theory* at least, for an efficacious
distance from the overwhelming, apocalyptic consequences of our violence.
It is, however, an *empirical* question whether that distance is sufficient for

fear and anger to be released or discharged, rather than forestalled by the prevailing recognition that our situation is objectively apocalyptic. Here again we come to the weakness of a Girardian analytic of sacrificial violence: its structural indifference to matters of praxis and ameliorative transformation. That indifference renders the establishing of cathartic balance or "optimal distancing" (Scheff) only an implicit, intuitive achievement in Girardian theory, rather than an explicit systematic feature. To vary the terms: Girardian theory lacks a performance orientation. Here the analytic approach to apocalyptic themes—what Christopher Collins calls "critical interpretation" of texts—needs to be complemented by a performative approach—an "enactive interpretation" of texts. Now such a performative or enactive reading of Apocalypse is implicit (albeit unrealized) in Adela Collins's cathartic interpretation of the text. Like Christopher Collins's iconic reading that plays or reenacts the text on the screen of the imagination, Adela Collins's readers project the drama of Apocalypse onto a cosmic screen of good and evil.

Most remarkably Adela Collins's interpretation offers an ameliorative or "antistructural" form of catharsis, in contrast to Girard's theory of normative or "structural" (Victor Turner) catharsis. Normative catharsis consists in the (nonameliorative) release of emotions that operates in acts of sacrificial violence—acts that effectively reestablish and maintain the status quo. By comparison Collins postulates a catharsis that subverts or counters the normative operation of vengeful violence in society, in place of a catharsis that merely facilitates or mediates the cycles of violence.[27] In this regard we see how each author, Adela Collins and René Girard, needs what the other can provide. Collins needs Girard's analytic of violence; Girard needs Collins's hypothesis of ameliorative catharsis. (And each needs Scheff's new theory of catharsis, alongside Christoper Collins's performance theory of iconic reading.) Were Girard's suggestively balanced reading transposed to a performance mode, in which a "fear-anger drama" (Scheff) could be enacted with suitably rendered cues for emotional discharge, we might reasonably expect to see dramaturgical moments of optimal distancing and effective catharsis.

The basis for such a transposition is evident in Christopher Collins's distinction between, and correlation of, critical and enactive interpretation. These two modes of interpretation are each necessary for an adequate theory of performative reading. A literate culture requires both critical forms of "mediate exegesis"—as in scholarly commentary—and enactive forms of "immediate construal"—as in poetic performance. The goal here is not to suggest one interpretive mode rather than the other; rather, these two modes of interpretation reciprocally inform and advance each other. "A critical interpretation is only as informative as its enactive interpretation was skillful . . . an enactive interpretation is only as skillful as its critical interpretation was informed."[28] Having been informed by Adela Collins's and René Girard's critical interpretations of the Book of the Apocalypse, how do we shift to the skills of enactive interpretation?

Providence

*The American Negro has always felt an intense personal interest in
discussions as to the origins and destinies of races.*

W.E.B. Du Bois, "The Conservation of Races"[29]

We are now in a position to examine the criteria for achieving ritual and
cathartic efficacy in an area more central to the interests of this study. With
his disavowal of preaching "the end of the world," Girard inadvertently
evokes a little known nineteenth century Afro-American work, *The End
of the World* (1888). This text is representative of a genre of black American
apocalyptic literature called "race histories." Its author, Theophilus Gould
Steward, was an articulate theologian of the African Methodist Episcopal
church writing at the end of the nineteenth century. His work is useful
here for recontextualizing the foregoing discussion, in order to take into
account black Christian literary and performative modes of conjuring apocalypse. An element of agonized conflict is immediately evident in the
expulsory tone of the subtitle, "Clearing the Way for the Fullness of the
Gentiles." Mimetic rivalry is also evident, appearing immediately in the
ethnic references of the introduction.

> The world is made for the Saxon, who is its lord, and all the other tribes are
> to clear it up for him. Surely this God is very good to the Saxon, although
> so very cruel to the Indian, Negro, Chinaman and the rest of mankind. . . .
> Again, I say, What a very useful God the Saxons have! Oh, that each of the
> other races had one just as good! As these theological carpenters and sculptors
> have made this one, perhaps some genius will do as much in the future for
> the other races. It is an old saying that what has been done may be done.[30]

Steward's declaration, "what has been done may be done," scarcely
veils his intention to emulate those white "theological carpenters and
sculptors." He does so by means of a theological reconstruction of apocalyptic that 'clears' them out of the way of true Christianity. Steward indeed
mimicks the ethnic aggrandizement of his white Christian antagonists, yet
he does so by means of interpretive feats involving great ludic skill. He
first countervalorizes dark-skinned and African peoples within an apocalyptic perspective that includes the central text of Ethiopianism, Psalm
68.31. "After God shall have scattered the fierce nations who delight in
war, and who have carried the destructive science to a height never before
conceived of, then 'Princes shall come out of Egypt, and Ethiopia shall
stretch out her hands unto God.' "[31] Second, within this apocalyptic context Steward claims the operation of mundane and historical (as distinguished from supernatural) agencies of nemesis. Indeed, like Girard he
closes the distance between his audience and the terrors of the Apocalypse
by attempting to render his contemporary situation objectively apocalyptic.
"The collapse of the dominant civilization of the world," he declares, "will

be the result of laws which are already at work, physical, social and moral."
Nevertheless, Steward also insists, the mundane nature of this apocalyptic
transformation

> will not at all detract from its character as a visitation from the Almighty, nor
> prevent its being accepted as the fulfillment of prophecy.... The work of
> destruction may be accomplished by secondary causes, and those who are
> smitten may be too blinded by sin and worldliness to see in it the hand of
> the Divine smiter; but it will be Jehovah's doings nevertheless."[32]

In this nimble manner the author asserts both the human agency of the
"work of destruction" and the conventional apocalyptic element of divine
judgment.

A third aspect of Steward's reading also parallels the Girardian reading:
specifically the distancing feature by which the authors commonly disavow
immediate cataclysm. For Steward too the objective reality of the Apocalypse does not necessitate a literal or immediate "end of the world."
Rather, the Apocalypse entails the end of this age only: the age of white
Christian domination. "We never read in the Scripture of the end of the
kosmos; but we do read of the end of the aion. Hence we may dismiss
from our thoughts the alarming ideas of physical catastrophe, which have
so long clustered around this subject ... and may say in advance, that the
end of this age does not necessarily involve the destruction of all things."[33]
The age that is ending is characterized by a distorted form of white su-
premacist Christianity, which Steward described as a "clan." The term
connotes European ethnocentrism as a source of inauthentic Christian
culture. Purging the religion of "this clan principle," Steward claims, will
allow a new age of genuine Christian history to ensue.

> The white races of the earth have formed a mighty confederation with the
> Anglo-Saxon leading, and ... they have modified the Christian idea to an
> alarming extent by this clan principle, so that it has become a white man's
> religion, and is so recognized by the darker races ... [who] instead of being
> blessed by Christian intercourse have been cursed by it.... [T]he rule holds
> that white races taken as a whole, have much more of clan than of Christ,
> and the clan spirit repels more generally than the Christ-spirit attracts....
>
> The writer thinks that the indications point about as follows: A great
> crash of the Christian nations, and a liberation of the Christian idea from the
> dominance of the principle of clan, and a consequent purification of the
> Christian Church.[34]

What Steward's 'ethnological' reading lacks, of course, is the deep
structural recognition that we find in Girard among others: that the true
nemesis of human felicity is humanity itself, not one ethnic group or some
particularly diabolical historical development, institution, or culture.[35] It
is unclear whether Steward's reading simply constitutes his own ethnic
animus (despite his disclaimer)—thus inciting the sacrificial impulse toward the primitive sacred in the form of a divine nemesis. For, despite
his portrayal of anonymous historical processes producing "a great crash

of the Christian nations," those processes may represent Steward's own ethnocentric will to constitute and expel new victims for the sake of a sacralized purpose: "clearing the way for the fullness of the Gentiles." As we know too well, such an expression—"clearing the way" (literally, for the fulfillment of the kingdom of God among all peoples) all too readily comes to mean the expulsion of victims for the sake of 'righteous' persecutors. In this regard Steward's exposition is ambiguous, and perhaps for that very reason is suspect. Nonetheless it does evince an awareness that the end of victimization is the critical factor, a *conditio sine qua non*, of the kingdom of God. "The end is not far. The bloody wave will soon have spent its force, and then shall the end come—the end of war and oppression; the end of insolence of white pride and black contempt, and the ushering in of a new era in which righteousness shall prevail, and the peaceful, loving spirit of the Lord Jesus Christ shall reign over all the earth."[36] This is essentially the messianic vision of the Book of Isaiah, with its prophecy of a 'peaceable kingdom' in which "they shall not hurt or destroy in all my holy mountain, for the earth shall be full of the knowledge of the Lord as the waters cover the sea" (Isaiah 11.9; cf. 65.25). Nonetheless it falls short of an explicit statement of ethnic reconciliation, such as we find in Martin Luther King's famous declaration: "I have a dream that one day on the red hills of Georgia, sons of former slaves and sons of former slave-owners will be able to sit down together at the table of brotherhood."[37]

The rise of black peoples in a third age

In a second race history we find formulated a little known aspect of this genre: the expectation of a 'third age' configuration in the providential destiny of black peoples. Perhaps the most lucid explication of this view was provided by the AME Bishop James Theodore Holly. In his 1884 work, "The Divine Plan of Human Redemption, in Its Ethnological Development," Bishop Holly offered a theological rationale for the demise of white supremacy and the elevation of black humanity. That rationale depended on a tradition of biblical typology derived from Genesis, chapter 10, in which the sons of Noah constitute the prototypes for human ethnic groups after the great flood that depopulated the earth. The repeopling of the earth by Noah's sons and their descendants—the sons Shem, Japheth, and Ham—figuratively accounts for the differing ethnic divisions (respectively) of Semites or Semitic peoples; Japhetic, or Caucasian peoples; and Hamitic, or black peoples. (It is evident that this typology leaves out of account Asian, Oceanic, and aboriginal peoples.) With this typological framework in mind, Holly espoused a providential or 'dispensational' schema in which the conversion and missionary labors of black peoples consummate the kingdom of God on earth:

> In the development of the Divine Plan of Human Redemption the Semitic race had the formulating, the committing to writing and the primal guardi-

anship of the Holy Scriptures during the Hebrew dispensation. The Japhetic race has had the task committed to them of translating, publishing and promulgating broadcast the same Holy Scriptures.... But neither the one nor the other of those two races have entered into or carried out the spirit of those Scriptures. This crowning work of the will of God is reserved for the millennial phase of Christianity, when Ethiopia shall stretch out her hands directly unto God.... [Both Semitic and Japhetic races] alike await the forthcoming ministry of the Hamitic race to reduce to practical ACTION that spoken word, that written thought.[38]

Providential construals of black experience and destiny possess a mixed lineage. The effort to derive a "divine plan of human redemption" that includes not only the peoples of the Bible, and of European church history, but also the darker peoples of the world preceded these efforts by Bishop Holly. St. Clair Drake has indicated that the earliest source of providential thought in black religious tradition was the missionary theology of northern Quakers and antislavery leaders. At first it was predominantly white religious leaders who wanted to send black Christians to Africa as teachers and missionaries. As a rationale they "elaborated the doctrine of 'Providential Design' to give sanction to their plans—'God, in his inscrutable way, had allowed Africans to be carried off into slavery so that they could be Christianized and civilized and return to uplift their kinsmen in Africa.'"[39] We can now recognize the inherently sacrificial and, indeed, prejudicial content of such a doctrine. It is a theologically dubious and spiritually pernicious (even diabolical) rationalization of the Atlantic slave trade. For it posits the menace of a deity who permits a historical process that wastes thousands of lives, in order to 'save' and civilize the resulting survivors and their psychologically maimed descendents. Evident in that doctrine is a view of divine nature deriving from a 'primitive sacred', in which the blood of victims is required for the god to save, protect and deliver. We may well ask, moreover, what value such teachings assign to the humanity of its prospective converts.

It was inevitable that many black religious leaders would internalize the subordinate view of African humanity deriving from white missionary efforts to rationalize Christianity with the slave trade. It is understandable, therefore, that "the combination of Ethiopianism with the doctrine of Providential Design also stereotyped Africa as 'heathen', 'dark', and 'benighted' ... [and that] in general, Negro missionaries were as censorious of African customs as white missionaries were."[40] Eventually, however, black intellectuals began to recast providential thought in terms consistent with the survival and human integrity of African peoples. Indeed, this providential revisionism supplied the spiritual motivation behind black nationalism in the United States and Pan-Africanism abroad. A variant of the doctrine of "providential design" emerged in which the exaltation of Africa was one pole, and the nemesis of the West was the other. As stated by Wilson Jeremiah Moses, the Ethiopian prophecy in Psalm 68.31 "came to be interpreted as a promise that Africa would 'soon' experience a dramatic political, industrial, and economic renaissance." But this interpretation was also capable of conveying a parallel feature: ethnic elevation on the one hand would result in ethnic nullification on the other. "Others have insisted that the real meaning of the scripture is that some day the black man will rule the world. Such a belief is still common among older black folk today."[41] Thus Ethiopianism was the source of this notion of a third age in which African peoples will gain the ascendancy over their former oppressors.

The fall of Babylon America and the end of white rule

*The Black Panthers absorbed much of the Black Muslim vocabulary,
particularly the idea of the Beast—which Elijah Muhammad identified
from the Book of Revelation as Babylon America. These names were used
to describe the cruelty and animality of white devil America. They
became essential to my own word list; and when Lee Lockwood
interviewed me in Algiers [he asked] . . .*

L: *Why did you say "Babylon"?*

C: *Because of all the symbols that I've ever run across to indicate a
decadent society, I find the term Babylon, which I take from
Revelation in the Bible, to be the most touching. . . .*

L: *It's an analogue?*

C: *It's an analogue. The United States of America is described in
Revelation. I'm not being a prophet. I'm just saying that I dig that.*

Eldridge Cleaver, *Soul on Fire*[42]

As in preceding chapters, I am concerned not to limit this treatment to the biblical figural tradition, or to Christian modes of conjurational spirituality. It is particularly fitting to extend our view of these phenomena to extra-Christian social-historical configurations—particularly with this treatment of the Apocalypse figure. For that figure bears major import for iconic formations in another African American religious culture: black Muslims, or the former Nation of Islam. Despite the centrality of the Koran as the sacred text of all Islamic cultures, Muslims also revere the Hebrew Bible and the Christian Scriptures as supplemental texts. Indeed, the three religions descended from Abraham—Judaism, Christianity, and Islam—share the Hebrew Scriptures as a foundational text. Significantly, each of them adds another sacred text or texts as canonical resources for interpreting and even supplanting that foundational text (respectively: Mishnah, Talmud, and other rabbinic literature for Judaism; the New Testament with or without the Apocrypha, creeds, and conciliar and magisterial dogma for Christianity; and the Qur'an, Sunna, and Hadith for Islam).

Islam, like Judaism and Christianity, is also a "religion of the book" in this regard. Yet the preeminence of its holiest book does not entirely eclipse the foundational stature of the Hebrew Bible, and allows for some degree of legitimacy for the Christian Scriptures as well. This secondary or tertiary role of the Bible has been even more pronounced in the heterodox Islam of African Americans, since the inception of the Nation of Islam in the 1930s. The inevitable closeness and complex interrelationship of black Muslims and black Christians in North America have made the Hebrew and Christian Bibles to some degree as familiar as the Qur'an.[43] That familiarity has resulted in similar traditions of figural interpretation and parallel iconic appropriations of biblical themes, in particular the theme of Apocalypse.

The roots of the Nation of Islam are historically grounded in the Moorish Science Temple of Noble Drew Ali, founded in Newark, New Jersey, in 1913 and later extended to other northern cities (Harlem, Philadelphia, Pittsburgh, Chicago, and Detroit) and to some southern cities. In 1930 a rival in the Detroit temple, Wallace D. Fard, formed a new group, which eventually deified him after his mysterious disappearance in 1933. The group subsequently became the Nation of Islam under the leadership of Messenger Elijah Muhammad. Elijah Muhammad's theology and teachings were indebted not only to Fard, his predecessor and mentor. More generally, the doctrine of this new Islamic community emerged from the esoteric framework of the Moorish Science Temple, in which Noble Drew Ali taught "a synthesized version of orthodox Islam, Garveyism, Christianity and various extractions from oriental philosophy."[44] A principal innovation contributed by Elijah Muhammad, however, was the doctrine that Islam is the true religion of all black people, in which Allah (God) is also black and black men are representative of Allah. By contrast Christianity is the religion of an enemy people—that is, white people, who are also (by a logic of inversion) representative of the devil despite their claims to be a divine, "chosen people."

An elaborate myth of origins accompanied this dualist theology, in which the domination and enslavement of darker peoples, by Europeans in the slave trade and during colonialism, are represented figurally. These are the deeds of "devils," of people of the "serpent" (Genesis 3) or the "beast" (Revelations 13), who will be overthrown by Allah and his (black) people at the end of the age.[45] Not incidentally, this theology and its mythic elements accompanied a vigorous social agenda that economically rehabilitated some urban communities. Black Muslim theology also spiritually revitalized black people who were neglected or disaffected by traditional black churches. In this connection the prison ministry of Elijah Muhammad and the Nation of Islam was uniquely successful. To that ministry we owe the reclamation of Malcolm X from his preconversion life-style as drug pusher and pimp, his eventual emergence as a black nationalist leader, and now, since his martyrdom in 1965, his valorization as a contemporary icon of black power.

Elijah Muhammad was clearly conversant in the hermeneutic tradition of biblical typology, as evinced by his use of the expression "the type of Israel."[46] In another of his figural constructions, "Babylon," the consummate wicked city in the Book of the Apocalypse, becomes a type of the

United States or "America." The focal point for this construction, however, is not merely the identity of the type (Babylon) and its antitype (the United States), but rather their identical destiny—their common destruction and fall. In fact, we find in the following statement the kind of 'figural prophecy' (Auerbach) that characterizes the Christian hermeneutic tradition of typology: "We can easily and truthfully," the Messenger declared, "liken the fall of America to the prophetic symbolic picture given in the (Bible) Revelation of John (18:2). The name Babylon used there does not really say whether it is ancient Babylon or a picture of some future Babylon.... [The description] given to the Babylon by the Prophets compares with the present history and people of America and their fall."[47]

The question most pertinent to this chapter is whether black Muslim figuration of 'white America' duplicates, or rather obviates, the sacrificial construal that we find in black Christian figuration of a third age. To answer this question we turn to Elijah Muhammad's version of the kind of race histories presented by Theophilus Gould Steward and Bishop James Theodore Holly. From personal revelations granted him by Allah, the Messenger taught that "6,600" years ago, after many thousands of years of black people's inhabiting the earth alone and creating great works of civilization, "our nation gave birth to another God whose name was Yakub [Jacob]." This Yakub was a scientist who created over the course of six hundred years a new race, white people, who were destined to rule "the original black nation" for the ensuing six thousand years. But for two thousand years these people were forcibly exiled by black people in "West Asia (now called Europe)," lacking civilization, industry, and literature. "After that time, Musa (Moses) was born: the man whom Allah would send to these exiled devils to bring them again into the light of civilization." Thus Moses was sent as a savior to white people by this other "God" born to the original nation— a God referred to variously as "Allah," the "One God," the "Supreme Being," and the "Mahdi," but also as the "Christ, the "second Jesus," and the "Son of Man." After the tyrannical rise of white people this complex, interfaith 'anthropotheism' will result in a final victory and vindication of black people, when Allah "will remove and destroy the present, old warring wicked world of Yakub (the Caucasian world) and set up a world of peace and righteousness, out of the present so-called Negroes, who are rejected and despised by this world."[48]

The genocidal destruction of "the Caucasian world" in this schema is certainly typical of sacrificial social systems. Such systems invariably require the expulsion, mutilation, or outright murder of some class of "wicked" victims in order to achieve the unanimity and felicity of a favored community. Elijah Muhammad's race history implies as well the destruction of other peoples of color, whom the evil scientist (Yakub), operating here as a trickster figure, created through a process of gradual deracination: "After the first 200 years, Mr. Yakub had... [all brown babies]. After another 200 years, he had all yellow or red.... Another 200 years, which brings us to the six hundredth year, Mr. Yakub had an all-pale white race

... made devils [who] were really pale white, with really blue eyes; which we think are the ugliest of colors for a human eye."[49] In this last comment we find a definitive indication of the motivating force driving this race history. Its inversions of white dominance and black destruction are designed to be curative or antidotal: in this instance, to supply an antidote to the perceived African American pathology of an internalized beauty standard, in which blacks are attracted to ("unlike") white-skinned and blue-eyed people and are repelled by their own ("like") dark-skinned and dark-eyed people. Indeed, speaking explicitly against the Christian doctrines of 'brotherly love' and 'love of enemies', the Messenger declared that white people are not naturally related to black people: "they are not your brothers . . . [make] brotherly love and friendship with your own kind first."[50] It is evident that this antidotal strategy for promoting ethnic solidarity among black people is chauvinistic and virulent toward white people and leads eventually to the genocidal trajectory of Elijah Muhammad's race history.

Precisely that antagonized aesthetic of blacks preferring black-skinned people and rejecting white-skinned people was transcended in the remarkable transformation of Elijah Muhammad's former disciple, Malcolm X, during his pilgrimage to Mecca (the Hajj) in the last years of his life. In 1964, after his breakup with Elijah Muhammad as his spiritual father and mentor, Malcolm X undertook the Muslim obligation of the Hajj. The break with Muhammad, on one side of that experience, and his assumption of a new name—El-Hajj Malik El-Shabazz—on the other, encompassed Malcolm's stunning turn from the race antipathies of the Nation of Islam. Again and again in *The Autobiography of Malcolm X* we find his transformation expressed in the aesthetic terms articulated: in terms typical of Elijah Muhammad's rhetoric of distinctions, in which the value of human beings is determined on the basis of differences in skin, eye, and hair color. In retrospect Malcolm acknowledged that his gradual (perhaps never fully consummated) repudiation of those terms began with the airplane trip to Jedda: "Packed in the plane were white, black, brown, red, and yellow people, blue eyes and blond hair, and my kinky red hair—all together, brothers! All honoring the same God Allah, all in turn giving honor to each other."

After arrival and upon reflection on the day's events the next morning, "I first began to reappraise the 'white man'. . . I first began to perceive that 'white man,' as commonly used, means complexion only secondarily; primarily it described attitudes and actions . . . toward the black man, and toward all other non-white men." The climax of his dawning realization occurred, however, during this pilgrim's ritual performances of circumambulating the Ka'ba, making the seven-times run between the hills of Mt. Al-Safa and Al-Marwah, and during prayers in the holy city of Mecca, in Mina, and on Mt. Arafat. "There were tens of thousands of pilgrims, from all over the world. They were of all colors, from blue-eyed blonds to black-skinned Africans. But we were all participating in the same ritual, displaying a spirit of unity and brotherhood that my experiences in America had led me to believe never could exist between white and non-white." With a kind of intoxication or euphoria that characterizes similar experiences of communitas (Victor Turner), Malcolm seemed never to exhaust his sense of amazement and surprise in discovering the fundamental humanity of white people. Again, that discovery was mediated through the communal and ritual practices of his religious tradition.

> I have eaten from the same plate, drunk from the same glass, and slept . . . on the same rug—while praying to the *same God*—with fellow Muslims, whose eyes were the bluest of blue, whose hair was the blondest of blond, and whose skin was the whitest of white. And in the *words* and in the *actions* and in the *deeds* of the "white" Muslims, I felt the same sincerity that I felt among the black African Muslims of Nigeria, Sudan, and Ghana.[51]

Malcolm's metamorphosis as El-Hajj Malik El-Shabazz implicitly repudiated the anti–white race hatred and ethnic chauvinism of Elijah Muhammad. The new leader did not live long enough to consummate his transformation in institutional forms. Some indication of the direction he might have taken is available, however, in the ongoing transformation of the African American Muslim community since the death of Elijah Muhammad in 1975. The succession of his son, Imam Wallace Muhammad, has resulted in new ventures of outreach to white Americans, on the one hand, and to the worldwide Islamic community on the other. That dual outreach has resulted in a major reversal in character from the former Nation of Islam, which one commentator has described as the shift from "religious Black nationalism to Americanism and orthodox Islam."[52] Not all African American Muslims are reconciled with this shift, however, notably the communities gathered under the alternative leadership of Minister Louis Farrakhan. Farrakhan resists both the rapprochement with white Americans and the new emphasis on orthodox Islam, to the degree that these developments detract from the ethnocentric or chauvinist emphases of Messenger Elijah Muhammad, and from his continued valorization as the great spiritual leader of Islam in America.

The alternative trajectory of El-Hajj Malik El-Shabazz may be discerned, as a final contrast, in his ameliorated version of black America's race history as expressed at the end of *The Autobiography of Malcolm X*. Note here the conditional versus the typically predestined feature of a third age, that coming age in which the white Christian West will be judged for its sins against peoples of color. "I believe that God now is giving the world's so-called 'Christian' white society its last opportunity to repent and atone for the crimes of exploiting and enslaving the world's non-white peoples." We even find a note of redemptive mission in Malcolm's statement that "sometimes, I have dared to dream to myself that one day, history may even say that my voice—which disturbed the white man's smugness, and his arrogance, and his complacency—that my voice helped to save America from a grave, possibly even a fatal catastrophe." The transformation of Malcolm X evinced here is remarkable, in that he opens himself to the possibility of a messianic vocation as a 'savior' in relation to a society which he elsewhere caricatured as 'white Devil America.' In this regard he converges in suggestive ways with his erstwhile Christian antagonist, the redemptionist Martin Luther King, Jr.[53] Moreover he testifies to the antisacrificial capacities available in Islam as a world religion, for overcoming expectations of—and inducements toward—a racial Armageddon as a necessary, predestined feature of Apocalypse.

In conclusion, we may observe a singular black theological effort to articulate the termination of white rule in a manner that does not require the exclusion of Western or European peoples. For this reason I return to the "ethnological" schema of peoples in the race history of Bishop James Theodore Holly: the 'table of nations' in which (1) the Semitic peoples received God's first dispensation (revelations to Israel), (2) the Japhetic peoples received the second dispensation (Caucasian rule and propagation of Christianity), and (3) the Hamitic peoples will receive God's final dispensation (African people's fulfillment of divine intention). In the following passage Bishop Holly directly addressed the issue of the exclusion of Caucasian peoples in a future divine dispensation: "Can it be," he queried, anticipating the third age of the divine plan,

> that [the Japhetic nations] shall be excluded from the active instrumentalities that God will employ during the millennial phase of the Christian dispensation, because they have so dreadfully perverted the peaceful spirit of the Gospel by the warlike disposition which they have manifested and the predatory incursions they have carried on during its apostolic phase which was committed to their ministry?

Opinions will differ whether the bishop's answer lacked the integrity of his office, or rather evinced a kind of theological maturity and wisdom, not 'to go where angels fear to tread'. Instead of boldly declaring either the exclusion or the inclusion of white peoples in the climactic age of divine fulfillment, Bishop Holly abstained from answering the question.

> Here it is better to imitate the silences of Holy Scripture than to attempt to answer a question that would perhaps lead us into an irreverent intrusion into the inscrutable designs of Almighty God upon which He has not spoken to us.[54]

A thoroughly nonsacrificial figuration of Apocalypse is perhaps not available in any African American traditions of biblical typology, whether Muslim or Christian. The nearest approach is this notable restraint of Bishop Holly, not to advance down the path of prophesied exclusion taken by his fellow black Christians such as Theophilus Gould Steward—and by his future coreligionist—Messenger Elijah Muhammad. In any case we still await a figural appropriation of the Book of the Apocalypse that is able to affirm the cosmic defeat of evil, without also therein sanctioning the genocidal elimination, or the subordinating exclusion, of other human beings. The question remains whether the lack of such a nonsacrificial rendering of Apocalypse indicates the inherent virulence of the text and the figure which encodes it. It may alternatively indicate our current incapacity for figurations that are not contaminated by our collective rivalries and vengeance. Perhaps the final answer is as much a matter of praxis as of theology.

There is, however, one ancient resource for a nonsacrificial rendering of Apocalypse. It comes to us from the Apocrypha (Greek: hidden, obscure), that collection of ancient Christian literature which is excluded from both the Catholic and Protestant canons of New Testament texts (an ironic exclusion, in the context of this discussion). Like a hidden and well-preserved tonic, this fragment lies waiting for formulation by some inspired future figuralist, one able to make efficacious for our earthly existence a dissenting effort to *include* and reclaim the tormented souls enduring their apocalyptic judgment.

> And unto them, the godly, shall the almighty and immortal God grant them another *boon*, when they shall ask it of him. He shall grant them to save men out of the fierce fire and the eternal gnashing of teeth: and this will he do, for he will gather them again out of the everlasting flame and remove them elsewhither, sending them for the sake of his people unto another life eternal and immortal, in the Elysian plain where are the long waves of the Acherusian lake exhaustless and deep bosomed.[55]

Notes

1. Donald G. Mathews, *Religion in the Old South* (Chicago: University of Chicago Press, 1977), p. 231.

2. Some of the most notable, recent exceptions are historical and theological works that appeared within five years of Mathews's statement. Historical accounts of slave beliefs concerning the Apocalypse, particularly the divine judgment against white Americans and slavery, can be found in Lawrence Levine's *Black Culture and Black Consciousness, Afro-American Folk Thought from Slavery to Freedom* (New York: Oxford University Press, 1977) and in Albert Raboteau's *Slave Religion: The "Invisible Institution" in the Antebellum South* (New York: Oxford University Press, 1978), each reviewed subsequently. Theological treatment of 'black eschatology' can be found in James Cone's writings, particularly in *The Spirituals and the Blues: An Interpretation* (New York: Seabury Press, 1972) and in Gayraud Wilmore's *Last Things First* (Philadelphia: Westminster Press, 1982), each referred to subsequently.

3. Mathews, *Religion in the Old South*, p. 231.

4. Ibid.

5. Levine, *Black Culture and Black Consciousness*, p. 43.

6. Greek *logos*: word. Cf. the first verse of the Gospel according to John: "In the beginning was the Word, and the Word was with God, and the Word was God."

7. Wilmore, *Last Things First*, pp. 90, 95.

8. George Shepperson, introduction to *The African Diaspora: Interpretive*

Essays, ed. Martin Kilson and R. Rotberg (Cambridge, Mass.: Harvard University Press, 1976), p. 8.

9. Here a slave woman, "Aggy," responds to her master's beating of her daughter. Raboteau, *Slave Religion*, p. 313.

10. Sacvan Bercovitch, *The American Jeremiad* (Madison: The University of Wisconsin Press, 1978), p. 132f.

11. Wilson Jeremiah Moses, *Black Messiahs and Uncle Toms: Social and Literary Manipulations of a Religious Myth* (University Park: The Pennsylvania State University Press, 1982), p. 33f.

12. James H. Cone, *The Spirituals and the Blues: (An Interpretation* (New York: Seabury Press, 1972).

13. Adela Yarbro Collins, *Crisis and Catharsis: The Power of the Apocalypse* (Philadelphia: Westminster Press, 1984), pp. 152–53.

14. Ibid.

15. Ibid., pp. 170–71.

16. Ibid.

17. René Girard, *Things Hidden Since the Foundation of the World*, with J.-M. Ourgoulian and G. Lefort (Stanford: Stanford University Press, 1987), p. 188.

18. Ibid., p. 186.

19. Collins, *Crisis and Catharsis*, p. 161.

20. Girard, *Things Hidden*, p. 195. With regard to demythologization, Girard makes this criticism of Bultmann's modernist project: "Critics like Rudolf Bultmann, whose theories dominated German theology after the Second World War, still have attempted to 'demythologize' the Gospels. Bultmann simply cut out what could no longer be contained in the sacrificial interpretation and what could not yet be given a nonsacrificial reading. In consequence, he invited his readers to forget the theme of Apocalypse, referring to it as an old Jewish superstition that has nothing whatsoever to offer the modern mind. Indeed Bultmann, like Albert Schweitzer, always saw the Apocalypse in terms of the vengeance of God, a reading that has no basis in the Gospels themselves." Ibid., pp. 259–60.

21. T. J. Scheff, *Catharsis in Healing, Ritual, and Drama* (Berkeley: University of California Press, 1979), pp. x–xii, 48, 53, 60.

22. Ibid., pp. 59–61.

23. Ibid., p. 63.

24. Ibid., pp, 141, 142, 148.

25. Girard, *Things Hidden*, pp. 260–61.

26. Ibid.

27. On catharsis in Girard's theory see Girard, *Violence and the Sacred*, pp. 30, 81, 290–92, 296. For Victor Turner's treatment of structure and antistructure in social transformation, and his notion of normative communitas as distinguished from a communitas that is authentically equalitarian and antistructural, see Turner, *The Ritual Process: Structure and Anti-Structure* (Ithaca, N.Y.: Cornell University Press, 1969), p. 131ff., and related passages in Turner, *Dramas, Fields and Metaphors: Symbolic Action in Human Society* (Ithaca, N.Y.: Cornell University Press, 1974).

28. Christopher Collins, *Reading the Written Image: Verbal Play, Interpretation, and the Roots of Iconophobia* (University Park: The Pennsylvania State University Press, 1991), p. 104.

29. The full, first sentence of Du Bois's 1897 essay, "The Conservation of Races," reads, "The American Negro has always felt an intense personal interest

in discussions as to the origins and destinies of races: primarily because back of most discussions of race with which he is familiar, have lurked certain assumptions as to his natural abilities, as to his political, intellectual and moral status, which he felt were wrong." *W.E.B. Du Bois*, The Library of America (Cambridge, England: Press Syndicate of the University of Cambridge, 1986), p. 815.

30. Theophilus Gould Steward, *The End of the World; or, Clearing the Way for the Fullness of the Gentiles; with an Exposition of Psalm 68.31* (Philadelphia, A.M.E. Church Book Rooms, 1888), pp. 74–76. But note this sardonic disclaimer: "The writer asks exemption from the charge of partiality should the views herein contained be found to favor especially those races of earth with which he feels especially identified. It would hardly seem necessary to make this plea, however, since the rule of the age is to interpret almost everything in science, history and religion to the advantage of that form or type of man which for the present holds ascendency." Ibid., pp. 2–3.

31. Ibid., pp. 123–24. Note also this passage from the commentary appended to Steward's text, "Exposition of Psalm 68.31 by James A. Handy, D.D.": "Its meaning, plain and simple, is that the land of Mizraim and Cush [Africa]—the heads of her numerous tribes—princes, her petty Kings—shall be converted to God and lead on the host. . . . In the fulness of time, appointed by unerring wisdom, Mizraimites and Cushites from centre to circumference of Africa, shall unite in one Christian faith and fellowship, led on by Jesus Christ, the captain of our common salvation with her Egyptian princes in the van and her millions following, they shall be conducted into the militant mansions of bliss to possess the everlasting enjoyments of God on earth . . . together with the church triumphant. Amen." Ibid., pp. 149–151.

32. Ibid., pp. 3–5.

33. Ibid., pp. 14–15.

34. Ibid., pp. 120–21, 123.

35. The "Christian realism" of the theologian Reinhold Niebuhr (1892–1971) is especially insistent on this point. The refusal to ascribe the sources of human evil to humanity itself constitutes what Niebuhr called the "easy conscience of modern man," which attributes evil to history, to religious or sociopolitical institutions, to economic systems, to deficient education, or (as in this case) to a particular culture. "[Modern man] considers himself the victim of corrupting institutions which he is about to destroy or reconstruct, or of the confusions of ignorance which an adequate education is about to overcome . . . in such pathetic contradiction with the obvious facts of his history." Reinhold Niebuhr, *The Nature and Destiny of Man*, vol. 1 (New York: Charles Scribner's Sons, 1964), pp. 94–95. Cf. also Reinhold Niebuhr, *Moral Man and Immoral Society* (New York: Scribner's Sons, 1960).

36. Ibid., pp. 69–71.

37. *A Testament of Hope: The Essential Writings and Speeches of Martin Luther King, Jr.*, ed. James M. Washington (New York: HarperCollins Publishers, 1991), p. 219.

38. James Theodore Holly, "The Divine Plan of Human Redemption, in Its Ethnological Development," *A.M.E. Church Review* 1 (October 1884): 84–85. Compare this version by Edward W. Blyden:

> The all-conquering descendants of Japheth have gone to every clime, and have planted themselves on almost every shore. By means fair and unfair, they have spread themselves. . . . The Messiah—God manifest in the flesh—was of the tribe of Judah. He was born and dwelt in the tents of Shem. The promise to

Ethiopia, or Ham, is like that to Shem, of a spiritual kind. It refers not to physical strength, not to large and extensive domains, not to foreign conquests, not to wide-spread dominions, but to the possession of spiritual qualities, to the elevation of the soul heavenward, to spiritual inspirations and divine communications. 'Ethiopia shall stretch forth her hands unto God.' Blessed, glorious promise! Our trust is not to be in chariots or horses, not in our own skill or power, but our help is to be in the name of the Lord. (Quoted in Gayraud S. Wilmore, *Black Religion and Black Radicalism: An Interpretation of the Religious History of Afro-American People*, 2nd ed. (Maryknoll, N.Y.: Orbis Books, 1983), pp. 119–20, and in Howard Brotz, ed., *Negro Social and Political Thought, 1850–1900, Representative Texts* (New York: Basic Books, 1966)), pp. 121–22. See also Thomas V. Peterson, *Ham and Japheth: The Mythic World of Whites in the Antebellum South* (Metuchen, N.J.: Scarecrow Press, 1978).

39. St. Clair Drake, *The Redemption of Africa and Black Religion* (Chicago: Third World Press, 1970), p. 41. We find a version of this doctrine even in the words of Edward W. Blyden, writing as a major ideologue of Ethiopianism and an early architect of Pan-Africanism in the midnineteenth century. Blyden discerned divine providence operating in the New World experience of Africans:

First, by suffering them to be brought here and placed in circumstances where they could receive a training fitting them for the work of civilizing and evangelizing the land whencce they were torn, and by preserving them under the severest trials and afflictions. Secondly, by allowing them, notwithstanding all the services they have rendered to this country, to be treated as strangers and aliens, so as to cause them to have anguish of spirit, as was the case with the Jews in Egypt, and to make them long for some refuge from their social and civil deprivations. Thirdly, by bearing a portion of them across the tempestuous seas back to Africa, by preserving them through the process of acclimation, and by establishing them in the land, despite the attempts of misguided men to drive them away. Fourthly, by keeping their fatherland in reserve for them in their absence. (Wilmore, *Black Religion and Black Radicalism*, pp. 118–19.

40. Ibid., pp. 41, 53.

41. Wilson Jeremiah Moses, "The Poetics of Ethiopianism: W.E.B. DuBois and Literary Black Nationalism," *American Literature* XLVII:3 (November 1975): 412.

42. Eldridge Cleaver, *Soul on Fire* (Waco, Tex.: Word Books, 1978), p. 92. Elsewhere Cleaver did indeed employ the Babylon figure for explicit prophetic purposes: "If we understand ourselves to be revolutionaries, and if we accept our historic task, then ... there will be a new day in Babylon, there will be a house-cleaning in Babylon." Cited in William L. Van Deburg, *New Day in Babylon: The Black Power Movement and American Culture, 1965–1975* (Chicago: University of Chicago Press, 1992), p. 5., and quoted from Cleaver, *Eldridge Cleaver: Post-Prison Writings and Speeches*, ed. Robert Scheer (New York: Random House, 1969), n.p. Although in the first quotation Cleaver attributes his revolutionary employment of the term "Babylon" to black Muslims in North America, we should also recall the prominent use of the term in Jamaica's Rastafarian movement, as highlighted above in the discussion of iconics, Chapter 4.

43. See Elijah Muhammad's treatment of this matter in his *Message to the Blackman in America* (Chicago: Muhammad Mosque of Islam No. 2, 1965),

pp. 86–99, under the subheadings "The Bible and Holy Qur-An: Which One Contains Words of God?" and "The Bible and Holy Qur-An: Which One Is Right?"

44. Maulana Karenga, "Black Religion," in Gayraud S. Wilmore, ed., *African American Religious Studies: An Interdisciplinary Anthology* (Durham, N.C.: Duke University Press, 1989), p. 292.

45. Muhammad, *Message to the Blackman*, pp. 103–26.

46. Ibid., p. 268.

47. Ibid., p. 276.

48. Ibid., pp. 110–11, 118.

49. Ibid., pp. 115–16. The trickster features of Yakub are evident in his fabrication of a presumably unnatural principle of repulsion and attraction: "Yakub was the founder of unlike attracts and like repels." Elijah Muhammad anathematized this principle in its social operation between black and white people: Yakub's deceit and mischief consisted in creating "an unlike [white] human being, made to attract [black] others, who could, with the knowledge of tricks and lies, rule the original black man—until that nation could produce one greater and capable of overcoming and making manifest his race of tricks and lies, with a nation of truth" (p. 112).

50. Ibid., p. 122.

51. Malcolm X, with Alex Haley, *The Autobiography of Malcolm X* (New York: Grove Press, 1965), pp. 323, 333, 340.

52. Karenga, "Black Religion," p. 297.

53. Malcolm X, *The Autobiography*, pp. 370, 377. Cf. "Movement Toward Martin" in James H. Cone, *Martin and Malcolm and America: A Dream or A Nightmare* (New York: Orbis Books, 1991), p. 192f. Note also the "complementary" relationship between Malcolm X and Martin Luther King, Jr., as represented by Wilmore, *Black Religion and Black Radicalism*, pp. 189–91.

54. Holly, "The Divine Plan of Human Redemption," pp. 82–83.

55. Montague Rhodes James, trans. and ed., *The Apocryphal New Testament*, 1924 reprint (London: Oxford University Press, 1972), p. 524. The editor concludes with the following addendum, however: "Some artless iambic lines of uncertain date are appended here, which show what was thought of the doctrine: 'Plainly false: for the fire will never cease to torment the damned. I indeed could pray that it might be so, who am branded with the deepest scars of transgressions which stand in need of utmost mercy. But let Origen be ashamed of his lying words, who saith that there is a term set to the torments.' " These texts come from the Second Book of the Sibylline Oracles (lines 330–38) and are evidently found in the Apocalypse of Peter.

Conclusion: Diaspora

The Biblical image which has been at the heart of the black
[American's] faith in the eventual appropriation of the American myth
must be replaced. . . . My own very untested suggestion about a possible
new image is that of an African Diaspora based on the Biblical story of
the Babylonian Exile and the final Jewish Diaspora. It is to the end of
the Biblical history of Israel that black America must look rather than to
the beginning.

Charles Shelby Rooks, "Toward the Promised Land"[1]

In 1973 the black theologian Charles Shelby Rooks proposed to abandon the now "compromised" figuration of Promised Land, along with its tarnished secular derivative of "the American dream." Rooks's tentative suggestion was to reconfigure contemporary black experience as a recapitulation of the Babylonian Captivity and the Jewish Diaspora. But what Rooks proposed speculatively at the beginning of the 1970s seems increasingly applicable as the century comes to a close. In contemporary black culture the configurations of Captivity and Diaspora appear most compelling. "The idea that America is the Promised Land is compromised almost beyond repair. Injustice, war, ecological devastation, runaway technology, etc., have served to tarnish the dream, perhaps forever."[2] What may remain viable after the atrophy of this particular figure, however, is the black figural tradition itself with its modes of incantatory and conjurational performance. As the theologian Robert Bennett admits, "the black experience in America is not the Jewish Christian experience in ancient Palestine." Nonetheless, he adds, "the same hermeneutical process which confronts us with the message from Scripture also suggests those categories by which we can deal creatively with the word being spoken by the black experience."[3] As demonstrated throughout this study, a major source of creativity in the engagement of African Americans with biblical hermeneutics has been their conjurational spirituality.

The term "African Diaspora" has become popular since the 1960s black consciousness movement. However, the figural correspondence between the worldwide dispersal of Jews and that of African peoples has

been recognized at least since the early nineteenth century. The word
"diaspora" itself derives from the Greek word for dispersion and was
typically applied to the 'scattering' (as in Nehemiah 1.8) of the Jews among
Gentile nations beginning with the Fall of the Northern Kingdom to
Assyria in 721 B.C.E. The dispersal of Jews in the Hellenistic world of the
Roman empire sets the scene for the appearance of the word in the Christian Scriptures (for example, John 7.35). But the historian George Shepperson reminds us that Diaspora has extended applications as well.

> This process of Jewish migration from their homeland into all parts of the
> world not only created a term which could be applied to any other substantial
> and significant group of migrants, but also provided a concept which could
> be used to interpret the experience (ofter very bitter experiences) of other
> peoples who had been driven out of their native countries by forces similar
> to those which had dispersed the Jews: in particular, slavery and imperialism.[4]

But like the other configurations, Diaspora inherits certain religious
or spiritual implications, by virtue of its figural embeddedness in biblical
narrative. It may seem unlikely that Christian spiritual dynamics are attached to the Diaspora figure as it is currently employed in black culture.
The term may appear incontrovertibly secular now, because of the ostensible social-political orientation of its usage. In this regard we can note
its political utility in providing a post-Ethiopianist conceptual framework
for Pan-Africanism, as "the latter-day ideology of the African diaspora"
(Shepperson). Nonetheless the derivation or "ethnogenesis" (Sollors) of
political configurations is not inconsequential. Unavoidably a transcendent
dimension coinheres in Diaspora, not through some obscure mysticism
but because the cultures that employ it (notably here Jewish and African
American religionists) are also hermeneutic communities that maintain
traditions of biblical self-identification. Can it really be coincidental that
a biblical figure is increasingly accepted by literate Afro-Americans, as a
designation for their contemporary experience and realities?

Can a compelling symbol emerge rootless in such a culture, severed
from its origins in a sacred narrative? On the contrary, as Stephen Crites
maintains in his essay on the narrative quality of human experience: "A
people's mundane stories are implicit in its sacred story, and every mundane
story takes soundings in the sacred story."[5] Crites's comment applies in a
peculiar manner to a diasporan people like Afro-Americans. It is remarkable that, in a characteristically oral and ecstatic culture, a people "of the
spirit" have also become a people "of the book." In that process they have
incorporated the spirituality of their Jewish, Muslim, and Christian cor-
eligionists, thereby participating in the religions of Abraham. They have
adopted the sacred texts of their diasporan host cultures. So intensive has
been this hermeneutic transference—this process of intratextuality—that
the Bible has come to serve as a surrogate sacred text for an ethnic community lacking indigenous texts (or estranged from its ancestral oral and
epic productions). Indeed, the 'surrogacy' of biblical narrative for black

America means that the culture inscribes its experience in the world of Scripture as an extension of that world—as if the Bible were its own literary record of divine encounter.

The self-inscription of cultures into the framework of biblical narrative is a precise, even paradigmatic, instance of intratextuality. Biblical narrative draws communities so profoundly into its worlds that, as Erich Auerbach has declared, the effect is autocratic. The Bible is "tyrannical," to repeat; "it excludes all other claims.... All other scenes, issues, and ordinances have no right to appear independently of it, and it is promised that all of them, the history of all mankind, will be given their due place within its frame."[6] As the theologian George Lindbeck has pointed out, however, this intratextual quality of the Bible derives from a hermeneutical ortho-doxy in which "it is important to note the direction of interpretation. Typology does not make scriptural contents into metaphors for extras-criptural realities, but the other way around. It does not suggest, as is often said in our day, that believers find their stories in the Bible, but rather that they make the story of the Bible their story."[7] According to this view of intratextuality, postbiblical typological communities do not adapt the Scriptures to their circumstances. Rather, they 'read' their his-torical and contemporary experience into the texts of biblical narrative in the terms provided by its coded, figural universe. According to this view the final figure to be examined in this study, Diaspora, configures a people's eschatological (end of the age) dispersal from every earthly homeland.

The figure of Diaspora, as a permanent condition of peoplehood, is eloquently captured in the image of Abraham as the archetypal nomad or pilgrim (the "wandering Aramean" of Deuteronomy 26.5). Indeed, per-manent Diaspora provides the focus for this Christian celebration of the patriarchs' homelessness (Hebrews 11.8–10, 13–16):

> By faith Abraham, when he was called, obeyed by going out to a place which he was to receive for an inheritance; and he went out, not knowing where he was going. By faith he lived as an alien in the land of promise, as in a foreign land, dwelling in tents with Isaac and Jacob ... for he was looking for the city which has foundations, whose architect and builder is God.... All these died in faith ... having confessed that they were strangers and exiles on the earth.... Therefore God is not ashamed to be called their God; for [God] has prepared a city for them.

In this text Abraham's nomadic existence constitutes a model of Diaspora. Abraham himself is represented as a nomadic exemplar, and through this exemplary figure Diaspora becomes not only a form of social-historical existence, but one of eschatological or transhistorical existence. On the one hand such 'confessing exiles'—like Abraham and other nomadic pa-triarchs and matriarchs—"make it clear that they are seeking a country of their own": that is, they appear thoroughly grounded in the circumstances and challenges of their social-historical existence. On the other hand they also continue to confess their identity as strangers who cannot find such a city in this world. "And indeed if they had been thinking of that country from which they went out, they would have had opportunity to return. But as it is, they desire a better country, that is a heavenly one" (vss. 14–16). Because their "desire" transcends the possibility of fulfillment by any temporal country, the writer concludes, desire must seek fulfillment beyond history. Only one kind of citizenship can satisfy such extraordinary desire, therefore "God is not ashamed to be called their God" and to vouchsafe them an eschatological, end-time city.

> This *displacement* of desire, from attachment to every earthly home-land, is the mode in which the Diaspora figure encodes the transhistorical trajectory of typological communities. In the context of this study that trajectory extends from the Bible's nomadic exemplar, Abraham, to the nineteenth century 'sojourning' model of Sojourner Truth, and thence to today's diasporan African American communities of faith and hope. In each instance Diaspora replaces the desire for a temporal homeland with an orientation and yearning toward a world-transcending citizenship. If we inquire into the content of that citizenship, it too is encoded by the Bible. Here we come to the figure at the end of the Book of the Apocalypse, the New Jerusalem of Revelation 21 and 22. In that end-time city a remnant have finally been freed from entanglement in worldly systems of injustice and oppression; they have been liberated from the world's "principalities and powers" (Ephesians 6.12). Liberation from "the Powers" (Wink)[8] is apocalyptically prefigured where, as history ends with the death knell of Babylon the oppressive city, the remaining people of God are summoned out of their long exile, out of their last captivity: "Come out of her, my people," the heavenly voice rings out, "lest you take part in her sins, lest you share in her plagues" (Revelation 18.4).

Here I conclude that African Americans are coparticipants in a contemporary, global configuration of all Abrahamic communities currently experiencing Diaspora.[9] In the diverse configurations of those several communities, Jewish, Christian, Islamic. and others, we can observe both traditional and renewed yearnings for the fulfillment of the Bible's prophetic vision of social justice, universal well-being, and enduring peace. (Consider here Dr. Martin Luther King's triad of antiracism, antipoverty, and antiwar ideals, elaborated as the most radical challenges facing the global community.[10]) Whether they regard themselves as rooted or captive in their host cultures, diasporan communities are undergoing the challenge of myriad forms of social transformation and human solidarity. I especially note the operation in such communities of various expressions of activism in society, solidarity with victims, and resistance against all forms of prejudice and oppression. Most consequential, however, may be a common awareness of being summoned or called by the possibilities of transcendence: transcending the excesses of allegiance to various forms of nationalism and transcending the oppressive conditioning of their host cultures in order to acknowledge, support, and advance the humanity of peoples elsewhere and everywhere.

Diasporan communities are uniquely situated to experience transcendence as prophetic distance from their nations of residence, and as a summons to "come out . . . lest you take part" in national sins and plagues. Not as summons to retreat or withdraw from society, but instead to adapt and 'update' the prophetic transformations of their past, African Americans and others may find themselves improvising a biblical vision and version of Diaspora. They may help 'pharmacognize' an antidote for nationalism, for example, out of the pharmacopeic wisdom of their ancestral heritage (on the one hand), and out of the conjurational biblicalism of their New World heritage (on the other). A diasporan approach to the toxic ideologies that plague national homelands, for instance, may be possible from the perspective of a conjurational culture. In such a culture it is axiomatic, in terms of understanding "the two sides of herbal power," that one seeks an optimal balance in all relationships and human structures, since "the same thing that can cure you, if used improperly can kill you."[11] The scope of this study has allowed us only to hint at such possibilities here, however. (Elsewhere I have described such possibilities as attempts to reconvene or "convene the ritual cosmos."[12]) It must suffice to suggest that new forms of transformation and negation, that restructure social-historical processes according to a rational-ritualistic and 'pharmaconic' understanding of the world, could be devised.

One crucial requirement for these developments is a praxis that intrinsically incorporates forms of self-negation. A praxis is needed that is capable of inhibiting a community from sedimentation into its own forms of ideological rigidity.[13] Reminiscent of Thomas Jefferson's recurrent revolution, such a praxis would be conducive to a permanent diaspora in which one is always undergoing dispersion from monocultural structures to more inclusive formations. Thereby it would effectively subvert and retard sedimentation—whether sedimentation of thought or of social structure (e.g., of ethnicity, class, and nationality). As in *Where Do We Go from Here?* where Dr. Martin Luther King presents the metaphor of the "world house," such a praxis would serve to disperse communities continually beyond themselves to engage "black and white, Easterner and Westerner, Gentile and Jew, Catholic and Protestant, Moslem and Hindu."[14] Through apprehending an inclusive, global communitas, such communities would enter the postmodern period with a consciousness that transcends ethnocentrism by virtue of its grounding in the diverse, composite heritage of multiple cultures.

We have seen some of the rich rewards that derive from such a multicultural heritage in North America. We have seen the fruits of both ritual improvisations and biblical hermeneutics—both Afro-American and Euro-Christian dynamics—operating in African American heritage. But it remains to be seen whether, in future configurations of black culture, its emancipatory movements and social-historical formations will cooperate in synergy with biblical narrative. If so then the power of biblical narrative—from Genesis to Apocalypse—to circumscribe black experience will be augmented by a rich conjurational heritage. It is a heritage that, like the indigenous music of black North Americans (the spirituals and the blues, gospel music and jazz), is capable of encompassing both sacred and secular developments in religion and culture. It is likely that diverse forms of *conjuring diaspora* will provide the next sphere of operation, and a more global context, for the continued vitality of that heritage.

Indeed, a more explicitly theological formulation of that prospector possibility can be stated in conclusion. 'Conjuring culture' through the incantatory use of biblical figures like Exodus and Promised Land, requires a specifically theological, rather than a merely rhetorical or literary, figural reading. By "theological" here I mean that biblical figures are employed in synergy with a Deity who cooperates in the concrete historical realization of such figures. We have seen that in African American theological perspective, for example, Exodus becomes historical in the emancipation of the slaves in the 1860s, and in the civil rights movement of the 1960s, under the historical supervision of a provident God. It remains to be seen, however, whether this tradition of figural-theological reading and synergetic reenactment will continue to flourish in black America. It remains to be seen, that is, whether the culture will continue to let the biblical text absorb the world, "rather than the world the text."[15] Yet whatever the future brings, conjuring culture with biblical figures now holds a distinctive place among contemporary spiritualities in North America.

Notes

1. Charles Shelby Rocks, "Toward the Promised Land: An Analysis of the Religious Experience of Black America," *The Black Church* II:1 (September 1973): p. 8.

2. Ibid.

3. Robert A. Bennett, "Black Experience and the Bible," *Theology Today* 27 (January 1971): 423, 433.

4. George Shepperson, introduction to *The African Diaspora: Interpretive Essays*, ed. Martin Kilson and R. Rotberg (Cambridge, Mass.: Harvard University Press, 1976), pp. 2–3, 8.

5. Stephen Crites, "The Narrative Quality of Experience," *Journal of the American Academy of Religion* 39 (September 171): 296.

6. Erich Auerbach, *Mimesis: The Representation of Reality in Western Literature*, trans. Williard Trask (Garden City, N.Y.: Anchor Press/Doubleday, 1952), p. 15.

7. George A. Lindbeck, *The Nature of Doctrine: Religion and Theology in a Post-Liberal Age* (Philadelphia: Westminster Press, 1984), p. 135.

8. See Walter Wink, *Engaging the Powers: Discernment and Resistance in a World of Domination*, vol. 3 of 3 (Minneapolis: Fortress Press, 1992).

9. Cf. Scott Heller, "Worldwide 'Diaspora' of Peoples Poses New Challenges for Scholars," *The Chronicle of Higher Education* XXXVIII: 39 (June 3, 1992): A7–9.

10. Martin Luther King, Jr., *Strength to Love*, in *A Testament of Hope: The*

Essential Writings of Martin Luther King, Jr., ed. James M. Washington (New York: HarperCollins Publishers, 1991), pp. 621–29.

11. Gary Edwards and John Mason, *Black Gods—Orisha Studies in the New World* (Brooklyn, N.Y.: Yoruba Theological Archministry, 1985), pp. 29–30. This quotation is discussed further in the section "Pharmacosm," Chapter 1.

12. Thee Smith, "W/Riting Black Thelogy," *Forum* 5:4 (December 1989): 50–51.

13. Consider this critique of dogmatic Marxism by Karl Mannheim: "Socialist thought, which hitherto has unmasked all its adversaries' utopias as ideologies, never raised the problem of determinateness about its own position. It never applied this method to itself and checked its own desire to be absolute." Karl Mannheim, *Ideology and Utopia* (N.p., n.d.), p. 225; as quoted in Reinhold Niebuhr, *The Nature and Destiny of Man* (New York: Charles Scribner's Sons, 1964), pp. 196–97. A programmatic intention, to negate the endemic tendency to self-absolutization, is the pharmaconic praxis envisioned here. That intention has been lucidly elaborated by the Marxist feminist scholar Erica Sherover-Marcuse as a principal feature in "a practice of subjectivity" that seeks to retard dogmatic and ideological formations even (or especially) in emancipatory projects:

> The emancipatory *intent* of a subjective practice cannot guarantee that its own activity in the service of liberation will be free from domination. This practice cannot escape its embeddedness in the historical context of domination. There is no external vantage-point from which a subjective practice could claim an immunity to the influences of the oppressive society against which it itself is directed.... A practice of subjectivity thus faces a permanent risk of being 'contaminated' with the toxins of domination.... An emancipatory subjective practice would thus have to struggle continuously against its own reification, against the incremental sedimentation of liberatory processes into fossilized procedures, against the distortions of domination which ingress into all attempts at liberation. It could only do so if its own praxis nourished and encouraged in individuals a critical intelligence and a sense of self-worth *in the context of a developing solidarity*.... Therefore an emancipatory practice of subjectivity must posit as its goal not the immediate realization of "the (given) self," but the *emergence* of a "self-in-solidarity." Erica Sherover-Marcuse, *Emancipation and Consciousness: Dogmatic and Dialectical Perspectives in the Early Marx* (New York: Basil Blackwell, 1986), pp. 141–42.

14. King, *Strength to Love*, p. 617.

15. Lindbeck, *The Nature of Doctrine*, p. 118. The African American New Testament scholar Vincent Wimbush sounded the alarm in this regard with his article questioning the future of biblical literacy—overagainst biblical literalism and fundamentalism—in black America. See Vincent Wimbush, "Rescue and Perishing: The Importance of Biblical Scholarship in Black Christianity." *Reflection* 80:2 (1983), 9–11. See also Henry Mitchell, *Black Preaching* (New York: Harper & Row, 1979): "Black dependence on scripture is not slavish or literal" (p. 113) but rather free or improvisational. "The riff or improvisation on the melody, so characteristic of the Black jazz instrumentalist or vocalist, is Black spontaneity at its best. The same freedom applied to the melodic line in Black gospels or religious Soul music is the very trademark of Black culture" (p. 198; cf. pp. 142ff, 202).

Coincidentally, the present study offers a figural and realist alternative to literalist appropriations of the Bible; an alternative, for example, to the recurrent and (most recently) 'Afrocentric' insistence on a genetic black (vs. Hebraic) identity for Moses, Jesus, and other biblical figures. The conjurational construal of black America's biblical hermeneutic is consistent, I judge, with George Lindbeck's and Hans Frei's concerns for narrative realism on the one hand, and with Vincent Wimbush's fears and Henry Mitchell's admonitions against narrow literalism on the other.

For a concluding point of corroboration see the recent, related discussion by James H. Evans, Jr., *We Have Been Believers: An African-American Systematic Theology* (Minneapolis: Fortress Press, 1992). "The hermeneutical process developed by the slaves centered around what one might call the'divinization' or the 'conjuring' of the [biblical] text." Such diviners were (and are) "moral analysts" and "cultural hermeneuts." They were practitioners intent on disclosing the "secret sense" of experience on the basis of an African mystical sensibility, and practitioners for whom "conjure also referred to an act of interpretation" (p. 48). Yet immediately following this insight Evans stops short of connecting such "imaginative" interpretations to his subsequent discussion of the Bible as realistic or "history-like" narrative (Hans Frei). Precisely that connection provides the focus of the present study. Still, while drawing upon Erich Auerbach's concept of *figura* Evans usefully inaugurates an indigenous, emancipatory christology on the basis of "the 'figura' of Christ in African-American experience" (p. 77f.). In this regard both Evans's efforts and the theological perspectives of the present work address Lindbeck's concern, stated a decade ago, when he lamented "much talk at present about typological, figurative, and narrative theology, but little actual performance. Only in some younger theologians does one see the beginnings of a desire to renew in a posttraditional and postliberal mode the ancient practice of absorbing the universe into the biblical world. May their tribe increase" (*The Nature of Doctrine*, p. 135).

Selected Bibliography

Ahlstrom, Sidney. *A Religious History of the American People*. New Haven: Yale University Press, 1972.

Anyanwu, Chukwulozie K. *The Nature of Black Cultural Reality*. Washington, D.C.: University Press of America, 1976.

Aptheker, Herbert, ed. 2 vols. *A Documentary History of the Negro People of the United States*. New York: Citadel Press, 1974.

———. *Nat Turner's Slave Rebellion*. New York: Humanities Press, 1966.

Armah, Ayi Kwei. *The Healers*. Nairobi, Kenya: East African Publishing House, 1978.

Asante, Molefi K. *The Afrocentric Idea*. Philadelphia: Temple University Press, 1987.

———, and Kariamu W. Asante, eds. *African Culture: The Rhythms of Unity*. Westport, Conn.: Greenwood Press, 1985.

Auerbach, Erich. *Mimesis: The Representation of Reality in Western Literature*. Translated by Williard Trask. Garden City, N.Y.: Doubleday/Anchor Books, 1952.

———. *Scenes from the Drama of European Literature*. Foreword by Paolo Valesio. edited by Wald Godzich and Jochen Schulte-Sasse, Minneapolis: University of Minnesota Press, 1984.

Badejo, Diedre L. "The Yoruba and Afro-American Trickster: A Contextual Comparison." *Présence Africaine* 147:3 (1988): 3–17.

Baer, Hans A. *The Black Spiritual Movement: A Religious Response to Racism*. Knoxville: University of Tennessee Press, 1984.

Baker, Houston, Jr. *Blues, Ideology and Afro-American Literature: A Vernacular Theory*. Chicago: University of Chicago Press, 1984.

Baldwin, James. *The Fire Next Time*. New York: Dell, 1963.

————. "Many Thousands Gone." In *Black Expression: Essays by and about Black Americans*, ed. Addison Gayle, Jr. New York: Weybright and Talley, 1969.

Baratz, Joan, and Stephen Baratz. "Black Culture on Black Terms: A Rejection of the Social Pathology Model." In *Rappin' and Stylin' Out*. edited by Thomas Kochman (Urbana: University of Illinois Press, 1972), pp. 3–16.

Barrett, Leonard E. *Soul Force: African Heritage in Afro-American Religion*. Garden City, N.Y.: Anchor Press/Doubleday, 1974.

Bell, Michael Edward. *Pattern, Structure, and Logic in Afro-American Hoodoo Performance*. Ann Arbor, Mich.: University Microfilms International, 1980.

Bellah, Robert N. *The Broken Covenant: American Civil Religion in Time of Trial*. New York: Seabury Press, 1975.

————, and Richard Madsen, William M. Sullivan, Ann Swidler, Steven M. Tipton. *Habits of the Heart: Individualism and Commitment in American Life*. New York: Harper & Row, 1985.

Bennett, Lerone Jr. *Confrontation: Black and White*. Baltimore, Md.: Penguin Books, 1965.

————. *The Negro Mood and Other Essays*. Chicago: Johnson Publishing Co., 1964.

————. *What Manner of Man: A Biography of Martin Luther King* Chicago: Johnson Publishing Co., 1964.

Bennett, Robert A. "Biblical Theology and Black Theology." *Journal of the Interdenominational Theological Center* 3 (1976): 1–16.

————. "Black Experience and the Bible." *Theology Today* 27:4 (1971): 422–433.

Bercovitch, Sacvan. *The American Jeremiad*. Madison: The University of Wisconsin Press, 1978.

————. *The Puritan Origins of the American Self*. New Haven: Yale University Press, 1975.

————, ed. *Typology and Early American Literature*. Amherst: University of Massachusetts Press, 1972.

Bernal, Martin. *Black Athena: The Afroasiatic Roots of Classical Civilization*. Vol. 1, *The Fabrication of Ancient Greece 1785–1985*. New Brunswick, N.J.: Rutgers University Press, 1987.

Bernard, Jacqueline. *Journey toward Freedom: The Story of Sojourner Truth*. New York: Feminist Press of the City University of New York, 1990.

Blassingame, John W. *The Slave Community*. New York: Oxford University Press, 1972.

————. *Slave Testimonies*. Baton Rouge: Louisiana State University Press, 1977.

Blauner, Robert. *Racial Oppression in America*. New York: Harper & Row, 1972.

Bondurant, Joan V. *Conquest of Violence: The Gandhian Philosophy of Conflict*, Rev. Ed. Berkeley: University of California Press, 1965.

Bontemps, Arna. *Free at Last: The Life of Frederick Douglass*. New York: Dodd, Mead, 1971.

Bracey, John H., August Meier, and Elliott Rudwick, eds. *Black Nationalism in America*. Indianapolis: Bobbs-Merrill, 1970.

Breiner, Laurence A. "The English Bible in Jamaican Rastafarianism." *The Journal of Religious Thought* 42:2 (Fall–Winter 1985–86): 30–43.

Brown, David. "Conjure/Doctors: An Exploration of a Black Discourse in America, Antebellum to 1940." *Folklore Forum* 23:1/3 (1990): 26ff.

Brown, Sterling. "The Spirituals." In *The Book of Negro Folklore*, edited by

Langston Hughes and Arna Bontemps. New York: Dodd, Mead, 1958, pp. 279–89.

Bruce, Calvin E. "Black Spirituality and Theological Method." *Journal of the Interdenominational Theological Center* 32 (Spring 1976): 65–76.

———. "Black Spirituality, Language and Faith." *Religious Education* LXXI:4 (July–August 1976): 363–376.

———, and William R. Jones, eds. *Black Theology II: Essays on the Formation and Outreach of Contemporary Black Theology*. Lewisburg, Pa.: Bucknell University Press, 1978.

Brumm, Ursula. *American Thought and Religious Typology*. New Brunswick, N.J.: Rutgers University Press, 1970.

Burke, Kenneth. *The Philosophy of Literary Form: Studies in Symbolic Action*. New York: Vintage Books, 1957.

Butler, Jon. *Awash in a Sea of Faith: Christianizing the American People*. Cambridge, Mass.: Harvard University Press, 1990.

Cameron, Vivian K. "Folk Beliefs Pertaining to the Health of the Southern Negro." Master's thesis, Northwestern University, 1930. Cited in Melville Herskovits, *The Myth of the Negro Past*. Boston: Beacon Press, 1958, pp. 239–42.

Cannon, Katie. *Black Womanist Ethics*. Atlanta: Scholars Press, 1988.

Carr, David. "Narrative and the Real World: An Argument for Continuity." *History and Theory*. XXV:2 (1986), 117–31.

Carter, Harold A. *The Prayer Tradition of Black People*. Valley Forge, Pa.: Judson Press, 1976.

Cherry, Conrad, ed. *God's New Israel: Religious Interpretations of American Destiny*. Englewood Cliffs, N.J.: Prentice-Hall, 1971.

Chestnut, Charles W. *The Conjure Woman*. 1899. Reprint. Darby, Pa.: Arden Library, 1978.

Chimezie, Amuzie. "Black Biculturality." *The Western Journal of Black Studies*. 9:4 (Winter 1985): 224–35.

Cleage, Albert. *Black Christian Nationalism: New Directions for the Black Church*. New York: William Morrow, 1972.

———. *The Black Messiah*. New York: Sheed & Ward, 1969.

Coan, Josephus R. "Daniel Coker: 19th Century Black Church Organizer, Educator and Missionary." *The Journal of the Interdenominational Teological Center*. III:1 (Fall 1975):17–31.

Coker, Daniel. "A Dialogue between a Virginian and an African Minister." 1810. Reprint. *Negro Protest Pamphlets*, edited by Dorothy Porter. New York: Arno Press/The New York Times, 1969.

———. "Journal of Daniel Coker." 1820. Reprint. *Paul Cuffee, Peter Williams, Daniel Coker, Daniel H. Peterson, and Nancy Prince*. Nendeln/Liechtenstein, Switzerland: Kraus Thomson Organization, 1970.

Collins, Adela Yarbro. *Crisis and Catharsis: The Power of the Apocalypse*. Philadelphia: Westminster Press, 1984.

Collins, Christopher. *Reading the Written Image: Verbal Play, Interpretation, and the Roots of Iconophobia*. University Park: The Pennsylvania State University Press, 1991.

Cone, Cecil W. *The Identity Crisis in Black Theology*. Nashville: African Methodist Episcopal Church, 1975.

Cone, James H. *Black Theology and Black Power*. New York: Seabury Press, 1969.
———. *For My People: Black Theology and the Black Church*. Maryknoll, N.Y.: Orbis Books, 1984.
———. *God of the Oppressed*. New York: Seabury Press, 1975.
———. *Martin and Malcolm and America: A Dream or a Nightmare*. MaryKnoll, N.Y.: Orbis Books, 1991.
———. *The Spirituals and the Blues: An Interpretation*. New York: Seabury Press, 1972.
———. "The Story Context of Black Theology." *Theology Today* XXXII:2 (1975): 144–50.
Conwill, Giles. "The Word Becomes Black Flesh: A Theoretical and Practical Paradigm for the Evangelization of Black Catholic Americans Based on Victor Turner's Concepts and Models of Dominant Symbol, Liminality, *Communitas* and Root Metaphor." Ph.D. dissertation, Emory University, 1986.
Costen, Melva W., and Darius L. Swann, eds. "The Black Christian Worship Experience: A Consultation." *The Journal of the Interdenominational Theological Center* XIV:1:2 (1986–1987): 1–256.
Courlander, Harold. *Negro Folk Music, U.S.A.* New York: Columbia University Press, 1963.
Craig, E. Quita. *Black Drama of the Federal Theatre Era*. Amherst: University of Massachusetts Press, 1980.
Crites, Stephen. "The Narrative Quality of Experience." *Journal of the American Academy of Religion* 39 (September 1971): 291–311.
Cruse, Harold. *The Crisis of the Negro Intellectual*. London: W. H. Allen, 1969.
Davis, Gerald A. *I Got the Word in Me and I can Sing It, You Know: A Study of the Performed African American Sermon*. Philadelphia: University of Pennsylvania Press, 1985.
Dennison, Sam. *Scandalize My Name: Black Imagery in American Popular Music*. New York: Garland Publishing, 1982.
Dickson, Kwesi, and Paul Ellingworth, eds. *Biblical Revelation and African Beliefs*. Maryknoll, N.Y.: Orbis Books, 1969.
Diner, Hasia R. *In the Almost Promised Land: American Jews and Blacks, 1915–1935*. Westport, Conn.: Greenwood Press, 1977.
Dixon, Christa K. *Negro Spirituals: From the Bible to Folk Song*. Philadelphia: Fortress Press, 1976.
Dixon, Vernon J., and Badi G. Foster. *Beyond Black or White: An Alternate America*. Boston: Little Brown, 1971.
Drake, St. Clair. *Black Folk Here and There: An Essay in History and Anthropology*. 2 Vols. Los Angeles: Center for Afro-American Studies, University of California, 1987.
———. *The Redemption of Africa and Black Religion*. Chicago: Third World Press, 1970.
Du Bois, W.E B. *Dusk of Dawn: An Essay toward an Autobiography of a Race Concept*. New York: Harcourt Brace, 1940.
———. *The Souls of Black Folk*. New York: New American Library/Signet Classic, 1969.
Dundes, Alan, ed. *Mother Wit from the Laughing Barrel: Readings in the Interpretation of Afro-American Folklore*. New York: Garland Publishing, 1981.

Dunston, Alfred G., Jr. *The Black Man in the Old Testament and Its World*. Philadelphia: Dorrance and Co., 1974.

Dvorak, Katherine L. "After Apocalypse, Moses." In *Masters and Slaves in the House of the Lord*, edited by John B. Boles. Lexington: University Press of Kentucky, 1988, pp. 173–242.

————*An African-American Exodus: The Segregation of the Southern Churches*. Brooklyn, N.Y.: Carlson Pub., 1991.

Edwards, Gary, and John Mason. *Black Gods: Orisa Studies in the New World*. Brooklyn: Yoruba Theological Archministry, 1985.

Eliade, Mircea. *Shamanism: Archaic Techniques of Ecstasy*. Princeton: Princeton University Press, 1964.

————. *From Medicine Man to Muhammad: A Thematic Source Book of the History of Religions*. New York: Harper & Row, 1974.

Elkins, Stanley. *Slavery*. 2nd ed. Chicago: University of Chicago Press, 1968.

Ellis, Carl F., Jr. *Beyond Liberation: The Gospel in the Black American Experience*. Downer's Grove, Ill.: Intervarsity Press, 1983.

Ellison, Ralph. *Shadow and Act*. New York: Random House, 1964.

Epstein, Dena. *Sinful Tunes and Spirituals*. Chicago: University of Illinois Press, 1977.

Essien-Udom, E. U. *Black Nationalism: A Search for Identity in America*. Chicago: University of Chicago Press, 1962.

Evans, James H., Jr. *Spiritual Empowerment in Afro-American Literature: Frederick Douglass, Rebecca Jackson, Booker T. Washington, Richard Wright, Toni Morrison*. Lewiston, N.Y.: Edwin Mellen Press, 1987.

————. *We Have Been Believers: An African-American Systematic Theology*. Minneapolis: Fortress Press, 1992.

Fauset, Arthur H. *Sojourner Truth: God's Faithful Pilgrim*. 1938. Reprint. New York: Russell Co., 1971.

Felder, Cain Hope. *Stony the Road We Trod: African American Biblical Interpretation*. Minneapolis: Fortress Press, 1991.

————. *Troubling Biblical Waters: Race, Class, and Family*. Maryknoll, N.Y.: Orbis Books, 1989.

Fisher, Miles Mark. *Negro Slave Songs in the United States*. New York: Citadel Press, 1969.

Foner, Philip S. *Black Socialist Preachers*. San Francisco: Synthesis Publications, 1983.

Foster, William Z. *The Negro People in American History*. New York: International Publishers, 1954.

Fox, Robert Elliot. *Conscientious Sorcerers: The Postmodernist Fiction of Leroi Jones/ Amiri Baraka, Ishmael Reed, and Samuel R. Delaney*. Westport, Conn.: Greenwood Press, 1987.

Frazer, James G. *The Golden Bough: A Study in Magic and Religion*. Vol. 1 of 2. New York: Macmillan and Co., 1911.

Frazier, E. Franklin. *The Negro Church in America*. New York: Schocken Books, 1964.

Frei, Hans W. *The Eclipse of Biblical Narrative: A Study in Eighteenth and Nineteenth Century Hermeneutics*. New Haven: Yale University Press, 1974.

Freire, Paolo. *Pedagogy of the Oppressed*. New York: Herder & Herder, 1970.

Frye, Northrop. *The Great Code: The Bible and Literature*. New York: Harcourt Brace Jovanovich, 1982.

Garnet, Henry Highland. "An Address to the Slaves of the United States of America." In *Black Nationalism in America*, edited by John H. Bracey, August Meier, and Elliott Rudwick (Indianapolis: Bobbs-Merrill, 1970) pp. 67–76.

Gates, Henry Louis, Jr. *Figures in Black: Words, Signs and the "Racial" Self.* New York: Oxford University Press, 1987.

———, ed. *"Race," Writing and Difference.* Chicago: University of Chicago Press, 1985/86.

———. *The Signifying Monkey: A Theory of African-American Literary Criticism.* New York: Oxford University Press, 1988.

Gayle, Addison, Jr, ed. *The Black Aesthetic.* Garden City, N. Y.: Anchor Press/ Doubleday, 1972.

———. *Black Expression: Essays By and About Black Americans in the Creative Arts.* New York: Weybright and Talley, 1969.

Geertz, Clifford. *The Interpretation of Cultures: Selected Essays.* New York: Basic Books, 1973.

Genovese, Eugene D. *Roll, Jordan, Roll: The World the Slaves Made.* New York: Pantheon, 1974.

Georgia Writers' Project. Works Project Administration. *Drums and Shadows.* Athens: University of Georgia Press, 1940.

Girard, René. *The Scapegoat.* Tarnslated by Yvonne Frecerro. Baltimore: The Johns Hopkins University Press, 1986.

———. *Things Hidden Since the Foundation of the World.* with J.-M. Ourgoulian and G. Lefort. Stanford: Stanford University Press, 1987.

———. *Violence and the Sacred.* Translated by Patrick Gregory. Baltimore: The Johns Hopkins University Press, 1977.

Glass, James M. "The Philosopher and the Shaman: The Political Vision as Incantation." *Political Theory* 2:2 (May 1974): 181–196.

González-Wippler, Migene, ed. *The New Revised Sixth and Seventh Books of Moses and the Magical Uses of the Psalms.* Bronx, N.Y.: Original Publications, 1991.

Grant, Jacquelyn. "Black Theology and the Black Woman." In *Black Theology: A Documentary History 1966–1979*, edited by Gayraud Wilmore and James Cone. Maryknoll, N.Y.: Orbis Books, 1979, pp. 418–33.

———. *White Woman's Christ and Black Woman's Jesus: Feminist Christology and Womanist Response.* Atlanta: Scholars Press, 1989.

Grant, Joanne, ed. *Black Protest: History, Documents, and Analyses, 1619 to the Present.* New York: Fawcett World Library, 1968.

Grossinger, Richard. *Planet Medicine: From Stone Age Shamanism to Post-Industrial Healing.* Berkeley, Calif.: North Atlantic Books, 1980.

Gustavo Gutierrez, *A Theology of Liberation: History, Politics, and Salvation* (Maryknoll, N.Y.: Orbis Books, 1973).

Hamerton-Kelly, Robert, ed., *Violent Origins: Ritual Killing and Cultural Formation.* Stanford: Stanford University. Press, 1987.

Hanigan, James P. *Martin Luther King, Jr., and the Foundations of Nonviolence.* Lanham, Md.: University Press of America, 1984.

Harding, Vincent. "The Acts of God and the Children of Africa." In *Shalom.* Philadelphia: United Church Press, 1973.

———. "The Uses of the Afro-American Past." In *The Religious Situation*, 1969, edited by Donald R. Cutter. Boston: Beacon Press, 1969, pp. 829–40.

Harrison, Paul C. *The Drama of Nommo.* New York: Grove Press, 1972.

Haskins, James. *Voodoo and Hoodoo.* New York: Stein and Day, 1978.

Hemenway, Robert E. *Zora Neale Hurston: A Literary Biography*. Foreword by Alice Walker. Urbana and Chicago: University of Illinois Press, 1980.

Henson, Josiah. *Father Henson's Story of His Own Life*. Introduction by Walter Fisher. New York: Corinth Books, 1962.

Hershberger, Guy F. *The Recovery of the Anabaptist Vision*. Scottdale, Pa.: Herald Press, 1957.

Herskovits, Melville. *The Myth of the Negro Past*. Boston: Beacon Press, 1958.

Heusch, Luc de. *Why Marry Her? Society and Symbolic Structures*. Cambridge and New York: Cambridge University Press, 1981.

Holly, James Theodore. "The Divine Plan of Human Redemption, In Its Ethnological Development." *A.M.E. Church Review* 1 (1884): 79–85.

Hopkins, Dwight N. *Shoes That Fit Our Feet: Sources for a Constructive Black Theology*. Maryknoll, N.Y.: Orbis Books, 1993.

———, George C. L. Cummings. *Cut Loose Your Stammering Tongue: Black Theology in the Slave Narratives*. Maryknoll, N.Y.: Orbis Books, 1991.

House, H. Wayne. "An Investigation of Black Liberation Theology." *Bibliotheca Sacra* 139 (1982): 159–76.

Hughes, Dennis E. "Shamanism and Christian Faith." *Religious Education* 71 (1976): 392–404.

Hughes, Langston, and Arma Bontemps, eds. *The Book of Negro Folklore*. New York: Dodd, Mead, 1958.

Hurston, Zora Neale. *Moses: Man of the Mountain*. Urbana and Chicago: University of Illinois Press, 1984.

———. *Mules and Men*. 1935. New York: Harper & Row, 1990.

———. *Tell My Horse*. Berkeley: Turtle Island Press, 1983.

Hyatt, Harry M., ed. *Hoodoo, Conjuration, Witchcraft, Rootwork*. 5 vols. Hannibal, Mo.: Western Publishing, Inc., 1970.

Idowu, E. Bolaji. *African Traditional Religion: A Definition*. Maryknoll, N.Y.: Orbis Books, 1973–75.

———. *Olodumare, God in Yoruba Belief*. London: Longmans, 1962.

Jahn, Jahnheinz. *Muntu: The New African Culture*. New York: Grove Press, 1961.

James, Robinson B. "A Tillichian Analysis of James Cone's Black Theology." *Perspectives in Religious Studies* 1:1 (Spring 1974):

Janzen, John M. *The Quest for Therapy in Lower Zaire*. Los Angeles: University of California Press, 1978.

———, and Wyatt MacGaffey *An Anthology of Kongo Religion: Primary Texts from Lower Zaire*. Lawrence: University of Kansas Press, 1974.

Johnson, Clifton H., and A. P. Watson, eds. *God Struck Me Dead: Religious Conversion Experiences and Autobiographies of American Slaves*. Philadelphia: Pilgrim Press, 1969.

Johnson, James Weldon. *The Autobiography of An Ex-Colored Man*. 1912. Reprint. New York: Hill and Wang, 1960.

———. *God's Trombones: Seven Negro Sermons in Verse*. New York: The Viking Press, 1969.

Jones, LeRoi. *Black Music*. New York: William Morrow & Co., 1963.

———. *Blues People*. New York: William Morrow & Co., 1963.

Jones, William R. *Is God a White Racist?* Garden City, N.Y.: Anchor Press/Doubleday, 1973.

———. "Theodicy: The Controlling Category for Black Theology." *Journal of Religious Thought* 30:1 (Spring–Summer 1973): 28–38.

Jordan, R. L. *Black Theology Exposed*. New York: Vantage Press, 1982.

Joyner, Charles. *Down by the Riverside: A South Carolina Slave Community*. Urbana: University of Illinois Press, 1984.

Kelsey, David H. *The Uses of Scripture in Recent Theology*. Philadelphia: Fortress Press, 1975.

Kent, George E. *Blackness and the Adventure of Western Culture*. Chicago: Third World Press, 1972.

Kilson, Martin, and R. Rotberg, eds. *The African Diaspora: Interpretive Essays*. Cambridge: Harvard University Press, 1976.

King, Martin Luther, Jr. *A Testament of Hope: The Essential Writings of Martin Luther King, Jr.* Edited by James M. Washington. New York: HarperCollins Publishers, 1991.

Kochman, Thomas. *Black and White Styles in Conflict*. Chicago: University of Chicago Press, 1981.

———, ed. *Rappin' and Stylin' Out: Communication in Urban Black America*. Urbana: University of Illinois Press, 1972.

Lampe, G.W.H. "The Reasonableness of Typology." *Essays on Typology*. Naperville, Ill: Alec R. Allenson, 1957.

Lanternari, Vittorio. *The Religions of the Oppressed: A Study of Modern Messianic Cults*. New York: New American Library, 1965.

Lee, Carlton. "Toward a Sociology of Black Religious Experience." *The Journal of Religious Thought* 29:2 (1972): 5–18.

Levine, Lawrence W. *Black Culture and Black Consciousness: Afro-American Folk Thought from Slavery to Freedom*. New York: Oxford University Press, 1977.

Levi-Strauss, Claude. *The Savage Mind*. Chicago: University of Chicago Press, 1966.

Lewis, I. M. *Ecstatic Religion: A Study of Shamanism and Spirit Possession*. 2nd ed. London and New York: Routledge, 1989.

Lincoln, C. Eric. *The Black Church Since Frazier*. New York: Schocken Books, 1974.

———, ed. *The Black Experience in Religion*. Garden City, N.Y.: Anchor Press/ Doubleday, 1974.

Lindbeck, George A. *The Nature of Doctrine: Religion and Theology in a Post-Liberal Age*. Philadelphia: Westminster Press, 1984.

Lipsky, Suzanne. "Internalized Oppression," *Black Re-Emergence* 2 (Seattle: Rational Island Publishers, 1977): 5–10.

Locke, Alain, ed. *The New Negro: An Interpretation*. New York: Albert & Charles Boni, 1925.

Loewenberg, Bert James, and Ruth Bogin. *Black Women in Nineteenth Century American Life*. University Park: Pennsylvania State University Press, 1976.

Long, Charles H. "Perspective for a Study of Afro-American Religion in the U.S." *History of Religions* 2 (August 1971): 54–66.

———. *Significations: Signs, Symbols, and Images in the Interpretation of Religion*. Philadelphia: Fortress Press, 1986.

———. "Structural Similarities and Dissimilarities in Black and African Theologies." *Journal of Religious Thought* 33 (Fall–Winter 1975): 9–24.

Lovell, John, Jr. *Black Song: The Forge and the Flame: The Story of How the Afro-American Spiritual Was Hammered Out*. New York: Macmillan Publishing Co., 1972.

Lowance, Mason I., Jr. *The Language of Canaan: Metaphor and Symbol in New*

England from the Puritans to the Transcendentalists. Cambridge, Mass.: Harvard University Press, 1980.

MacGaffey, Wyatt. "Complexity, Astonishment and Power: The Visual Vocabulary of Kongo Minkisi." *Journal of Southern African Studies* 14:2 (January 1988): 188–203.

Malcolm X, with Alex Haley. the *Autobiography of Malcolm X*, New York: Ballantine Books, 1973 (reprint).

Malinowski, Bronislaw. *Magic, Science and Religion and Other Essays*. Garden City, N.Y.: Doubleday and Co., 1954.

Marks, Morton. "Exploring *El Monte*: Ethnobotany and the Afro-Cuban Science of the Concrete." In *En Torno a Lydia Cabrera*, edited by Isabel Castellanos and Josefina Inclán. Miami: Ediciones Universal, 1987.

———. "Uncovering Ritual Structures in Afro-American Music." In *Religious Movements in Contemporary America*, edited by Irving Zaretsky and Mark Leone. Princeton: Princeton University Press, 1974.

Marty, Martin E. "America's Iconic Book." In *Humanizing America's Iconic Book*, edited by Gene Tucker and Douglas Knight. Chico, Calif.: Scholars Press, 1982.

———. "Ethnicity: The Skeleton of Religion in America." *Church History* 41 (1972): 5–21.

Mathews, Donald. *Religion in the Old South*. Chicago: University of Chicago Press, 1977.

Maultsby, Portia K. "Afro-American Religious Music: A Study in Musical Diversity." *The Papers of the Hymn Society of America* XXXV (n.d.): 2–19.

Mays, Benjamin E. *The Negro's God as Reflected in His Literature*. 1938. Reprint. New York: Atheneum, 1973.

Mbiti, John S. "An African Views American Black Theology." *Worldview* 17 (August 1974): 41–44.

———. "The Biblical Basis for Present Trends in African Theology." In *African Theology en Route*, edited by Appiah-Kubi and Torres. Maryknoll, N.Y.: Orbis Books, 1979.

———. *Concepts of God in Africa*. New York: Praeger Publishers, 1970.

———. *New Testament Eschatology in an African Background*. London: Oxford University Press, 1971.

McClendon, James Wm., Jr. *Biography as Theology: How Life Stories Can Remake Today's Theology*. Nashville: Abingdon Press, 1974.

———. *Ethics*. Vol. 1 of *Systematic Theology*. Nashville: Abingdon Press, 1986.

McFeely, William S. *Frederick Douglass*. New York: Norton, 1991.

McKinney, Don. "Brer Rabbit and Brer King: The Folktale Background of the Birmingham Protest." Paper presented at the Southeastern Commission for the Study of Religion, American Academy of Religion, March 10–12, 1989, Atlanta, Georgia.

Middleton, John, ed. *Magic, Witchcraft and Curing*. Garden City, N.Y.: The Natural History Press, 1967.

Miller, Keith D. "Composing Martin Luther King, Jr." PMLA: *Publication of the Modern Languages Association of America* 105:1 (January 1990): 7–82.

———. *Voice of Deliverence: The Language of Martin Luther King, Jr. and Its Sources*. New York: Free Press, 1991.

Miller, Perry. *Errand Into the Wilderness*. Cambridge, Mass.: Harvard University Press, 1956.

————. Introduction. *Images or Shadows of Divine Things*, by Jonathan Edwards. New Haven: Yale University Press, 1948. Mitchell, Faith. *Hoodoo Medicine: Sea Islands Herbal Remedies* N.p.: Reed, Cannon & Johnson Co., 1978.

Mitchell, Henry H. *Black Belief: Folk Beliefs of Blacks in America and West Africa*. New York: Harper & Row, 1975.

————. *Black Preaching*. Philadelphia: Lippincott, 1970.

————, and Nicholas C. Lewter. *Soul Theology: The Heart of American Black Culture*. San Francisco: Harper & Row, 1986.

Mitchell, Mozella G. *Spiritual Dynamics of Howard Thurman's Theology*. Bristol, Ind.: Wyndham Hall Press, 1985.

Mitchell-Kernan, Claudia. "Signifying." In *Mother Wit from the Laughing Barrel*, edited by Alan Dundes. New York: Garland Publishing, 1981, pp. 310–328.

Morrison, Roy D. II. "Black Philosophy: An Instrument for Cultural and Religious Liberation." *The Journal of Religious Thought* XXXIII:1 (1976): 11–24.

Moses, Wilson Jeremiah. *Black Messiahs and Uncle Toms: Social and Literary Manipulations of a Religious Myth*. University Park: Pennsylvania State University Press, 1982.

————. "The Poetics of Ethiopianism: W.E.B. DuBois and Literary Black Nationalism." *American Literature: A Journal of Literary History, Criticism, and Bibliography* XLVII:3 (November 1975): 411–426.

Mosley, Albert G. "Negritude, Magic, and the Arts: A Pragmatic Perspective." In *Philosophy Born of Struggle*, edited by Leonard Harris. Dubuque, Iowa: Kendall/Hunt Publishing Co., 1983.

Moyd, Olin P. *Redemption in Black Theology*. Valley Forge, Pa.: Judson Press, 1979.

Muhammad, Elijah. *Message to The Blackman in America*. Chicago: Muhammad Mosque of Islam No. 2, 1965.

Murray, Albert J. *The Omni-Americans: New Perspectives on Black Experience and American Culture*. New York: Outerbridge & Dientsfrey, 1970.

Murray, Pauli. "Black Theology and Feminist Theology: A Comparative View." In *Black Theology: A Documentary History 1966–1979*, edited by Gayraud Wilmore and James Cone. Maryknoll, N.Y.: Orbis Books, 1979, pp. 388–417.

Oduyọye, Modupẹ. *The Sons of the Gods and Daughters of Men: An Afro-Asiatic Interpretation of Genesis 1–11*. Maryknoll, N.Y.: Orbis Books, 1984.

Ong, Walter. *Orality and Literacy*. London: Metheun, 1982.

Osofsky, Gilbert, ed. *Puttin' on Ole Massa*. New York: Harper & Row, 1969.

Outlaw, Lucius T. "Black Folk and the Struggle in 'Philosophy'." *Radical Philosophers' Newsjournal* (1976): 21–30.

————. "Language and the Transformation of Consciousness: Foundations for a Hermeneutic of Black Culture." Ph.D. dissertation, Boston College, 1972.

————. "Language and Consciousness: Towards a Hermeneutic of Black Culture." *Cultural Hermeneutics* 1 (1974): 403–413.

Painter, Nell Irvin. *Sojourner Truth, A Life, A Symbol*. New York: W.W. Norton, 1994

Palmer, Richard E. "Phenomenology as Foundation for a Post-Modern Philosophy of Literary Interpretation." *Cultural Hermeneutics* 1 (1973): 207–223.

Paris, Peter J. "The Bible and the Black Churches." In *The Bible and Social Reform*, edited by Ernest R. Sandeen. The Bible in American Culture series, no. 6. Philadelphia: Fortress Press, 1982.

————. *The Social Teaching of the Black Churches*. Philadelphia: Fortress Press, 1985.

Paris, Walter E. *Black Pentecostalism: Southern Religion in an Urban World*. Amherst: University of Massachusetts Press, 1982.

Pelton, Robert D. *The Trickster in West Africa: A Study of Mythic Irony and Sacred Delight*. Berkeley: University of California Press, 1980.

Peterson, Thomas V. *Ham and Japheth: The Mythic World of Whites in the Antebellum South*. Metuchen, N.J.: Scarecrow Press, 1978.

Pipes, William H. *Say Amen Brother! Old-Time Negro Preaching*. New York: William-Frederick Press, 1951.

Pitts, Walter F. "Keep the Fire Burnin': Language and Ritual in the Afro-Baptist Church." *Journal of the American Academy of Religion* LVI:1 (1988): 77–97.

———. *Old Ship of Zion: The Afro-Baptist Ritual in the African Diaspora*. New York: Oxford University Press, 1993.

Plumpp, Sterling. *Black Rituals*. Chicago: Third World Press, 1972.

Poland, Lynn. "The Secret Gospel of Northrop Frye." Review of *The Great Code: The Bible and Literature*, by Northrop Frye. *The Journal of Religion* 64:4 (October 1984): 513–519.

Porterfield, Amanda. "Shamanism: A Psychosocial Definition." *Journal of the American Academy of Religion* 55:4 (Winter 1987): 721–739.

Pryse, Marjorie, and Hortense Spillers, eds. *Conjuring: Black Women, Fiction and Literary Tradition*. Bloomington: Indiana University Press, 1985.

Puckett, Newbell Niles. *Folks Beliefs of the Southern Negro*. Chapel Hill: University of North Carolina Press, 1926.

Raboteau, Albert J. "The Afro-American Traditions." In *Caring and Curing: Health and Medicine in the Western Religious Tradition*, edited by Ronald L. Numbers and Darrel W. Amundsen. New York: Macmillan Publishing Co., 1986.

———. "The Black Experience in American Evangelicalism: The Meaning of Slavery." In *The Evangelical Tradition in America*, edited by Leonard Sweet. Macon, Ga.: Mercer University Press, 1984.

———. *"Ethiopia Shall Soon Stretch Forth Her Hands": Black Destiny in Nineteenth-Century America*. Tempe: Department of Religious Studies, Arizona State University, 1983.

———. *Religion and the Slave Family in the Antebellum South*. Notre Dame, Ind: University of Notre Dame Press, 1980.

———. *Slave Religion: The "Invisible Institution" in the Antebellum South*. New York: Oxford University Press, 1978.

Rampersad, Arnold. *The Art and Imagination of W.E.B. Du Bois*. Cambridge: Harvard University Press, 1976.

Rawick, George P., ed. *The American Slave: A Composite Autobiography*. 19 vols. Westport, Conn.: Greenwood Press, 1972.

Redkey, Edwin S. *Black Exodus: Black Nationalist and Back to Africa Movements, 1890–1910*. New Haven: Yale University Press, 1969.

Reed, Ishmael, ed. *19 Necromancers from Now*. Garden City, N.Y.: Doubleday, 1970.

Reid, Stephen Breck. *Experience and Tradition: A Primer in Black Biblical Hermeneutics*. Nashville: Abingdon Press, 1990.

———. "Review of God of the Oppressed, by James Cone." *The Drew Gateway* 48:1 (1977): 53–56.

Richards, Dona. "The Implications of African-American Spirituality." In *African Culture: The Rhythms of Unity*, edited by Molefi Asante and Kariamu Asante. Westport, Conn.: Greenwood Press, 1985., pp. 207–31

Ricoeur, Paul. "Toward a Hermeneutic of the Idea of Revelation." In *Essays on Biblical Interpretation*, edited by Lewis S. Mudge. Philadelphia: Fortress Press, 1980.

———. "Temps biblique." *Archivio di Filosofia* 53 (1985): 29–35.

———. *Time and Narrative*. Vol. I. Chicago: University of Chicago Press, 1984.

Roberts, J. Deotis. *A Black Political Theology*. Philadelphia: Westminster Press, 1974.

———. *Black Theology Today: Liberation and Contextualization*. New York: Edwin Mellen Press, 1983.

———. *Liberation and Reconciliation: A Black Theology*. Philadelphia: Westminster Press, 1971.

Roberts, John W. *From Trickster to Badman: The Black Folk Hero in Slavery and Freedom*. Philadelphia: University of Pennsylvania Press, 1989.

Rooks, Charles Shelby. "Toward the Promised Land: An Analysis of the Religious Experience of Black America." *The Black Church* II:1 (1972): 1–48.

Rosenberg, Bruce. *The Art of the American Folk Preacher*. New York: Oxford University Press, 1970.

Sanders, Cheryl J. "Christian Ethics and Theology in Womanist Perspective." *Journal of Feminist Studies in Religion* 5:1 (Fall 1989): 83–91.

Savitt, Todd L. *Medicine and Slavery: The Diseases and Health Care of Blacks in Antebellum Virginia*. Urbana: University of Illinois, 1978.

Scheff, T. J. *Catharsis in Healing, Ritual, and Drama*. Berkeley: University of California Press, 1979.

Scherer, Lester B. *Slavery and the Churches in Early America 1619–1819*. Grand Rapid, Mich.: Eerdmans, 1975.

Schreiter, Robert J. *Constructing Local Theologies*. Maryknoll, N.Y.: Orbis Books, 1985.

Sernett, Milton C., ed. *Afro-American Religious History: A Documentary Witness*. Durham, N.C.: Duke University Press, 1985.

———. *Black Religion and American Evangelicalism: White Protestants, Plantation Missions and the Flowering of Negro Christianity, 1787–1865*. Metuchen, N.J.: Scarecrow Press, 1975.

Sewall, Samuel. *The Selling of Joseph: A Memorial*. Edited by Sidney Kaplan. Amherst: University of Massachusetts Press, 1969.

Shaw, Talbert. "Religion and AfroAmericans: A Propaedeutic." *The Journal of Religious Thought* XXXII:1 (Spring–Summer, 1975): 65–73.

Sharp, Gene. *The Politics of Nonviolent Action*, 3 vols. Boston: Porter Sargent, 1973.

Shepperson, George. "Ethiopianism and African Nationalism." *Phylon* XIV 1 (1953): 9–18.

———. Introduction. *The African Diaspora: Interpretive Essays*. Edited by Martin Kilson and R. Rotberg. Cambridge: Harvard University Press, 1976.

Sherover-Marcuse, Erica. *Emancipation and Consciousness*. New York: Blackwell, 1986.

Simpson, George Eaton. *Black Religions in the New World*. New York: Columbia University Press, 1978.

Simpson, R. B. *A Black Church: Ecstasy in a World of Trouble*. Ph.D. dissertation, Washington University, St. Louis, 1970.

Smith, Archie, Jr. *The Relational Self: Ethics and Therapy in a Black Church Perspective*. Nashville: Abingdon Press, 1982.

Smith, Luther E. *Howard Thurman: The Mystic as Prophet*. Washington, D.C.: University Press of America, 1981.

Smith, Theophus H. *The Biblical Shape of Black Experience: An Essay in Philosophical Theology*. Ph.D. dissertation, Graduate Theological Union, 1987.

———. "A Phenomenological Note: Black Religion as Christian Conjuration." *Journal of the Interdenominational Theological Center* XI:1 (1983): 1–18.

———. "The Spirituality of Afro-American Traditions." In *Christian Spirituality*, edited by Louis Dupré and Don E. Saliers. Vol. 18 of *World Spirituality: An Encyclopedic History of the Religious Quest*. New York: Crossroad Press, 1989.

———. "W/Riting Black Theology." *Forum* 5:4 (December 1989): 1–18.

———, and Mark I. Wallace, eds. *Curing Violence: Religion and the Thought of René Girard*. Sonoma, Calif.: Polebridge Press, 1994.

Smith, Timothy L. "Slavery and Theology: The Emergence of Black Christian Consciousness in 19th Century America." *Church History* 41 (1972): 497–512.

Smitherman, Geneva. *Talkin' and Testifyin': The Language of Black America*. Boston: Houghton Mifflin Co., 1977.

Sobel, Mechal. *Trabelin' On: The Slave Journey to an Afro-Baptist Faith*. Westport, Conn.: Greenwood Press, 1979.

Sollors, Werner. *Beyond Ethnicity: Consent and Descent in American Culture*. New York: Oxford University Press, 1986.

Spariosu, Mihai, ed. *Mimesis in Contemporary Theory*. Philadephia: John Benjamins, 1984.

Stepto, Robert B. *From Behind the Veil: A Study of Afro-American Narrative*. Urbana: University of Chicago Press, 1979.

Steward, Theophilus Gould. *The End of the World; or, Clearing the Way for the Fullness of the Gentiles; with an Exposition of Psalm 68.31*. Philadelphia: A.M.E. Church Book Rooms, 1888.

Stewart, Jimmy. "Introduction to Black Aesthetics in Music." In *The Black Aesthetic*, edited by Addison Gayle, Jr. Garden City, N.Y.: Anchor Press/Doubleday, 1972.

Stowe, Harriet Beecher. *The Key to Uncle Tom's Cabin: Presenting the Original Fact and Documents upon Which the Story Is Founded, Together with Corroborative Statements Verifying the Truth of the Work*. 1854. Reprint. New York: Arno Press, 1968.

———. "Sojourner Truth, The Libyan Sibyl." *Narrative of Sojourner Truth: With a History of Her Labors and Correspondence Drawn from her "Book of Life."* Edited by Olive Gilbert. 1878. Reprint. New York: Arno/The New York Times, 1968.

———. Uncle Tom's Cabin. New York: Penguin Books, 1981.

Stuckey, Sterling. *Slave Culture: Nationalist Theory and the Foundation of Black America*. New York: Oxford University Press, 1987.

Sullivan, Lawrence, ed. *Hidden Truths: Magic, Alchemy, and the Occult*. New York: Macmillan Publishing Co., 1987/1989.

Suttles, William C., Jr. "African Religious Survivals as Factors in American Slave Revolts." *Journal of Negro History* 56 (1971): 97–104.

Thomas, George B. "African Religion: A New Focus for Black Theology." In *Black Theology II*, edited by Calvin E. Bruce and William R. Jones. Lewisburg, Pa.: Bucknell University Press, 1978.

Thomas, H. Nigel. *From Folklore to Fiction: A Study of Folk Heroes and Rituals in the Black American Novel.* Westport, Conn.: Greenwood Press, 1988.

Thomas, Keith. *Religion and the Decline of Magic: Studies in Popular Beliefs in Sixteenth and Seventeenth Century England.* London: Weidenfield and Nicolson, 1971.

Thomas, Latta R. *Biblical Faith and the Black American.* Valley Forge, Pa.: Judson Press, 1976.

Thompson, Robert Farris. *Black Gods and Kings.* Bloomington: Indiana University Press, 1976.

———. *Flash of the Spirit: African and Afro-American Art and Philosophy.* New York: Vintage Books, 1984.

———, and Joseph Cornet. *The Four Moments of the Sun: Kongo Art in Two Worlds.* Washington, D.C.: National Gallery of Art, 1981.

Thurman, Howard. *Deep River and the Negro Spiritual Speaks of Life and Death.* Richmond, Ind.: Friends United Press, 1983.

———. *Jesus and the Disinherited.* Nashville: Abingdon-Cokesbury Press, 1949.

Tillich, Paul. *Systematic Theology.* Vol. 1. Chicago: University of Chicago Press, 1951.

———. *Theology of Culture.* London: Oxford University Press, 1959.

Tracy, David. "Ethnic Pluralism and Systematic Theology." *Ethnicity.* No. 101 of *Concilium: Religion in the Seventies.* Edited by Andrew M. Greeley and Gregory Baum. New York: Seabury Press, 1977.

Trulear, Harold Dean. "An Analysis of the Formative Roles of Ideational and Social Structures in the Development of Afro-American Religion." Ph.D. dissertation, Drew University, 1983.

———. "The Lord Will Make a Way Somehow: Black Worship and the Afro-American Story." *The Journal of the Interdenominational Theological Center* XIII:1 (1985): 87–104.

Truth, Sojourner. *Narrative of Sojourner Truth: With a History of Her Labors and Correspondence Drawn from her "Book of Life".* Edited by Olive Gilbert. 1878. Reprint. New York: Arno/The New York Times.

Turner, Victor. *Dramas, Fields, and Metaphors: Symbolic Action in Human Society.* Ithaca, N.Y.: Cornell University Press, 1974.

———. *The Ritual Process: Structure and Anti-Structure.* Ithaca, N.Y.: Cornell University Press, 1969.

Wahlman, Maude Southwell. "Religious Symbolism in African-American Quilts." *The Clarion* (Summer 1989): 36–44.

Walker, Alice. *In Search of Our Mothers' Gardens.* New York: Harcourt Brace Jovanovich, 1983.

Walker, David. *David Walker's Appeal.* Edited by Charles M. Wiltse. 1829. Reprint. New York: Hill & Wang, 1965.

Walker, Sheila S. *Ceremonial Spirit Possession in Africa and Afro-America: Forms, Meanings and Functional Significance for Individuals and Social Groups.* Leiden: E. J. Brill, 1972.

Walker, Wyatt T. *Somebody's Calling My Name: Black Sacred Music and Social Change.* Valley Forge, Pa.: Judson Press, 1979.

Wallace, Anthony F. C. "Revitalization Movements." *American Anthropologist* LVIII (1956): 264–281.

Washington, James Melvin. *Frustrated Fellowship: The Black Baptist Quest for Social Power.* Macon, Ga.: Mercer University Press, 1986.

————, ed. *A Testament of Hope: The Essential Writings of Martin Luther King, Jr.* New York: HarperCollins Publishers, 1991.

Washington, Joseph R. *Black Religion.* Boston: Beacon Press, 1964.

————. *Black Sects and Cults.* Garden City, N.Y.: Anchor Press/Doubleday, 1973.

————. *The Politics of God.* Boston: Beacon Press, 1967.

Weisbrot, Robert. *Freedom Bound: A History of America's Civil Rights Movement.* New York: Norton, 1990.

West, Cornel. *Prophesy Deliverance! An Afro-American Revolutionary Christianity.* Philadelphia: Westminster Press, 1982.

Whitten, Norman E., Jr. "Contemporary Patterns of Malign Occultism among Negroes in North Carolina." *Journal of American Folklore* 75:298 (1962): 311–325.

————, and John Szwed, eds. *Afro-American Anthropology: Contemporary Perspectives.* New York: Free Press, 1970.

Wilder, Amos Niven. *Theopoetic: Theology and the Religious Imagination.* Philadelphia: Fortress Press, 1976.

Williams, Delores S. "Black Theology's Contribution to Theological Methodology." *Reflections* 80:2 (January 1983): 12–16.

Williams, Delores S. *Sisters in the Wilderness: The Challenge of Womanist God-Talk.* Maryknoll, N.Y.: Orbis Books, 1993.

Williams, Melvin D. *Community in a Black Pentecostal Church: An Anthropological Study.* Pittsburgh: University of Pittsburgh Press, 1974.

Williams-Jones, Pearl. "Afro-American Gospel Music: A Crystallization of the Black Aesthetic." *Ethnomusicology* 19:3 (September 1975): 373–385.

————. "The Musical Quality of Black Religious Ritual." *Spirit: A Journal of Issues Incident to Black Pentecostalism* 1 (1977): 20–30.

Wills, David W., and Richard Newman, eds. *Black Apostles at Home and Abroad: Afro-Americans and the Christian Mission from the Revolution to Reconstruction.* Vol. 1 of 2. Boston: G. K. Hall, 1982.

Wilmore, Gayraud S., ed. *African American Religious Studies: An Interdisciplinary Anthology.* Durham, N.C.: Duke University Press, 1989.

————. *Black Religion and Black Radicalism: An Interpretation of the Religious History of Afro-American People.* Maryknoll, N.Y.: Orbis Books, 1983.

————. *Last Things First.* Philadelphia: Westminster Press, 1982.

————. "Revising the Color Symbolism of Western Christology." *Journal of the Interdenominational Theological Center* 2 (Fall 1974): 8–18.

————, and James H. Cone, eds. *Black Theology: A Documentary History 1966–1979.* Maryknoll, N.Y.: Orbis Books, 1979.

Wilson, Bryan R. *Magic and the Millennium: A Sociological Study of Religious Movements of Protest Among Tribal and Third-World Peoples.* New York: Harper & Row, 1973.

Wimberly, Edward P. and Anne Streaty. *Liberation and Human Wholeness: The Conversion Experiences of Black People in Slavery and Freedom.* Nashville: Abingdon Press, 1986.

Wimbush, Vincent. "Rescue the Perishing: The Importance of Biblical Scholarship in Black Christianity." *Reflection* 80:2 (1983): 9–11.

Wink, Walter. *The Bible in Human Transformation: Toward a New Paradigm for Biblical Study.* Philadelphia: Fortress Press, 1973.

————. *Engaging the Powers: Discernment and Resistance in a World of Domination.* Vol. 3 of 3. Minneapolis: Fortress Press, 1992.

Winks, Robin, ed. *An Autobiography of the Reverend Josiah Henson*. Reading, Mass.: Addison-Wesley, 1969.

Wintrob, Ronald M. "Hexes, Roots, Snake Eggs? M.D. versus Occult." *Medical Opinion* 1 (1972): 54.

Wood, Peter H. " 'Jesus Christ Has Got Thee at Last': Afro-American Conversion as a Forgotten Chapter in Eighteenth Century Southern Intellectual History." *Bulletin of the Center for the Study of Southern Culture and Religion*. 3:3 (1979): 1–7.

Yoder, John Howard. "Exodus and Exile: The Two Faces of Liberation." *Cross Currents* (Fall 1973): 297–308.

———. "Peace without Eschatology." *The Original Revolution*. Scottdale, Pa.: Herald Press, 1971.

Young, Josiah U. *Black and African Theologies: Siblings or Distant Cousins?* Maryknoll, N.Y.: Orbis Books, 1986.

Young, Robert Alexander. "The Ethiopian Manifesto." In *A Documentary History of the Negro People of the United States*, edited by Herbert Aptheker. Vol. 1. New York: Citadel Press, 1951.

Zeusse, Evan M. *Ritual Cosmos: The Sanctification of Life in African Religions*. Athens: Ohio University Press, 1979.

Index